MANGROVES AND MAN-EATERS

10 March 2011

To Anna and Stephen

With very best wishes

Dan

Dan Freeman.

MANGROVES AND MAN-EATERS

AND OTHER *wildlife* ENCOUNTERS

DAN FREEMAN

Whittles Publishing

Published by
Whittles Publishing,
Dunbeath,
Caithness KW6 6EG,
Scotland, UK

www.whittlespublishing.com

© 2010 Dan Freeman
Line Drawings © Robin Prytherch
ISBN 978-184995-009-1

Printed and bound in the UK by
J F Print, Ltd., Sparkford, Somerset.

For Becky, Michael, William and Edward, who stayed at home but were always with me...

Contents

FOREWORD

by *Tony Soper,* Kingsbridge

My first solo camping trip was to Dartmoor. They were great hiking days but I was unprepared for the bitter cold at night. Kitted out with a thin blanket and a couple of blanket pins, I ended up tearing out the pages from a Penguin book and covering my shivering body to no good effect. So I went back home and saved for a sleeping bag. Since then, I have eagerly grabbed every chance of any kind of expedition but it was some time before I became wise in the matter of choosing my companions. There are people who revel in coast-walking and the skills of finding a secret cliff ledge where you camp with the sounds of puffins and peregrines in your ears. And there are people who, not unreasonably, prefer visiting castles and mediaeval churches. Most people are happy enough to sample the local pub, so that was never a problem, but, on the whole, I've most enjoyed the company of those who go for the birds and the beasts. Which brings me to Dan Freeman.

I first met Dan when we found ourselves involved in the making of a six-part series, *Discovering Birds*, for the BBC in London. Neither of us could believe our luck as we drank coffee in the canteen and plotted safaris round the hottest birdspots of Britain, making sure that they took us to the most remote islands and wetland wildernesses. And it was while we were filming gannets on Bass Rock and avocets on the Tamar and swans in Dorset, sampling the full English breakfasts of a hundred B and Bs, that I enjoyed some of Dan's fascinating stories. Long before he joined the BBC, he had travelled the world on expeditions, meeting the cream of the world's birders and inevitably finding himself face-to-face with exotic animals. But he also seems to have had more than an average share of those sticky situations that need a firm hand and a healthy dose of luck, especially in the making of wildlife movies.

In this entertaining book, which is as much about people and places as it is about wildlife, you'll read of hair-raising encounters with giant crabs, lions and killer bees. But rest assured: with Dan Freeman you are travelling with the best of guides, the kind of guy who makes sure that after a long day with sandflies and piranhas, there's always a welcoming hostelry with a hot bath, a decent bed and a proper breakfast.

Tony Soper

PREFACE

There can't be many places as physically challenging to a human as a mangrove forest; and nor can there be many animals as intimidating as a man-eating tiger. Put the two together and you have one of the most daunting places on earth. You have the Sundarbans, a vast tidal forest shared between India and Bangladesh. It is here, in the Sundarbans, that the combined force of mangroves and man-eaters awaits thousands of people who must come to collect their wood, fish and honey.

My fascination with mangroves began in Australia in 1968, on a Natural History Museum expedition to study birds. I already knew about hundreds of mosquitoes, but not millions; and I knew, from family outings to Weston-super-Mare, something about mud; but not deep, cloying mud that clings to a latticework of tightly twisted roots blocking every way. I also knew about tides, but not about tidal races and the speed with which a mangrove forest can fill with water. Nor did I know that people could be killed by giant crabs. What I soon learnt, though, was that mangrove forests are a naturalist's paradise.

My fascination with man-eating tigers began when was I was eight or nine. My father read me bedtime stories from Jim Corbett's *Man-eaters of Kumaon*, and although I was a bit frightened of the dark I was always able to get to sleep. It must have been Jim Corbett's devotion to tigers radiating from the pages that overshadowed the potentially terrifying details of how he tracked and finally killed these big cats in the Himalayan foothills.

Thirty years later I was on a plane to Bangladesh. It was my turn to enter the Sundarbans, though not *despite* its man-eating tigers, but *because* of them. I was working on a BBC film looking at the problems facing the people of Bengal when they entered the forest to collect food and firewood. It was a humbling experience spread over two years, and although we were not able to film a single tiger, we felt their presence wherever we went.

Forest of Fear is just one of the many films I have worked on during 35 years of programme-making around the world. Each trip – and there were others for research, tourism and photography – offered different encounters with people and wildlife. Some were more challenging than others, though none came close to the physical and emotional difficulties posed by the Sundarbans, the tidal forest that combines two of my great interests. It is for this reason that I have chosen *Mangroves and Man-Eaters* to be the title of this book and I hope that you enjoy these pages and find them fascinating, perhaps even inspiring.

Dan Freeman

ACKNOWLEDGEMENTS

Mr E. J. Douglas, the headmaster at Hadham Hall School in Hertfordshire, was the first person outside my family to take my interest in birds seriously. Then, in 1960, Mary Ashton, a biology teacher and friend of my parents, took the 15-year-old me on my first expedition – a week on Bardsey Island off the coast of north Wales. The sound of Manx shearwaters calling at night was intoxicating. Around this time I was also devouring every *Zoo Quest* word published by David Attenborough and every man-eating tiger hunt written up by Jim Corbett. My fate was sealed.

Since those early days I have worked with and received help and inspiration from a great many other people. Among them are: Pat Hall (in particular), Derek Goodwin, Shane Parker, Graham Cowles, Peter Colston, Philip Burton, David Snow, Ian Galbraith, Dom Serventy, Allan McEvey, Charles Pitman, Con Benson, Alec Forbes-Watson, Don Turner, Fleur Ng'weno, Charles Wurster, Isobel Bennett and her sister Phyll, Brian Booth, Tony Hiller, Cliff Frith, Ralfe Whistler, Harry Butler, David Reed, Chris Parsons, Tony Soper, Robin Prytherch, Peter Jones, Nigel Ashcroft, Barry Paine, Keith Scholey, Ned Kelly, Pam Jackson, Sara Ford, Nigel Tucker, Dilys Breese, Paul Reddish, Hugh Miles, Tom Poore, Adrian Warren, Richard Matthews, Richard Brock, Pelham Aldrich-Blake, Brian Leith, Jeffery Boswall, John A. Burton, Bruce Coleman, Cyril Walker, Martin Brendell, Tony Allen, James Gray, Chris Catton, Ramon Burrows, John Cooke, Sean Morris, Dominic Crawford-Collins, Theo Cockerell, Chris Gomersall, Ken Marsland, Jonathan Kingdon, John Fa, Andy Copp, Nicholas Davies, Nigel Collar, Dale Brown, Simon Rigge, John Man, Bruce Coles, Julian Badcock, Somerville Anderson, Bernard (aka Jules B.) Davidoff, Peter Haley Dunne, Desmond Chaffey, Audrey Dickinson, Jan Aldenhoven and Glen Carruthers, Malcolm and Unity Coe, Michael and Penny Richards, Mike and Elaine Potts, Sarah and Stevie de Kock, Alan and Tessa McGregor, Wayne and Venessa Hinde, Mike and Juliet Wood, Richard and Julia Kemp, Simon and Catherine Bearder, Lesley and Alan Sunderland, Lindy and Peter Newton, Jim and Noreen Braund, my parents-in-law Cyril and Catherine Bagguley and my own parents, the painters Frank and Pamela Freeman, who set such a good example.

Special thanks go to Malcolm Coe, who gave me a place at St Peter's College, Oxford, to study zoology, long after I had left school. And to Sir Alec Cairncross, the then Master of St Peter's who, in 1976, the college's poor student fund being empty, personally lent me the airfare to travel to South Africa to work on bushbabies with Bob Martin and Simon Bearder.

Thanks go to the Bristol Evening Post for permission to reproduce their article in *Point Torment* and to Jan Aldenhoven and Glen Carruthers for their heartfelt contribution on filming doves in China.

Thanks are also due to those people who have read chapters and made helpful comments: my brother Oliver Freeman, Mic Cady, Rob Hume, John A. Burton, Polly Coles, Harriet and John Vice-Coles, Andrew Hornung, Pat Hall, Mike Potts, Michael Richards, Cliff and Dawn Frith, Ralfe Whistler, Brian and Anne Booth, Nick Blackman, Marcus Clapham and, in particular, Keith Whittles, who saw something in my encounters that encouraged him to make the commitment to publishing. I am more than grateful to Keith and to Sue Steven, Moira Hickey, Kerrie Moncur and Shelley Teasdale of Whittles Publishing for turning my typed pages into a proper book.

My wife Becky and my sister Sally read the entire manuscript and suggested changes which have improved the flow of the stories. Sons Michael, William and Edward helped with dates, maps and a baby blackbird who grew up to become our friend.

Tony Allen is another friend. We began work on our first film, in England, in 1980 and completed our last, in Zimbabwe, in 2000. Filming with Tony on five continents over 20 years – a mixed bag of piranhas, elephants, wolves, pandas, snakes and spiders – was a pleasure. In 2008 we joined forces with his business partner Andy Matheson (Panache Productions) and botanist Jenny Steel and her husband Alan Pottinger to make a DVD on Jenny's new wildlife garden in Shropshire.

Because *Mangroves and Man-Eaters* covers half a century, it is not surprising that many of the people mentioned in the text have died. I have not dwelt on this because there is, for me, a real sense in which they will never die. I am just grateful to everyone who has helped me fulfil my childhood dream of working with wildlife.

Which brings me to Tony Soper and Robin Prytherch. It has been a privilege to work with them on bird programmes for the BBC during the 1980s and to retain them today as friends. I consider myself fortunate to have met them. It is, therefore, an added bonus that they have contributed to this book, Tony with the foreword and Robin with the line drawings which introduce each of my wildlife encounters.

1

Bonding with Spiders

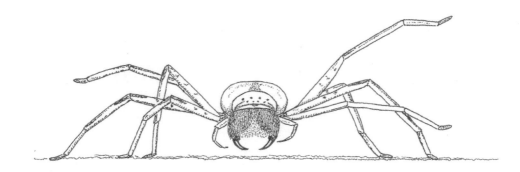

The bell rang for the third time in twenty minutes, announcing the arrival of yet another visitor. I got up from my desk and stepped out into the corridor, where the smell of mothballs from a million bird skins hung in the air. The stranger, swarthy and dark-haired, was striding towards me. And then, in an accent that was clearly American, 'The name's Bond, James Bond.' I blinked. Did he really think that having the same name as the fictitious MI5 agent was enough for him, the mere author of the *Birds of the West Indies*, to announce himself in this way? I forced a slightly embarrassed smile, but no more, as I shook his hand. It was hardly my place to engage these people in flippant banter, even if I had seen one or two of the early Bond films and could spot the intended humour. As a junior member of the Bird Room staff, I was in awe of the famous names passing through London's Natural History Museum on their way to the 1966 Ornithological Congress in Oxford. They were flocking from all over the world and I could hardly keep up with them.

That said, my response to this particular introduction might have been different had I known my Bond history. Apparently, the writer Ian Fleming thought the name of the person standing in front of me – a name he had first seen on the cover of a Caribbean bird book – would be perfect for the character he was creating. Soon, the special agent of book and film was launched with the identity of an island-hopping ornithologist. Little wonder that the real James

Bond, who met Fleming just once and who even looked like the actor Sean Connery, indulged his celebrity. I knew none of this that morning, not having had a proper briefing from Her Majesty's Secret Service!

The *Dr No* film had left me with the abiding memory of a large black spider crawling over 007's naked body in the dead of night: that bulge under the thin sheet moving purposefully across his skin, his fear, his sweat, his paralysis and that final muscular outburst as he leapt up and beat the spider to a pulp on the bedside rug. The sequence, of course, was heavily stage-managed. The spider didn't really die, however convincing Sean Connery's onslaught. I know that for a fact because, having met the real James Bond, I went on to meet the real spider. She was alive and biting long after the film was made. Her name was Amelia, and Doug Clark, the museum's own Spiderman, was her keeper. During the filming Doug was positioned off-camera to make sure everything went smoothly and that nobody, least of all his precious pet, got hurt.

Amelia was one of those creatures whose bark really was bigger than its bite. She certainly looked the part with her shiny, black upper body and large half-hidden fangs poised beneath a set of forward-staring eyes. She carried herself on eight thick, hairy legs spanning at least 10 cm, bunched and toe-tapping, a suppressed lethal energy. To us, though, she was docile, with venom no more painful than that of an everyday wasp in the garden. Her hunting technique in the tropical wild, down among the dead leaves or up on the shaggy trunk of a rainforest tree, would rely on stealth, a lady-in-waiting, ready to pounce and pierce the unfortunate insect or lizard that had blundered her way.

What is it about spiders, large or small, hairy or smooth, that bothers so many people? Psychoanalysts have had a field day, probing minds and linking the phobia to any number of innermost fears. In Freudian terms it is, or at least was, fashionable to link it to sexual experience. It is quite possible, though, that people are born with a genetically-determined fear of spiders, even though this fear is not shared equally by different peoples around the world. There are those in Africa, India and Australia, for example, who are not at all bothered by the poisonous spiders that sneak into their homes. If the fear of spiders is not inherited by children, then it must be learnt, either from personal experience or from the behaviour of the adults around them. Whatever its origin, there is no denying that the sudden appearance of a spider can trigger a violent and genuine response. Arachnophobia is not to be taken lightly.

The best spiders, if 'best' is the right word, are those that are safely confined to a web. A web neatly defines the limit of their movements and restricts them to a known place. Its fly-catching stickiness, reduces the need for excess venom. But if the spiders around me are the fleet-of-foot, non-web varieties, I feel uneasy, compelled to keep an eye on them until they have gone somewhere safe. The most difficult thing is to get to sleep knowing they are out there, tiptoeing around the walls in the dark, a feeling I was relieved to find shared by even the hardiest of arachnophiles. In South Africa, working on bushbabies with Simon and Catherine Bearder, we went one evening to the ranch house for a meal with Richard and Barbara Galpin. Casting an eye round their living room, I could not help noticing three large wolf spiders, perhaps 8 cm across their legs, walking purposefully around the walls. Richard casually picked one up, released it onto the back of his hand and tried to provoke it into attacking his finger, which it steadfastly refused to do. He stuck it back on the wall, a bit like a fridge magnet, explaining how good they were at mosquito control. Barbara agreed, though she admitted that she really didn't like them in the bedroom when they stomped across her face in the middle of the night on their prickly little toes.

The walls of the little wooden bush house I shared with Simon and Catherine for two months in 1976 were singularly lacking in spiders. That said, we were out most nights radio-tracking bushbabies from dusk until dawn in sub-zero temperatures and hardly had time to notice creatures around the house. What I did notice, though, was a large hole in the ground on the edge of the track just beyond the front door. For the first few weeks I assumed it to be the home of a small rodent. One evening, shining a torch onto it, my eye was caught by a small animal backing away from the light. Still thinking 'mouse', I went to have a closer look. Simon smiled knowingly and suggested I probe the hole with a piece of dead grass. I duly went fishing, got a bite first time and hauled the occupant to the surface: a baboon spider, a nocturnal wanderer a good 12 cm across its fat, hairy legs. It came halfway out into the open, released its hold on the grass and sat for a few minutes before slinking back into the darkness of its burrow. From that day on, I could not resist poking grass into almost every hole I came across, with the unsettling discovery that nearly all were occupied by these hairy heavyweights. It was some comfort to know that the freezing August nights of the Transvaal winter were too cold for them to be wandering around the woodland floor while we, our headlamps, eyes and radio antennae fixed firmly on the treetops, were concentrating on bushbabies as they bounced from branch to branch.

Twenty three years later, while filming at Imire Game Reserve in Zimbabwe with Tony Allen and Nigel Tucker, holes in the ground around our living quarters jogged my memory of bushbaby nights 800 km away to the south. Demonstrating the art of fishing for baboon spiders, which had everyone doing it for days, there emerged just one little difference between here and South Africa. It was still cold at night but because we were staying in a tourist lodge, there were lights along the paths to the rondavels where we slept. It seemed that all the winged and walking creepy-crawlies of Imire were attracted to these pools of brilliance. So, too, were their predators, the bats and the baboon spiders, the latter, no doubt, warmed by the heat of the bulbs whose brightness they tolerated in return for such an ample supply of food. Somehow, though, we had to deter these spiders from getting into our bedrooms. Early each evening we sprayed insect repellent round our shuttered doors and windows and not a single spider crossed the threshold.

One evening at Imire, the baboon spiders more or less under control, I went into Tony's room to ask for the air-freight details for his filming gear. The generator in the lodge had just gone down and the room was dimly lit by candles. Lazing on his bed after a long day filming buffaloes, Tony gestured towards a mound of equipment in the corner where I would find the details on the labels attached to the cases. In the shadowy, flickering light – or, more accurately, the shadowy, flickering dark – I leaned over and groped around in search of one of the labels. I grasped a living creature. Not a spider, but the smooth, sinuously twisting body of a snake. Had it been a baboon spider, I might have died from shock even if it hadn't bitten me. Being a snake – although it could easily have been one of the highly-poisonous boomslangs that lived in the tree canopy above our thatched roofs – it filled me with alarm, but not total dread. I still pulled back with a startled cry. Tony grabbed his torch and shone it beyond the cases. Firmly stuck to a pile of discarded gaffer tape was a metre-long snake. It had evidently been there for some time because it was exhausted by its efforts to escape. It was identified by the locals as a slightly venomous rat snake, a convenient term that seems to cover many nondescript snakes that come into people's homes all over the world to feed on their resident rodents. Later that evening, I made the mistake of telling Tony that I was lucky not to have been bitten. He was not sympathetic. The snake - had I forgotten already? - was actually in *his* room and, but for the gaffer tape, might have ended up in *his* bed.

Living through such memorable encounters is an occupational hazard for people working with wildlife. Studying birds in Australia in 1968, I often slept outside my stuffy tent in the fresh air. My camp bed had uprights at each corner, to which a mosquito net could be attached. Each morning, I would crawl out of the side and tuck the net back under the thin mattress, guaranteeing that nothing could get in during the day. One night, though, as I lay in bed looking up at the brilliant night sky, I noticed the shape of a large spider at the far corner above my right foot. I was not at all bothered by its presence on top of the mosquito net and was too tired and comfortable to get up and look any closer. I drifted off into my usual deep bush sleep.

In the morning, with the first wash of dawn giving shape to trees, tents and Land Rovers, the spider was still there, in exactly the same position. I lay looking at it, waking slowly, and as its body and legs came into focus, I realised that it was in fact on the inside of the netting. I went cold, not from worry over what it might do next but from realising that I had left myself vulnerable by falling asleep. For all I knew, it had walked all over me, returning to its own little corner to sleep out the approaching day. Sliding carefully out of the head-end of my camp bed, I tied the netting with string as tightly as I could below the spider, so that it was trapped. I then undid the net from its poles and removed my unwelcome guest to a very safe distance.

A few weeks later, on the same expedition, I went to unscrew the yellow lid on one of our black water-containers. Under the flickering flames of a night fire, images are not as clear as they are during the day, so it wasn't until my hand was about to make contact with the lid that I realised that a large spider was sitting on it, matching its colour almost perfectly. At the last minute, it raised and spread its front legs, bending back its head and thorax to expose its fangs as a warning. The unexpectedness of the moment made me jump back with surprise.

So what is it about spiders? They invade our homes, they scuttle, they hide, they are fast and secretive and they kill for a living. We treat them all as though they are equally poisonous. They have a kind of mystic power over us, causing knee-jerk reactions if they are encountered unexpectedly or move too quickly. Something about them can flick an internal switch of fear and I can't imagine that anyone will ever really know what it is. Perhaps our ancestors were blighted by super-venomous spiders that lurked in the dark recesses of caves. Perhaps they sneaked in unnoticed, snuggling down with unwary children wrapped in warm hides and grasses, their parents reacting violently, beating the spiders to death and instructing their offspring to do the same. Inherited or acquired, the fear of spiders is firmly lodged in many of us today. When Ian Fleming wrote Amelia into James Bond's bed, he knew exactly what he was doing.

Even David Attenborough, the grand master of natural history, can be emotionally hijacked by a spider. In *Life in the Undergrowth*, the 2005 six-part BBC series on animals without backbones, there was a programme on the importance of silk. Fittingly, it was about spiders and their webs. David sat on a small mossy bank in a Malaysian rainforest and poked provocatively at the tough web-lines radiating from a central covered burrow. He was demonstrating the connection between the lines of silk fanning out from the trap door and the spider hidden inside. Its feet, revealed by intimate filming beneath the camouflaged trap door, were placed on several of the tripwires, ready to respond to the vibrations of any small creature ambling across them. After enough twangs from David, the spider flipped its lid and launched itself at the stick to bite what could have been another meal. The speed of the spider's response, the eruption of the lid and the sudden rush towards him of a mass of fangs and feet had David recoiling in breathless surprise. Yet he retained his composure, completed the piece-to-camera with his usual aplomb and may even have sanctioned the use of his response in the finished film, knowing that his own, quite genuine reaction would be the same as that of many, if not most, of his viewers.

The Australian redback spider, a relative of America's infamous black widow, is considered one of the most poisonous spiders in the world. It is a small, black, shiny spider with a red body-stripe, attractive to look at but deadly. The larger female, even though she is no more than 2–3 cm long, spins an untidy but ingenious web in dark, dry hideaways in and around human dwellings. Outside lavatories, woodsheds, flowerpots and verandas are among her places of choice, and contact with people, particularly their carefree children, is inevitable. Redback females – the all-brown males are not big enough to be any threat to human skin – bite with small-fanged difficulty and yet more than 200 people are injected with their neurotoxins every year. Although a slow death from these bites was once a real threat, the widespread availability of antivenin has made fatalities a thing of the past.

Working for Partridge Films on a series about sex in the natural world, I set out for Brisbane with Tony Allen late in 1989. On our long filming list of creatures preparing for, engaging in or recovering from their animal intimacy was the redback spider. Once our studio was ready and escape-proof, all we needed was the spiders themselves. Geoff Monteith, from the Brisbane Museum, made it sound easy. He directed us to a nearby playground, suggesting we look beneath the little seats on poles that parents used while their children played. If we needed more spiders than we were sure to get there, we were to run a stick round the insides of the piled-up car tyres that acted as a kind of manual bouncy castle for the kids. To our horrified surprise, both suggestions worked. From under the bottoms of adults not in the least concerned about the proximity of either themselves or their children to potential death, we came away with ten female redbacks and four of their diminutive male partners.

Back in the studio, we gave each female spider its own special film set and waited a few days until they had spun their webs. We then introduced the males. Each went about its sexual imperative like a lamb to the slaughter. One by one they loaded their front palps with sperm and set off into a female web, twanging a coded rhythm on the silken strands to suppress her cannibal instincts. Creeping closer and closer, hesitating when she moved but driven by the powerful instinct to reproduce, they crawled between her legs until their loaded palps made contact with her receptive body. And then for the quick getaway before she stirred from her sexually-induced trance and turned hunter rather than hunted. Too late. A sudden long-legged embrace, a deadly bite and it was all over. One by one, the four little males were done for, wrapped and stored, ready to be sucked dry at a more convenient time. For each deceased male the life-job was done. But all was not lost. If being eaten had given the female the resources to make more eggs to be fertilised by his now safely-stored sperm, then there would be even more young redback spiders setting out into the playgrounds of Australia with his genes in their own little bodies.

My own father used to tell a story about how he once found me, aged three or four, sitting screaming on the edge of our lawn. To trace the problem to its source, he followed my outstretched arm to the end of an accusing finger and there, just in front of me, sitting on a leaf, was a spider a millimetre in diameter. It was a money spider, one of those arachnids which, if treated with respect, is said to bestow the hope of financial reward. What this means, and it was obviously something I still had to learn, is that a fear of some spiders can be suppressed, particularly by greed. Money spiders fail to promote fear in the hearts and minds of grown-ups and this is something that can easily be communicated to children, despite those terrifying first encounters.

At the opposite end of the spider spectrum is the outsized *Nephila*. You might need to be guaranteed a lottery win to overcome a distaste for one of these giants, even though, just like money spiders, they are quite harmless to people. They live in the tropics where the colourful females, whose bodies may be fully 5 cm long with a leg span three times that size, spin the largest individual webs of all known spiders. The comforting thing is that a web, however intimidating it may be to look at, anchors a spider to a fixed position and can usually be avoided. But *Nephila*'s web really *is* intimidating. Its support-lines may be 10 or 15 metres long, running from the tops of trees and lampposts to the ground. The actual web, the trap, can be several metres in diameter, its individual strands tough enough to hold a small bird flying into them at high speed. My first close encounter with one of these sticky frameworks was in Australia, when I first ventured into the wilds of the Kimberleys, in the northwest of the country, in 1968. That day in April showed more than any other my lack of experience in the tropics. With my attention focussed on something in the distance, I walked straight into one of these giant cobwebs and stuck fast. The harder I struggled, the more entangled I seemed to become. I just knew that, with a web that size, there must be a giant man-eater lurking nearby. But she remained hidden, for good reason, and I pulled myself free without mishap. I had learnt a lesson about concentrating on my immediate surroundings in a new and potentially hostile land.

It was on this same bird-studying expedition to Australia in 1968 that I had my own spider-bonding experience, one that, to my mind at least, ranks with the fictitious pairing of James and Amelia. It was October and we had set up camp just outside Oenpelli (now known as Gunbalanya) in Northern Territory. One of our tents was a laboratory where we prepared our birds as study skins, skeletons or spirit, by which the bird was preserved whole in formalin. The work tables were a general mess of cotton wool, bottles, dissecting kits, notebooks and data sheets.

The spider must have entered the tent at night, climbed the table leg and snuggled down in the cotton wool at my work space. That meant it was waiting for me. Since first light I had been out collecting more birds and now, fed and rested, I needed to process them and write up my notes during the heat of the day. Colleague Tony Hiller was standing at the entrance to the tent when I sat down and reached for the cotton wool. He said something and I looked up, opening the cotton wool as we spoke. The first thing I felt was the gentle touch of its feet as it climbed out of the cotton wool onto the back of my right hand. I sat and looked at it, detached, as its clustered eyes stared straight at me, its enormous fangs just a millimetre from my skin. Tony also saw the spider, and could see the panic mounting on my face. Being fearless, he casually offered the sensible advice of letting it go where it wanted. But the spider was on me, not him. It stopped at my elbow as though contemplating its next move.

Inside the humid tent, under the midday sun, I began to sweat even more than usual. A kind of paralysis set in and I found myself unable to do anything but watch as the spider tiptoed slowly towards my shoulder. I was all alone in my own little nightmare of nature. There was no Doug on standby, no director to cut the shot, no writer to turn the spider round. This was not acting. This was the real thing.

Before Tony had a chance to come to my rescue, my mind and body arrived at the same involuntary breaking point that Sean Connery had portrayed so convincingly on film. With a loud vocal outburst – I must have been holding my breath while all this was happening – I jumped up and, with a violent sweep of my left hand sent the spider flying. It hit the far wall of the tent, dropped to the floor, raised its upper body to expose its fangs and began a fierce kind of arachnid square dance. There was an audible crunch as it collided with a table leg and drove its fangs deep into the wood.

Both shaken and stirred, I sat and watched its antics with a growing sense of relief, taking long deep breaths. As the oxygen surged through my body, my head cleared and I just knew that the bond between me and that eight-eyed monster of a spider was, like diamonds, for ever. I was seized with a desire to keep it. But not alive. It would have no more licence to kill. Gripping the longest pair of forceps I could find, I approached the angry spider, clamped it between metal shafts and dropped it into a jar of 70% alcohol. Two days later, I lifted it out and laid it on the table. Quite incredibly, its legs began to move. It went straight into a large pot of formalin and from there, the following week, into a small bottle of alcohol.

Four months later, early in 1969, I handed my bottled trophy to museum Spiderman, Doug Clark, in London. It came back neatly labelled 'Neosparassus margareyi, female', an accompanying note advising that she was, or at least had been, fairly poisonous. She was a hunter, and although not one of those hairy bird-eating Amelia types, was still 16 cm across her outstretched limbs, which were thin and spiky, built more for speed and distance than for stealth and ambush.

Doug's Amelia lived in the Natural History Museum until she was at least fourteen. Her cast-off skins, shed annually as she grew, were mounted in a glass box and put on public display. My own spider sits on a shelf at home in the same bottle where it began its physical afterlife in 1968 – the permanent reminder of a moment of fear.

That fear, as deep-rooted in me as in anybody who doesn't like spiders, may have originated in our African ancestors a million years ago; a million years, that is, before the real James Bond published his *Birds of the West Indies* and Ian Fleming, still without a name for his special agent, fixed himself another martini.

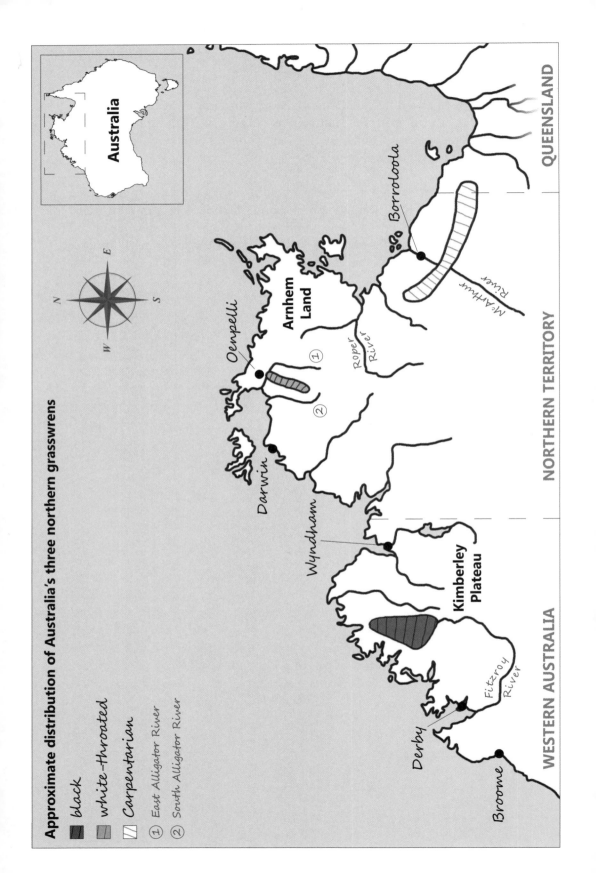

Approximate distribution of Australia's three northern grasswrens

black

white-throated

Carpentarian

① East Alligator River

② South Alligator River

Australia

N
W   E
S

QUEENSLAND

Borroloola

McArthur River

Arnhem Land

Oenpelli

①
②

Roper River

Darwin

Wyndham

NORTHERN TERRITORY

Kimberley Plateau

Derby

Fitzroy River

Broome

WESTERN AUSTRALIA

2
GRASSWRENS FROM THE NORTH

The black grasswren (*Amytornis housei*)

Richard Blythe rested the paper on his lap and turned towards us. 'You know, I've seen this before. Last year. Some people, I don't remember who, came up here looking for the black grasswren and they had exactly the same map.' He held up his hand to halt my protest. 'Dan, I recognise it, I remember these details,' his finger jabbing at the page.

But how could Richard, isolated up here in northern Australia, have seen my map before? It had been such an important, well-guarded secret, known to just four people in England, that the only explanation had to be that he was wrong. And yet he seemed so certain.

Early in 1967, I had traced the map from the 1902 original by F. S. Brockman. My copy had then been held under lock and key in London's Natural History Museum by my boss, J. D. Macdonald, until the bird-collecting expedition I was part of left England for Australia in March 1968. Just before we left, Mr Macdonald, who was Head of the Bird Room, had made a little ceremony of unlocking his safe and returning my precious map. He had wished us the best of luck in our search for the black grasswren, a bird that had not been seen for 67 years. Four months later, towards the end of June, Richard Blythe was telling us he had seen my map before.

Now, fast-forward two and a half years to a letter addressed to me and dated 29 December 1970. The letter was from Brigadier Hugh R Officer, an amateur ornithologist who had written books on Australian honeyeaters and flycatchers, and who had a life-wish to see all eight Australian grasswren species in the wild. At the top of his 'unseen' list was still the black grasswren, an enigmatic little bird confined to the sandstone heart of the continent's wild northwest.

Quite remarkably, Hugh's letter contained the following:

> 'You are probably unaware that I and three mates planned a search for *Amytornis housei* in October 1967. Thanks to Macdonald, I had Brockman's map and we did not anticipate any great difficulty. But how wrong we were. Over-century temperatures, due to our start being delayed by one of our party scalding his leg, plus the very rugged nature of that vertical sandstone country was too much for our simple little party for, of course, we had no outside help.
>
> In addition Richard Blythe, of Mt House, had told me that there was a track from Beverley Springs to the Charnley [River] which when we arrived at Beverley Springs, Barry Doherty said he was quite unaware of.
>
> And so our effort failed. We would have done better if we had gone in from the direction you did.'

So Richard had not been wrong. He *had* seen my map before, and this letter proved as much. It also revealed that Hugh Officer had failed to find the bird, which would explain why we had heard nothing of his little expedition before we reached Mount House eight months after him.

The Brigadier and Mr Macdonald had obviously been in touch with each other during 1967, but which of them, I wondered, had made the first move? Thinking back through what had happened, my hunch was Macdonald. He had known about the map, Hugh Officer had not.

Back to 1966. As a junior member of Mr Macdonald's Bird Room staff, I began the research for a bird-collecting expedition to the Kimberleys in Western Australia and to Arnhem Land in Northern Territory. This work entailed finding out how well the different birds in these areas were represented in museum collections and I soon came across three isolated species of grasswren, small secretive birds whose restricted ranges we would encounter. Although we would try and find all three species, the most exciting was the black grasswren from the Kimberleys, which had not been seen since its discovery in 1901. I soon realised that there had been several serious attempts to track it down and that their publicised failures had magnified the bird's mystery all the more. I decided to do a bit of my own research into the little bird which had acquired something of a 'missing-presumed-extinct' label, just to see if it was worth flirting with failure ourselves.

The black grasswren, *Amytornis housei*, was named after the ornithologist Dr F. M. House, who discovered the species in 1901. He was attached to a surveying party working through the Kimberleys under the leadership of F. S. Brockman. Initialled and numbered campsites, marked on the expedition map, plotted their slow but steady progress across a desolate sandstone plateau. Dr House wrote that he first encountered the new bird near camp FB25. He does not say where he encountered it again, although the expedition's route followed the River Charnley west for the next 80 km (camps FB26 – 29+).

Having unearthed a copy of Brockman's 1902 report in the British Museum library, I traced from its map the section of the Kimberleys surrounding camp FB25. The camp was in an interesting position, close to Rocky Mountain, the source of three rivers: the Drysdale, flowing north, the Charnley, flowing west and the Isdell, flowing south. I found FB25 clearly marked as being south of Rocky Mountain on the upper reaches of the Isdell, an important fact considering that, for nearly 70 years, people had more or less assumed this camp to have been northwest of Rocky Mountain, somewhere along the Charnley. Birds, admittedly, have wings and can cover great distances but Australia's grasswrens are not ordinary birds. They spend their lives down on the ground, scuttling like mice along rocky ledges, darting in and out of clumps of grass and gliding on small, rounded wings when they come to open areas to be crossed. They might even be in the long transitional phase of losing the power of flight altogether, so a mistaken locality of a few kilometres could make all the difference.

What struck me, though, was that previous expeditions to rediscover the black grasswren had spent their time and energy trying to reach Rocky Mountain from the west or southwest before beginning their search along the Charnley. Here, their progress and their well-being had been severely hampered by the 'very rough and stony' route taken by Brockman's little party.

Realising that camp FB25, the most positive location for the black grasswren and therefore the best place to start looking for it, had actually been some 20 km southeast of Rocky Mountain was enough to convince Major Brian Booth, the expedition leader, that we should take up the challenge. He agreed that we should follow the Isdell north, giving a wide berth to the difficult and possibly misinformed route towards and along the Charnley.

Armed with this plan of attack and my copy of the Brockman map, fresh from Mr Macdonald's safe, three of us left Southampton on the Greek ship *Ellenis* on 8 March 1968. The Major and I were accompanied by Cliff Frith, a colleague from the Natural History Museum's Bird Room and already something of a bowerbird freak in those early days. In Perth we would be joined by Western Australian naturalist, guide and cook, Harry Butler, by Ralfe Whistler, who would fly from England in 24 hours rather than spend an additional month away from his family on the ship, and by Tony Hiller, an experienced taxidermist who would come over from Sydney to join us at our Helena Valley campsite.

We were about two months into our journey through the Kimberleys when we arrived at Mount House Homestead, a cattle ranch named after the doctor who discovered the black grasswren in 1901. Richard and Rachel Blythe welcomed us into their isolated home, plying us with the creature comforts we had already forgotten. I particularly remember running water, ironed shirts, cornflakes and red wine. And, of course, the conversation over my map, when I showed it to Richard to see if he could recommend a good route up the Isdell River towards Rocky Mountain. Apart from the nonsensical idea that he had seen the map before, Richard made the intriguing comment that another waterway, Manning Creek, rose closer to Rocky Mountain than did the Isdell and that the two could even be confused because the name 'upper Isdell' was once used to describe the flow of wet-season water that extends the Manning above its dry-season source. Backing an immediate hunch that FB25 might, in fact, have been on Manning Creek and not on the Isdell, we bid the Blythes farewell and set off to establish our own base camp along the edge of the Barnett escarpment.

We headed north from Mount Barnett Homestead for 6 km before following Manning Creek away from its junction with Barnett River. The track gave way to ground so rough that even the four-wheel drive Land Rovers, down to first gear in their lowest ratios, were beginning to struggle. It was time to continue the journey on foot.

We chose an idyllic spot close to Manning Creek for our base camp, pitching our tents in an open sandy patch overlooking a deep pool surrounded by river gums. Purple-crowned fairy-wrens, the first we had seen, busied themselves in the low scrub as we set up camp. Under the warm glow of the setting sun, the first stage of our quest to rediscover the black grasswren ended with a refreshing swim, a drink, and another of Harry's 'guess-the-contents' stews. Tomorrow, 1 July, was the big day.

At eight in the morning, dressed up in our hiking gear and weighed down with provisions, Brian, Harry and I set out along Manning Creek towards the site of camp FB25. Ralfe, Tony and Cliff stayed behind to explore the local birdlife in our absence. Once gone, we had no way of communicating with them and the simple plan was that we would be back in four days. They would come and look for us on day five if we had not returned by midday.

Our course lay due north for the first 20 km. Following a direct compass bearing, we occasionally parted company from the meandering creek. At times we walked alongside it, admiring white egrets as they flapped lazily upstream or freshwater crocodiles that basked on islands of rock and heaved themselves into the water when we came too close. And then, we would be 30 metres above the crocodiles, where the voice of the sandstone shrike-thrush echoed jubilantly along corridors of red-and-black rock. Sheltered walls held silhouettes of hands or stylised paintings of animals and people. We were not alone. Of all the days, it was the most magical, the three of us held by the past as we pushed towards an unknown goal. We came across a spectacular waterfall which Harry said he would register as 'Halls Falls' after our expedition sponsor. Today it is Manning Falls, a popular tourist spot. By sundown, with 16 km behind us, we made camp at the water's edge. We swam and, with barely a thought for food, settled into a deep sleep round a small fire protecting us from mosquitoes.

We were up with the sun, bitten but refreshed, fortifying ourselves with tea, biscuits and cheese. Before breaking camp, we decided to make a quick 200 metre excursion to the top of the nearest sandstone outcrop. From here, we hoped to locate a northerly trig point, Joint Hill, and determine the course of the Manning ahead of us. Joint Hill was unidentifiable among the sea of trees and outcrops that stretched away to the horizon, but the creek – its flow revealed by the lush green of waterside vegetation – appeared to be bending to the northwest, towards Rocky Mountain. This was good news. We would be able to stay with it until midday. And then we would fill our waterbottles and begin our search for the bird that had evaded detection for so long.

As we turned to make our way back to the little creek-side camp, our plan was obliterated at a stroke. Harry yelled. He'd seen something, low down on the rocks. Brian and I froze and stared. A movement caught our eyes, and then, right in front of us, we saw them: black grasswrens, as excited by the sight of us, it seemed, as we were by them. There were probably five birds, but so active it was impossible to tell. They ran, heads and tails down, disappearing this way and that into small crevices and scuttling along ledges sheltered by clumps of needle-sharp spinifex grass, their feathers matching the black-and-brown sandstone so perfectly that they would vanish and reappear in an instant. Their flurry of activity was accompanied by grating, ticking alarm calls, impossible to locate, which certainly confused us but which no doubt kept them well in touch with each other. It was exhilarating, breathtaking and emotional, all rolled into one. But we also had a job to do. Within an hour we had come across another small group. Using guns that fired a fine spray of dust shot, we collected four specimens, a necessary proof-of-encounter for the scientific world.

Now that we had found our bird, it made sense to abandon any thought of reaching the presumed site of FB25. Instead, we would stay at our base on the creek, prepare our specimens – which meant skinning, preserving and stuffing them as study skins – and write up our observations while they were fresh in our minds. We would then have all the next day to observe the birds where we knew we would be able to find them.

We were in such high spirits that the first wave of mosquitoes took us completely by surprise. It was time to build another fire, a proper one with logs, while Harry applied his bushcraft to the creek and came back with large freshwater prawns threaded on a stick. They were delicious. Washed down with cups of sweet black tea and with the black grasswren literally in the bag, we settled into the perfect night's sleep round a blazing fire.

Again, we were up with the sun and setting off for the little outcrop to the east of the creek. At eight o'clock, Brian and I were sitting quietly among jumbled sandstone, half-hidden by an overhanging branch, when four black grasswrens suddenly appeared in front of us. This time, they did not seem to know we were there and their behaviour was much less frantic than before. One of them was a very young bird, accompanied by two adult females and an adult male. It was so young, in fact, that it was still begging to be fed, opening its yellow-gaped mouth, puffing its feathers and thrusting itself at the male who jumped on it with such force that both disappeared down a crack in the rocks. And came straight up again. All four birds chased each other, running in and out, up and down, never still, adult tails cocked or flattened against the rock, the male's shoulders gleaming chestnut in the morning sun. They were silent, too, uttering not a single sound until we made our move.

When we stood up, instantly triggering the alarm calls we had heard the day before, it was because we had decided to collect the youngster. Just a week or so out of its nest on 3 July, the egg it hatched from would have been laid in late May or early June, a clear sign that breeding could extend from the wet season into the beginning of the dry, when insect and plant food was still plentiful.

With seven birds collected, as well as the nest and eggs of the white-quilled rock-pigeon, we trekked south the following day, well pleased with our little trip. Ralfe, Tony and Cliff were waiting anxiously.

'Did you...?' one of them called as we came out of the scrub and into the clearing, 'Did you... find it?' 'Yes!' we said triumphantly, arms raised to the skies. 'Thank goodness for that!' said Cliff, because they, too, had found it, just 500 metres away, on yet another little outcrop of black-and-red sandstone. Unaware of our success, and with us not even considering theirs, they had taken four specimens, so now we had eleven – plenty for the world to see and, with luck, precluding the need for any more to be collected.

Our 1968 Harold Hall Expedition completed the set of five major bird-collecting expeditions carried out in Australia between 1962 and 1968 by London's Natural History Museum staff and invited scientists. It was a huge undertaking made possible by the generosity of Major Harold Hall, whose father had built a family fortune during the Australian gold rush years in the 19th century. There was opposition from scientists within Australia to our collecting in their country but our cause was championed at the highest level by the Australian ornithologist Dom Serventy.

J. D. Macdonald had also played an important role and it is now clear that he saw the success of the expeditions as critical to his own planned retirement to Queensland, where he would continue to work on birds. For him to have personally provided an Australian expedition with a map revealing a new location for camp FB25, perhaps leading to the rediscovery of a bird that vied with the night parrot as Australia's most enigmatic, would have been a big and colourful feather in his ornithological cap. Had he not been planning his own move to Australia – he was, in fact, already there by the time we returned to England for Christmas 1968 – I like to believe he would never have passed a copy of my map to Brigadier Officer. It is even possible that he contacted the Brigadier *because* he had the map, thereby initiating the failed 1967 expedition. I was in Brisbane in 1989 and I called round on the off chance of finding Mr Macdonald at home in Gleneagle Street. He was out shopping, a neighbour told me, and I was unable to return another day, so the conversation we might have had never did take place. It's too late now.

When my account of the rediscovery of the black grasswren was published in the Australian scientific bird journal *Emu* in September 1970, I wrote to Hugh Officer, enclosing a copy of the account . The only reason I did this was because my former London museum colleague and good friend, Shane Parker, who was by then working on birds in Australia, had told me about Hugh Officer's life-wish to see all Australia's grasswrens. Shane, without knowing anything about my map, mentioned this to me in one of our frequent letters and I contacted Hugh Officer purely out of goodwill, hoping that the details of our 1968 success might be helpful if he ever went to the Kimberleys himself. His reply, in December 1970, came as a shock. As well as explaining Richard Blythe's reaction to seeing my map, it added extra meaning to the delight shown by Mr Macdonald when I had handed him my copy of the map for safekeeping.

In October 1967, Hugh Officer and his three companions had set out for the Kimberleys to find the black grasswren. The map on which their hopes of success rested was a photocopy of the one I had traced early in 1967. It must have come from Mr Macdonald, who evidently wanted to be the instigator of their success rather than of ours. Hugh Officer's failure was down to any number of factors, the least of which may have been not reaching camp FB25. Our own experience showed that it wasn't necessary to get that far to find the black grasswren, which is locally distributed across appropriate sandstone country. I doubt anyone knows the full extent of its range even today.

The tone of Hugh Officer's December 1970 letter to me was so friendly that I warmed to him at once. Having failed with the black grasswren, he wrote that he would try to find one last grasswren before calling it a day. Sometime in

1971, he confided, he would mount an attempt on another grasswren isolated in northern Australia, the Carpentarian grasswren, *Amytornis dorotheae*. In 1968, we had had our own plans to search for this bird four months after our July success with the black grasswren in the Kimberleys. But before that, early in October, we had our second grasswren to look for (one the Brigadier *had* seen), the white-throated *Amytornis woodwardi*.

The white-throated grasswren (*Amytornis woodwardi*)

With the August departures of Harry and Ralfe behind us, Brian, Cliff, Tony and I set out from the Blue Dolphin Motel near Darwin in two overloaded Land Rovers. We followed a dirt road east through what is now Kakadu National Park, heading for Oenpelli (today known as Gunbalanya), an Aboriginal mission tucked away on the northwestern edge of the Arnhem Land plateau. Crossing the crumbling concrete causeway of the East Alligator River – Cahill's Crossing – and barely noticing the warning not to swim with man-eating crocodiles, we entered another magical world, a world that for 50,000 years has belonged to the people who came down from the north in their simple wooden boats. Two or three kilometres beyond the river, we set up camp in the shade of gum trees overlooking an enormous floodplain. The plain, mostly dry at the end of September, was surrounded by sheer escarpments of red sandstone that rose 40 metres above the eucalypts and glowed with the early evening sun. A distant honking heralded the first wave of magpie geese coming in to roost on the open lagoon. The stars were brighter and more noticeable than ever.

Brian decided that he and I should walk up the East Alligator River where it cut through the thick sandstone mass 3 km from camp. It looked ideal grasswren country, and to increase our chances of success we reckoned on being away for two nights. We had a map that showed the river's dry season pools we could rely on before our water bottles ran dry.

The white-throated grasswren, while restricted in its distribution, was not quite another 'missing' bird. Discovered in 1903 by an expedition working along the South Alligator River, about 100 km southwest of where we were camped, it had been found again in 1948 by Herbert Deignan, the ornithologist on a combined Australian–American scientific expedition to Arnhem Land. Of importance to us when we turned up 20 years later was that Deignan had found the grasswren on the sandstone outcrops close to Oenpelli Mission, only a 12 km drive from our base camp. By walking deep into the plateau in the opposite direction, Brian hoped we might extend the bird's range even further.

Early on the morning of 2 October, Tony drove Brian and me across the East Alligator floodplain to the entrance of the river gorge. From here, we set out on foot. The first kilometre took us along the edge of the dry river bed, a gentle walk through green vegetation drawing its water from deep below ground. With temperatures that would soar beyond 40°C by mid-afternoon, we followed the shade, questioning slightly the absence of standing water but confident we would soon come to the first of several pools marked on the map. The reliance we put on the accuracy of this rather aged map meant we were drinking a little too freely. Within two hours we had emptied three of our four water bottles and it was clear that something was wrong. This might have been the moment to turn round and head back down the gorge to Tony and Cliff.

We decided against doing this because we did not have enough water for the return journey in the heat of the day, and when we did reach the entrance to the gorge there would be no Land Rover to drive us back across the open plain to camp. So we agreed, instead, to keep going, optimistic that we would find a pool before long. If we failed, we would make camp, turn back at first light and walk in the relative cool of morning. Neither of us questioned our ability to survive.

The gorge became narrower and steeper. There were no pools and no hint of standing water anywhere. And then we were climbing, at first over broken boulders but then up through narrow ravines, forcing our way through bushes swarming with angry green tree-ants as we disrupted their delicate leafy nests. By the time we reached the top, drenched in sweat, scratched, bleeding and bitten, it had taken five hours to cover the last kilometre. We drained our last water bottle and took stock. Just ahead, but now at least 60 metres below us, was a large plain studded with trees. If we could get down to that before nightfall, we would, surely, find a supply of water.

And then, a black cloud bubbled up from behind the trees to darken the sky. The northern wet season had chosen the perfect moment to break on our parched little world. Not wishing to be exposed to lightning, we squeezed in beneath an overhanging shelf and waited. As my eyes adjusted to the shadows, the silhouettes of hands emerged on the ceiling just above my face. Someone had been here before – who knows how many thousands of years before? – also lying on their back but with black and red ochre paints to pass the time. As if to embrace another moment from the past, a rare white-lined honeyeater whistled its clear, tuneless notes. The first flash of lightning and the first crack of thunder silenced the bird and heralded a tropical downpour of immense power. Our billycan was overflowing in less than a minute, a vital gift from a storm that moved on as quickly as it had arrived. We now had a little over two hours to get to the bottom of the gorge. Scanning the lightly treed plain through binoculars, we could just make out a herd of water buffaloes to our left. We should head their way. With luck, they would be gathered round fresh water.

The rocks were steaming and slippery and another storm was brewing. Lowering himself into a narrow and thickly vegetated gully, Brian suddenly stopped and raised his right hand, his eyes fixed in front of him. It was not a grasswren but a white-lined honeyeater, and it had just flown up from its deeply cupped nest, a structure unknown to science. We sat quietly and watched. The first large drops of rain had already fallen when one of the adult birds slipped down through the branches and onto its gently swaying nest. Certain of its identification, we collected the nest, with its single youngster, and carried on down the boulder-strewn escarpment. Fortunately, the raindrops did not develop and our final hour-long scramble to level ground was not hampered by another violent downpour.

Walking easily through a grove of tall eucalypts, we found the buffaloes wallowing in a small mud-filled swamp. As they snorted their disapproval and squelched reluctantly out of our way, it was clear we were going to have to make do with the best water we could find right here. Daylight was fading fast and we had had enough. With 30 displeased buffaloes moved to a safe distance, Brian gathered wood while I took the billy into the mud.

There was no standing water to be scooped up. Instead, I was pushing the billycan into the hoof- and body-prints left by the buffaloes and waiting for it to fill with a dark orange liquid. This we boiled furiously over the fire, reducing it by half to convince ourselves we had killed any gremlins that might be lurking within. A sprinkling of tea leaves made the colour more inviting, and the addition of two heaped spoonfuls of sugar and a pinch of concentrated orange powder gave us a refreshingly hot, bitter-sweet drink. Collecting more water under the light of a blazing branch stuck into the mud, we boiled again, opened our army ration packs and conjured a piping-hot stew.

I remembered the stew four months later, back in England, when I went to our family doctor in Hampshire complaining that I had several hard lumps, each a centimetre or so across, on the calf muscle of my right leg. He raised an eyebrow when I told him they were moving around and immediately telephoned London's Hospital for Tropical Diseases which, in turn, alerted our local General Hospital. The date was set. Two scalpel-wielding experts would come down and investigate. But imagine their disappointment when, on the morning of the biopsy, I telephoned the hospital to say that the lumps had vanished overnight. Their train journey to Alton wasted, the experts advised me to contact them if anything reappeared, adding casually that either my blood had been too rich and the parasites had died and been absorbed, or they had gone to ground, probably in my liver, and would wait until I returned to the tropics before completing whatever kind of cycle they were on. I went back to the tropics three years later and was relieved to find that nothing happened. Brian had no similar body invasion to report.

The stew eaten, we spent the rest of that mosquito-ridden night under the glowing eyes of the displaced buffaloes, before packing up and allowing them to return to their wallow. They were as eager to be back as we were to leave. As they sniffed and plodded through the lingering smells that had invaded their space, we retraced our own steps, soon passing the deep gully we had climbed down the previous afternoon. Two hundred metres further on, we found the real East Alligator River, whose crystal-clear pools now beckoned us back towards camp rather than deeper into the unknown. We set off, twice breaking out from the gorge to search for *Amytornis* on nearby sandstone outcrops. We slept during the heat of the day and in the late afternoon walked the final 6 km to camp in one enjoyable go.

The gorge we had followed the day before was not marked on our map and, throughout the whole of that punishing afternoon, we had been just a few hundred metres from a never-ending supply of drinking water. It was our mistake, of course, though we would not, in hindsight, have swapped a single drop of that precious liquid for the first-ever nest of the white-lined honeyeater.

Despite this unintended success, we had still not found the white-throated grasswren. We decided to work the area where Herbert Deignan had succeeded in 1948, so we set out for a long day on the sandstone ranges between the East Alligator River and Oenpelli Mission. Tony and Cliff worked one side of the track from one Land Rover, Brian and I the other. After a ten-hour day which again peaked at around 40°C, we had our bird, officially flushed from the rocks for the first time in 20 years. As with other grasswrens, encounters like these seem to require luck as well as judgement. The birds are locally abundant but restricted to what appear to be vast areas of suitable habitat. Because they keep such a low profile among the spinifex-clad rocks, they are very easy to miss. Nowadays, though, with these edge-of-Arnhem-Land outcrops open to tourist and mining operations, the white-throated grasswren is on the list of birds that even a casual visitor might hope to see, just a short walk from the road.

The Carpentarian grasswren (*Amytornis dorotheae*)

Two grasswrens out of two meant we were optimistic about completing the set of three. A few weeks later, we drove out along the McArthur River towards the Gulf of Carpentaria and made camp at Caranbirini waterhole, 30 km from Borroloola. We were now on the southern edge of the sandstone escarpment where the Carpentarian grasswren, *Amytornis dorotheae*, had been collected in 1914 by H. G. Barnard a year after it had been recorded for the very first time. The little bird had been neither seen nor searched for again – hence its low press rating compared to the black grasswren – when, 54 years later, our little expedition drove up and pitched its tents in the heat and humidity of early November. Storms were now breaking regularly. Our days in the north were numbered.

During what was meant to be our final camp, we decided that nobody should be on the sandstone range with a loaded gun after 9.30 in the morning. In the continuing build-up to the wet season, with its sporadic, violent storms,

21

humidity was approaching saturation and the daily maximum temperature was 45°C. In such conditions, judgement can be impaired, inviting unnecessary risk and danger.

One morning, after a 4.30 start, Brian and I had walked a long way from the Land Rover in our quest to find our third grasswren and were only halfway back by 11 o'clock. I definitely remember feeling a bit woozy, seeing Brian ahead of me picking his way down into a dry river bed, and then standing at its edge myself. Without realising, I had walked out onto an overhanging ledge, which suddenly collapsed under my weight. Down I went with my 12-bore shotgun, which I somehow threw forward so that it was sliding with its barrel pointing straight up at me. It was loaded and, not having a safety catch, its firing pin was resting against the cartridge. I sensed what was coming and twisted to get out of the line of fire before it hit the rocks at the bottom of the creek. With a deafening roar, the gun discharged the contents of its single barrel in a tight cluster that smashed into the rock next to me and peppered me with shot. Brian was hit by the force of the ricochet that carried pellets 30 metres across the creek. Lying there among the rocks while he stumbled back to me, the proverbial smoking gun by my side, I felt as though I had been hit in the back by a giant sledgehammer studded with red hot needles. Engaging his military know-how, Brian stayed in front of me, calming me down with words and eye contact, ignoring his own wounds before moving round to inspect the damage to my back. With relief, he announced that, while there was a lot of blood, my injuries appeared to be no more than superficial.

The following day, Tony drove Brian 600 km to the hospital in Katherine to have a piece of lead shot removed from his right eye. While they were away, Cliff worked on me with scalpel, tweezers and a bottle of Mercurochrome to stave off infection. At night we were visited by dingoes. Increasingly bold in their search for food, one broke into the kitchen tent while others dipped into our rubbish pit and removed whatever scraps they could find. The smoking gun soon sent them on their way.

A few days later, with the wet season at bursting point and our return to Perth already delayed, we set out across country once more. We made an unscheduled camp just south of the junction of Clyde Creek and the McArthur River, close to where the Carpentarian grasswren had been collected in 1914. Despite the storms that threatened to isolate us from the main road, we were going to give the bird one final try. We could spare two days.

Dark clouds rumbled as Cliff and I set out along the mostly dry McArthur in blistering heat and humidity. We passed the empty corridor of the Clyde and headed towards distant outcrops that held great grasswren promise. It was slow going. As the evening storm beat down, we tucked into the top of the river bank and tried to sleep. It rained like it had never rained before, crashing down with a violence of thunder and lightning that roared and brightened the heavy, starless night.

We crawled out of our shelter at first light, cramped and soaked to the skin. By now immune to the pain of discomfort, we were determined to make the most of what would be the last of 232 days in the bush. One hundred metres from the river, the sandstone country proper rose up through the eucalypts. As we set foot on its loose, roughly-piled rocks, we heard grasswren-like calls coming from deep within the rocks. We waited. The callers would not show themselves. We waited even longer, desperate for the sounds to be embodied as birds. But we decided, eventually, that as our mystery voices were going nowhere and we had no time to spare, we had been deceived, probably by bats. We left them twittering in their hidden, sandstone world.

For the next four hours we worked half a dozen outcrops along the McArthur without success. Enough was enough. Our quest for grasswren number three, *Amytornis dorotheae*, known affectionately and conveniently as 'Miss Dorothy', was over. We accepted defeat and force-marched the handful of kilometres downriver to join the others by nightfall.

Eight days later, after a 5,500 km drive away from the northern rains, Brian, Cliff and I were back in the sunshine and cotton-wool snow of Perth, a city setting its sights on what it considered another white Christmas. Tony, who had left us at Port Augusta on a Sydney-bound train, was heading towards an uncertain future, both in professional and personal terms. I was sorry to see him go. It would be 21 years – the year of that missed opportunity to see Mr Macdonald – before we would meet again. He was by then happily re-married, to Katie, just as bearded, and running his own butterfly-breeding farm on Mount Glorious, not far from Brisbane.

I passed the details of our final grasswren experience to Hugh Officer when I replied to his letter of December 1970. His delightful reply, in September 1971, contained the following lines:

> 'In your letter of 20 April 1971 you put a p.s.: "You <u>must</u> see the Dorothy Grass-wren!" Well, we are just back from the McArthur River and we did just as you ordered for we succeeded in seeing Dorothy and what a thrill it was. We still feel elated.'

Hugh Officer and two friends had combined their own and our experiences with information supplied by Harry Morris, a local mining manager to whom we had also spoken in 1968. Making their own decisions, and with limited time, they walked up Clyde Creek away from its junction with the McArthur River – the main watercourse Cliff and I had followed to the squeaking bats – to search the sandstone outcrops there. They had, in effect, gone back to the place where the birds were first discovered. Two further lines from his letter said it all:

> '*Setting off at 0630 we climbed up a sandstone feature across the creek from our camp and at 0720 we put up a party of wrens. Quick work and pure luck!*'

Maybe, but it was the sort of luck that he and his companions thoroughly deserved. They were committed, they were adventurous and they went back to the first known locality, a basic ploy that had stood us in such good stead in the Kimberleys but which, for some reason, we had overlooked here. They succeeded with the Carpentarian grasswren where we, with our advantages of youth and recent encounters with two other grasswrens, had failed.

With so little precise information to guide either of us in our respective searches all those years ago, we don't really know why Hugh Officer had not taken the same 1971 Arnhem Land 'luck' with him to the Kimberleys in October 1967. He may have walked a different route from us (despite having the same map) at a different time of year, but it is still possible that he was right on top of the birds he so desperately sought without actually seeing them. All we do know is that he failed by an unknown margin to rediscover *Amytornis housei*, in the same way, perhaps, that we failed to rediscover *Amytornis dorotheae*.

But the details are not important. It is enough to consider that, between us, we had tracked down three of the world's least-known birds, three Australian grasswrens that had scuttled, undetected and unconcerned, among the sandstone rocks of the continent's rugged north for such a very long time.

3

REEF ENCOUNTER

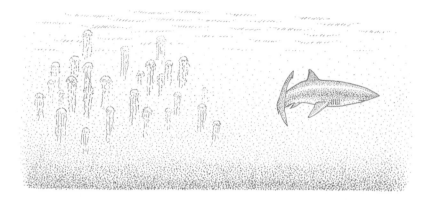

Sharks were the problem and Tony Hiller's instruction was as clear as the Timor Sea that sparkled in front of us: I was to shoot first and if that didn't work, it would have to be the spear. I wasn't sure how effective either tactic would be but, in a life or death situation, I would obviously do my best. Tony was going to dive on the deep side of the reef when the tide was out, and despite my reservations, I could not let him go on his own.

From our little camp under the trees it was a walk of about 200 metres to the exposed low-tide reef. The first landmark was a row of metal spikes planted during the Second World War to repel Japanese landing craft. Now, in 1968 and with the Japanese invasion of northern Australia more about trade and tourism than military possession, these spikes were corroded with rust and horribly out of place.

Beyond the spikes, a steep slope marked the limit of normal high tides, and beyond that ridges of sand ran parallel to the coast. They were pleasant to walk: cool, moist and firm underfoot, the crest of each exposing us to a refreshing onshore breeze.

We crossed three of these peak-and-trough ridges before coming to the reef. It was a long, flat-topped wall of coral with shallow pools containing sea creatures able to withstand the harsh tropical sun at low tide. We knew enough to be careful of stonefish, beautifully camouflaged and with a spike long and sharp enough to penetrate a shoe and poison the foot inside. Beyond the reef, the water was dark and menacing, a gently swelling void. Tony fitted his snorkel, adjusted the knife on his belt and waited while I loaded the 12-bore shotgun. As I settled the bags and positioned the spear by my feet, he was gone, shimmering down the steep wall to search for life on the ocean side of the reef. He had entered another world, leaving me all alone in mine, with its comforting sunshine, mewing gulls and gently lapping waves. I scanned the water for threatening shapes but none came. From the safety of the reef-top it was hard to imagine them even existing. At regular intervals Tony surfaced, filled his lungs with fresh air and turned to continue his journey along the hidden coral wall.

I didn't realise the tide had turned until the first small wave spilled onto the reef, forcing me to move our bags. I glanced back towards the shore and saw that the trough of the ridge nearest to us was filling with water. The advancing sea must have broken through the coral barrier some distance away and was now racing between the sand ridges on the landward side of the reef. My first thought was of being cut off from the shore, my second that here was a perfect way in for sharks. We were in the 21°C water belt where sharks feel so comfortable, where their attacks on people, statistically at least, are most likely to happen. Tony surfaced to find me clutching the bags, gun and spear, anxious to get out while the going was good. 'Just one more dive,' he said cheerily, 'and then we'll be off.'

The thought of wading into the sea where sharks could be running with the tide was not at all comforting. Two dives later Tony, whose arm was now bleeding from a scrape with the razor-sharp coral, grabbed his bag and the spear and, with a 'Well, there's no point hanging around here,' jumped off the reef and into the rapidly filling trough. I set off in pursuit, staying close but overlooking the blood flowing from his arm into the sea, not thinking that this is exactly how sharks can be attracted to their prey.

'Just keep the gun dry and be prepared to use it if you see anything coming towards us' was Tony's slightly gung-ho advice as we set off into the shallows. Thirty metres later we were walking across the summit of the first ridge. The water rose above our waists as we plunged into the second trough. I was feeling decidedly nervous, having no control over what might be happening below the surface. It was easiest to walk with my arms held aloft, but I couldn't do that. I felt too vulnerable. Abandoning any thoughts of firing the gun, I removed its cartridge and plunged it into the water, holding it in front of me like a metal bar for protection. I felt better for a step or two, but then the light dancing on the surface of the sea began to play tricks with my mind. Shadows and shapes were coming, going, circling. I began to feel sick and weak, a million butterflies playing havoc with my stomach. Another sandy peak loomed. It was a relief to wade up it, to stand mostly clear of the water. But I had to tear myself from this delusion of safety. The sea was still rising. The mind-sharks were still coming.

Halfway; and now wading through deep water, feeling the pull of the tide, not knowing what it might be bringing our way. Thirty metres is an eternity. Of the two of us, Tony was the more resilient, less bothered, less fearful of nature's surprises. He pushed through the water ahead of me, still leaking the blood that could be our downfall. But then we were climbing once more.

The top of this third ridge was covered by almost a metre of water and my feet somehow scrambled across it to the far, landward side. We stopped for just a moment, saw the dry beach 40 metres away, took deep breaths and waded in. The water was above head height. We had to swim. I held the gun out in front and paddled hard with my legs, fixing my eyes on a tree above the sand, blotting out thoughts of the sea and its ghoulish residents, pushing through the relentless tide and the heavy water that seemed reluctant to release us from its hold.

We reached the high-tide shelf. Our feet dragged and slipped in the sand that rose to meet us. And then, with one final effort we were free, scrambling past the now welcome spikes of war, onto the hot dry sand in front of the trees and our little camp. We were safe. Ten minutes that had felt like a lifetime. Looking back from the comfort of the beach, the top of the reef was just visible. White horses cavorted along its leading edge and the water between us was once again smooth and inviting.

Even on dry land we had our tormenters, but our campsite enemy was not imagined. It was the sandfly, a small two-winged creature with the sharpest blood-sucking bite. For three successive nights they had plagued us in their thousands, small and persistent enough to penetrate mosquito netting and enter our sleeping bags, piercing our skin from head to toe. We deserved a night of comfort before heading east to the Arnhem Land plateau and the floodplain of the East Alligator River.

Two travel-worn Land Rovers hauled themselves away from the beach towards Nightcliff's Blue Dolphin Motel. Showered, refreshed and clutching cold beers on the verandah after a wonderful seafood meal, we talked with the manager about camp life and the five months we had already spent roughing it in northern Australia. At one point he asked, quite casually, if we knew there was a red alert, that swimming in the sea around Darwin was banned because of a coastal invasion of sea wasps, box jellyfish whose long trailing tentacles are so poisonous they can kill anything they touch.

'And,' he added, as if to confirm their potency, 'When the sea wasps come to town, even the sharks stay away.'

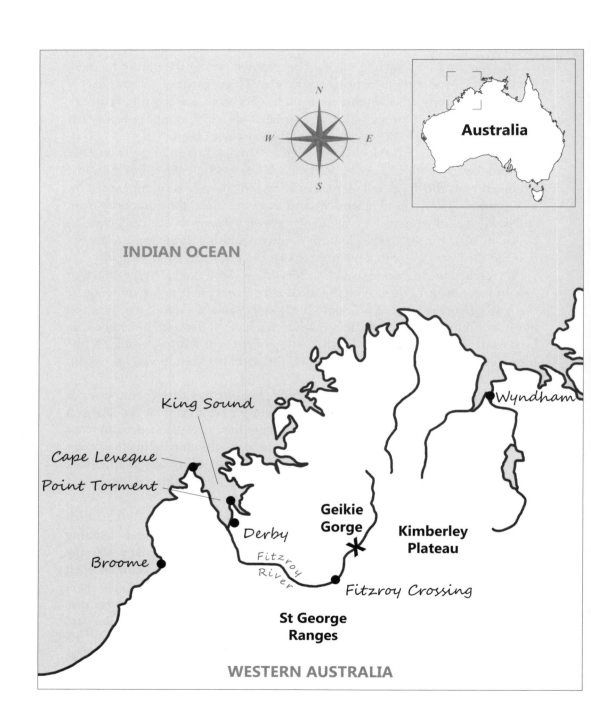

4
POINT TORMENT

'Seriously, mate, there's crabs in there the size of dinner plates,' the mechanic drawled over a mug of hot tea. He had just driven 15 km from Derby in Western Australia with news on the spare parts for the Land Rovers and was taking a lively interest in our plans. 'Mind you,' he continued, 'they're bloody good to eat, but the trick is to get them before they get you. They're that bad, I've heard some people even use 'em as guard dogs. If you're really goin' up there, you'd better warn yer mates.'

I had been all on my own in camp, holding the fort while the others were busy elsewhere. Cliff Frith, Ralfe Whistler and Tony Hiller were out collecting birds locally while Brian Booth and Harry Butler had gone to find the best way into Point Torment, a peninsula jutting out into King Sound, a sheltered bay on Australia's northwest coast.

Point *Torment*? None of us had given the name a second thought but, as we were to discover, it definitely held something to think about. And it had nothing to do with crabs the size of dinner plates. When they returned in the evening, Brian and Harry told how they had been bitten to pieces by millions of mosquitoes. It was a shock, even for those two seasoned campaigners, and a clear warning to the rest of us. Point Torment was the first location for our survey of mangrove birds around the Kimberleys in Western Australia and Arnhem Land in Northern Territory.

It was near Derby that the 17th-century seafarer William Dampier, one of the first English people to visit Australia, had observed from his ship that the locals lived in mud huts. All he had really seen were curious Aborigines peering at him from around the giant termite mounds that are such a feature of northern Australia's wooded grassland.

After Dampier, it was another 140 years before HMS *Beagle*, fresh from returning Charles Darwin to England after his defining wildlife journey around the world, arrived to carry out a survey of the Kimberley coast. From the ship forever associated with Darwin's 1859 book on natural selection and the origin of species, Lieutenant John Lort Stokes ventured into these mangroves. He suffered so badly from mosquitoes and sandflies that he immediately named the peninsula Point Torment. And now it was our turn, maybe just the third British expedition between the nicely juxtaposed dates of 1698 and 1968 to enter this outpost of small obnoxious flies.

The birds of Australia's mangroves are fascinating, not just because there are a lot of them but because many of them are found nowhere else on the continent. For a while, scientists assumed that the ancestors of these birds must have invaded Australia from the tropical north and spread round the lushly-vegetated coast, giving rise to new species. A more recent thought is that primitive birds were already 'on board' the island when it split from the southern super-continent of Gondwanaland as much as 160 million years ago and that they retreated to the lush coastal regions as the centre of the continent dried out. This all happened over such a long period, and there have been so many climatic changes within Australia as it drifted towards its present position, that the precise way its birds evolved may never be fully understood. Our contribution would be to collect a series of specimens from the mangroves around the northern coast and then, back in the Natural History Museum in London, find any differences between them that might show how and where their evolution into new species was taking place. The mangrove robin, mangrove golden whistler, white-breasted whistler and lemon-breasted flycatcher were the sort of birds we would be looking for in particular.

The distance round the adjoining coasts of the Kimberleys in Western Australia and Northern Territory is roughly 3,000 km. During our eight-month journey from Derby on the Indian Ocean to Borroloola on the Gulf of Carpentaria, near the Queensland border, we planned to sample the mangrove birds at five or six different places. Point Torment was to be our first encounter with a habitat that is as hostile as it is magical. The man from the garage had lived in Derby all his life and yet even he knew the place only by reputation. He finished a third mug of tea, climbed into his wagon and wished us the best of luck.

Mangrove trees thrive in salt water, their roots and trunks flooded and exposed by the rise and fall of twice-daily tides. They dig deep into the margins of the land, stabilising fertile silt washed down to the coast by inland rivers. The shallow, warm, rich and protected waters of mangrove swamps act as nurseries for a great many species of fish, some of which are commercially important. To destroy mangroves, as has happened around the tropical world, is to court disaster. We may not be able to live in them but this does not mean we can live without them. Once mangroves are removed, for firewood, furniture or development, offshore fisheries can collapse, or the force of cyclonic tidal waves and winds, no longer absorbed by a thick stand of vegetation, is felt further inland where people have set up their homes behind what was once a protective wall of trees. Left to grow in peace, mangroves can form life-saving barriers. The following article, which relates to Chapter 9, *Mangroves and Man-Eaters*, appeared in the *Bristol Evening Post* newspaper at the end of May 1985:

Trees save TV crew in cyclone

By Martin Powell

A WILDLIFE film crew from Bristol survived the cyclone and tidal wave that killed thousands in Bangladesh.

BBC television producer Dan Freeman flew out of Bangladesh as the cyclone moved in last Friday.

He left a cameraman and a sound-recordist in a boat in the Ganges delta.

By the time he arrived back in Britain the tidal wave had struck and he feared for the lives of the two.

But they were in a mangrove forest and missed the worst of the storm.

Cameraman Alan McGregor and sound-recordist Tessa Woodthorpe-Browne were today safely in a hotel in the Bangladesh capital of Dhaka.

Mr Freeman said today: "It's been a very anxious couple of days trying to find out if they were safe.

"I think they were so far into the Sundarbans mangrove forest that it absorbed the tidal wave.

"Ironically, I left them with instructions to film a storm. They didn't film that particular one because it happened at night.

They were in a boat 30 miles inland in the Ganges delta and were so far away from habitation that they were safe."

It was the fourth trip to Bangladesh by the freelance crew, filming for the BBC Natural World programme, which is produced by the Natural History Unit in Whiteladies Road, Bristol.

The two members still in Bangladesh will fly home within the next week.

31

Logging destroys coastal mangroves, exposing inland areas to cyclones, Sundarbans, Bangladesh (Author's collection)

Mangroves and mosquitoes go hand in hand. My first serious encounter with mosquitoes had been five months prior to leaving England for Australia in March 1968. It was in the Camargue in the south of France where the Rhône forms its own delta before emptying into the Mediterranean. Walking around the Tour du Valat Biological Research Station with John Walmsley, one of its researchers, we were set upon by clouds of whining mosquitoes. John, who had worked in the Camargue for a number of years, hardly noticed them, but I suffered dreadfully. Each bite on my face, hands and legs – it was still warm enough in October to be wearing shorts – reacted viciously and it was impossible not to scratch, making them swell, bleed and hurt all the more. So when Brian and Harry came back from their Point Torment recce and warned us to 'Be prepared!', that is exactly what I knew I would be.

There was, however, to be a short respite, a kind of lull before the storm, even though it was an actual storm that brought it about. We were still, at the end of May, not clear of the northern wet season and torrential downpours had left Brian and Harry's hard-won route to the mangroves impassable. We would need a week of dry weather before we could return.

While we waited, we decided to go inland, to familiarise ourselves with the birds around Fitzroy Crossing and Geikie Gorge. We would keep a weather eye on the coast. After a few days around the wonderfully orange-and-white layered walls of the gorge, Brian decided we had time to drive south into semi-desert and set mist nets over isolated waterholes near the St George Ranges. His faint hope was that we would be visited in the dead of night by the incredibly rare night parrot, if indeed the bird still existed. Ralfe and I staked out our nets and huddled under a cloudless sky. It was bitterly cold and we had nothing to report beyond some loud midnight twitterings that passed rapidly overhead. Brian and Tony fared no better. When it was light enough to dismantle our nets, we returned to Harry and Cliff, empty-handed and frozen to the core. They were not surprised by our lack of success. The elusive and widely-distributed parrot has since been found, in fact, a few thousand kilometres away in southeast Australia. I remember a 1970s newspaper report that told how pleased Australian ornithologists were to hear that the remains of a night parrot had been found in the radiator grille of an outback lorry, because, until then, they had thought the bird extinct.

It had stopped raining on the coast so we decided to return to Derby after one more inland camp. I was out with Harry the next morning when we realised we were being watched from the trees by a very old Aborigine man. We had a permit to shoot a kangaroo each day for food and, when I returned to the Land Rover with one slung over my shoulder, the old man shuffled forward, almost barring my way into the back of the vehicle. I could see his interest in the kangaroo and waited for his next move. Harry approached and the old man looked at us in turn, and then back to the dead kangaroo. He fixed us with eyes that seemed to convey some deep inner need. Slowly, he mumbled, 'Day after, same place, more same meat.' And then he trudged back to the trees.

Harry realised that the old man wanted us to bring another dead kangaroo to the same place, at the same time, the next day. We duly obliged, though slightly wondering if we would ever see him again. But then out he came, stopping by the kangaroo that lay in the grass next to the Land Rover. Harry gestured with his head and eyebrows and the old man kneeled, pulled out a large knife and began stabbing the kangaroo in the neck. He smeared blood on his arms and chest, and even took one of its feet and scraped the claws across his cheek and shoulder. Harry showed him where the bullet had entered the kangaroo's body and he dug around with his knife to remove it. Then, with a big grin, he stood up, pocketed the knife, heaved the bloody corpse onto his shoulder and strode back to the trees. We watched him fade into the distance, suspecting that we had just been party to an elaborate ruse to restore a man's ailing tribal prowess. 'Well, good luck to him,' shrugged Harry. 'His problems will return when the meat has all gone and he is expected to go out and get some more!'

We could avoid the persuasive influence of the sun no longer. The coastal rainclouds had dispersed and we returned to Derby after our pleasant interlude, leaving the old man to his short-lived revival of fortune. The following day we would go to Point Torment.

My diary entry, 12 June 1968:

> 'Up early and speedy preparation for trip to mangroves on Point Torment. Simply clad in pants, two pairs socks, pair shorts, pair jeans, pair thick trousers, polo-neck jersey, shirt, combat jacket, gloves, hat and headnet plus binoculars, two cartridge belts, water bottles, collecting bag and twelve-bore [shotgun]. [The] Idea being to keep out the expected (others reported millions) sandflies and mosquitoes. Sun shining strongly and I may lose 4 lbs [nearly 2 kilos] in sweat but it will be worth it.'

We drove 80 km from Derby into Point Torment. Following tracks that were still not properly dry, we stopped at the edge of an enormous mudflat. Our goal was a distant line of shimmering green. The tide was three-quarters out, giving us plenty of time. The walk to the mangroves was unpleasant. The more I sweated, the more I envied the others. In their one or two layers of clothing they were so much more comfortable under the scorching sun. Until, that is, we were among the trees. Despite the welcome shade, the humidity increased and the insects, unhindered by the stiff onshore breeze that had kept them at bay out in the open, had us, or should I say them, at their mercy. There were millions of them, humming, swarming, piercing and sucking. I was bitten, of course, but only on my hands, which I had to expose to use my binoculars, gun, notepad and pencil. The rest of my body, mummified in layers of sweaty, claggy clothing, was safe. Even peeing through a strategically placed length of plastic tubing was possible without being bitten.

Deep inside one of these mangrove swamps at low tide, with its thick canopy, twisted trunks, tight latticework of roots and carpet of short 'breathing' stems growing up through the mud like a bed of stubby nails, progress is slow. Identifying, shooting and retrieving birds is not easy, and neither is the vital chore of recording their soft-part colours, those taxonomically important bits like eyes, legs and patches of bare skin that fade so quickly after death. Within earshot of each other's guns, we waded around a 2 km stretch of swamp, each in his own insect-battered world.

I climbed out through a mass of tangled roots onto a small promontory. The tide had just turned and I was three or four metres above a sinister whirlpool, where two bodies of muddy water merged and flowed to the open sea. It was unpleasant to think of being down there. Even if there weren't any mosquitoes or pythons, there would be crocodiles and sharks. If you weren't eaten, or you didn't drown, you might be swept out to sea knowing that the next landfall beyond Cape Leveque was an Indonesian island 1,000 km away.

But it was bliss to stand there on that little jetty of mud, the open water in front, the dense trees behind and the mosquitoes again held off by the refreshing wind. Ten or fifteen minutes passed before I decided to retrace my footsteps through the mud. Why I chose that moment I will never know but, as I turned, I saw what looked like a large crack between me and the thick wall of trees about 5 metres away. I just stared at it but then realised that it actually was a crack and that it was running from left to right and opening up fast. The whole promontory, its balance tipped by my added weight or the erosion of one final tide, was about to slide down into the river. I scrambled towards the trees, threw my gun and bags as far as I could, caught hold of an overhanging branch and hauled myself over the widening gap. Grabbing my gear, I waded a further 5 metres into the trees before turning to see what was happening. Peering out through the latticework of branches that partially obscured my view, I was just in time to see several tons of mud slide down into the river, dragging saplings and bushes into the deadly whirlpool. I was left standing close to a sheer, ragged drop into the sea.

Gathering my muddy belongings and ignoring the renewed attacks of the mosquitoes, I took a large swig of water and set off with a still-pounding heart. The comforting sound of guns was not too far away and I decided to make for these, feeling in need of a bit of company. But when I stopped squelching through the mud to climb over a tangled mass of roots, I heard what I took to be a faint cry drifting through the trees. I listened carefully and there it was again: a human cry. I scrambled a further 20 metres and listened once more. A definite cry for help. I called back. A response, somewhere to my left, not far beyond the dark wall of roots, trunks and leaves. And then I found him, kneeling over a low stump, his arm extended, his hand mostly hidden by a thick, twisted root. It was Cliff, lost in a cloud of mosquitoes, his mind already abandoned to the pain and the thought that we might not find him before the evening tide crept his way.

'There,' he whispered, 'under the root', where his hand was lost in such a thick grey slime that it was impossible to see what he meant. Until it moved its legs. The crab, exactly one of those the mechanic from Derby had warned us about, was backing in beneath the tree. Cliff's muddied and barely discernible right hand was held by two giant pincers.

35

I could just make out the shell of his tormentor. It was at least 30 cm across, with massive forearms held up in front of its face. One pincer was closed round Cliff's middle finger and the tips of the other had him by the thumb, penetrating both sides, through the nail, and meeting in the middle. The only move Cliff could make with his hand was towards the crab, to release the tension, whereupon it took up the slack and dug itself a little deeper into the mud. Like the python that tightens its suffocating coils when its victim breathes, the crab was killing by stealth. And waiting for the rising tide.

I sat down next to Cliff and gave him a drink from my water bottle. Only one course of action was open to us. I would have to shoot the crab. Loading a solid .22 bullet into the 'under-and-over' gun that Cliff was using (a gun with two barrels, one on top of the other, the top barrel of .22 calibre, the lower of .410, giving a specialised collector's gun able to fire two different-sized cartridges), I rested the barrel on a root and moved it slowly towards the crab, making sure that Cliff's hand was out of the line of fire. Then, at the point of contact, I pulled the trigger, trusting the bullet would pass through a vital part of the crab's anatomy and down into the mud. It did exactly that. With an involuntary spasm that had Cliff gasping one final time, the crab relaxed its hold.

It was easy enough to unhook the claw holding his middle finger, but the one piercing his thumb was a problem. It was locked shut and could not be worked on without inflicting severe pain on Cliff, who by now was close to fainting. I had to cut it away from the shattered shell. With no thought of mosquitoes, or even of salvaging the crab's body for a feast, we hobbled out of the mangroves and across the mudflat to the distant Land Rovers. Harry immediately radioed Derby Hospital to warn them that a serious case of crab bite was coming in. On the way there, Cliff did his best to convince us that he had shot a small bird and that both he and the crab had gone for it at the same time, the crab winning a bigger trophy than it had intended. I think we all knew that Cliff's taste buds had got the better of him. But he had our sympathy – there would have been enough meat on that single crab to feed all five of us.

The nurses could hardly contain themselves. It wasn't the bite, they said, it was the word 'serious'. When the radio message came in from Point Torment, they had prepared an operating theatre and had called in the local anaesthetist. The surgeon was scrubbed up and ready to go somewhere important. But when our casualty hopped out of the Land Rover and neatly sidestepped the waiting stretcher-bearers, we could hear the groans of disappointment.

While Cliff was being de-pincered, cleaned and bandaged, the nurse told us about the man who manoeuvred one of these giant crabs into a bag, slung it over his shoulder and walked out of the local Derby mangroves towards his car. He never made it. The crab found his backbone through the flimsy sacking and went clean through, killing him instantly.

'That,' she said, 'is what *we* call serious!'

Forty one years later, Cliff's emailed recollection of that day pointed out that I first had to find and retrieve his gun from the mud where he had dropped it in the shock of the crab attack. He added that the little bird he had collected, a male red-headed honeyeater, 'fell to land directly in front of the bloody crustacean'. So there had been a bird, after all. And he added, 'You know, Dan, I have told this story [the one about the man who was killed by a crab] several times – but all I get is looks of utter disbelief.'

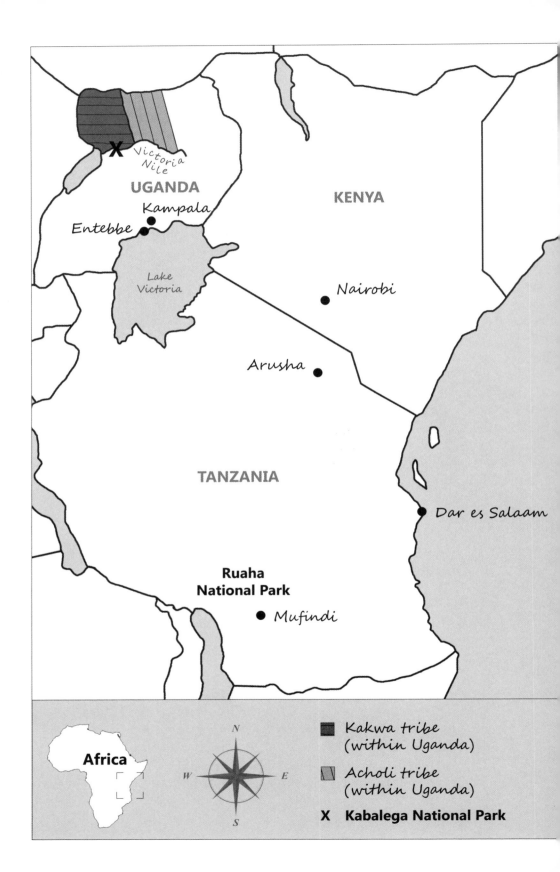

Victoria
Nile

UGANDA

KENYA

Kampala

Entebbe

Lake
Victoria

Nairobi

Arusha

TANZANIA

Dar es Salaam

**Ruaha
National Park**

Mufindi

Africa

N

W   E

S

Kakwa tribe
(within Uganda)

Acholi tribe
(within Uganda)

X   Kabalega National Park

5

A SAFARI TO REMEMBER

'Achtung! Achtung!' yelled the German tour leader, fearful of losing his group in the crowd at Entebbe Airport.

'Spitfires!' came the reply from the ageing Brit propped against a corner wall; a moment of humour in a country suffering its own Hitleresque tragedy. For this was 1972, the year 60,000 Asians were ordered out of Uganda by Idi Amin. Even now, several months before their November deadline, they were leaving in their droves, many to the makeshift camps that would be their first homes in Britain and other countries. The keys to a brand-new Mercedes could be bought in the airport car park for as little as £5, offered by families desperate to begin their new lives with whatever sterling they could muster.

1972 was a sad and destructive year for the Asians forced out of their adopted homeland. But the silver lining to their unfortunate cloud was that they would at least be escaping the clutches of a tyrant. For those Ugandans who were African by ancestry, there was nowhere else to hide. Thousands were killed as their President countered every perceived threat to his authority with violent blood-letting. And no tribe seemed to suffer more than the Acholi in the north of the country.

I met an Acholi man at Murchison Falls National Park (now called Kabalega) while working as a guide on a bird-watching safari early in 1972. We were staying at Paraa Lodge, overlooking the River Nile. His name was Alfred

Labongo and he was the Park Warden. He and Idi Amin, who came from the neighbouring Kakwa tribe, had known each other since childhood. But Amin had growing concerns about Alfred - he was intelligent and he was an Acholi, the President's hated tribe. In his favour, though, Alfred's role in the Park meant that he was in touch with tourists from overseas, and this suited Amin's desire for the outside world to know that Uganda's new regime meant business. Amin's Death Squads – euphemistically the Bureau of State Research or Military Police – dumped the bodies of their victims inside National Parks so that tourists, following vultures to what they assumed to be a regulation lion kill, would see for themselves what was happening. Park people like Alfred would apologetically supply the details, which would then travel out of Africa to the rest of the world. For this reason, Alfred, the messenger, was not worth shooting.

But the evening I met Alfred, a few days after that visit to Entebbe Airport, he drank far too much and threw caution to the wind. Through an alcoholic haze, he talked loudly and pointedly about Idi Amin, waving his arms and swearing that one day he would rid his people of this monster, raise the Acholi tribe to political prominence and become their leader himself. Even, one day, President of a Uganda where everyone lived in peace. I was new to Africa and, while others more knowledgeable bundled him away to his bed, I felt sorry for Alfred and the desperation that consumed him. Three months later, in Nairobi, where I was living, I heard that Alfred had continued his drunken outbursts. One night not long after our visit, he simply disappeared, his usefulness to Amin exhausted or outweighed. He would now be dead, butchered and fed to the vultures like so many of his people, powerless to halt the juggernaut of death that rode among them.

Later in the year, at about the time of the Asian departure deadline, I travelled to southern Tanzania to spend a fortnight in the isolated and wild Ruaha National Park with an enterprising group of elderly Americans. Not for them a zebra-striped minibus flogging the crowded Serengeti circuit. They wanted something like the trip they had had with Don Turner's East African Ornithological Safaris the previous year, when they were caught up in severe storms and flooding along the Tana River in central Kenya. They wanted, as they had said when booking Don to lead them once more, another 'safari to remember'.

I drove down through Tanzania, from Arusha to Iringa, with helper Emily Ford. We were a few days ahead of Don and the Americans, giving us time to finalise the tented camp. Don and George Dove, who ran the Serengeti's Ndutu camp, had chosen an idyllic spot where the Great Ruaha River meandered slowly through exposed sandbars, several metres below its wet-season level. We pitched the dining-room tent in the shade of a large acacia tree on a bank overlooking the river.

Setting up a well-appointed camp in the middle of nowhere is no easy task, but slowly our little canvas village took shape. We were quite alone in Ruaha because the new Park Warden had yet to take up office. His predecessor, who had been there for many years, had flown out in his light aircraft the week before we arrived.

The highlight of each day was when the elephants came by on their way to drink and bathe in one of the remaining dry-season pools. They were below us, but so close that we could have leaned out from the top of the bank and scratched their leathery hides. They filed past with little more than a glance in our direction.

On our third day, things were different. A delivery of unripe fruit was now stacked in the storeroom tent and one of our local helpers, Martin, was inside, connecting gas cylinders to the refrigerators. The elephants duly arrived on their morning trek to water and I stopped to watch them as usual. But this time, the last one in the procession halted immediately below the large acacia and raised its trunk enquiringly. It must have picked up the scent of the fruit because it turned towards the bank and began an amazing feat of mountaineering. It placed its front legs on the exposed roots of the tree about a metre up the bank, leant forward, pushed its tusks into the soft earth and held its weight on these while its back feet gained a foothold just behind its front ones. Settling its weight and holding this position, it eased its tusks out, dug them in higher up, leaned forward again and scrambled its front feet even higher up the bank. Alarmed by its improbable progress but half-expecting to see it topple over backwards, I called for the others to be on their guard.

The elephant was determined to complete its assault course. With one final push from its tusks and a scrabbling of knees and feet, it hauled itself up to our level and stood next to the tree. Now, as it towered above us, its immense size was unnerving, a distinct physical threat. As though totally aware of this, it strolled between two half-assembled tents and made for the one containing Martin and the fruit.

Thinking 'noise', I picked up an empty jerry can and starting bashing it with a hammer. The ringing, metallic echoes must have carried for kilometres but the elephant took not a scrap of notice. It walked slowly round the tent, hesitated and then began to push in under the awning at the front. Shouting a quick warning, while fully expecting Martin to unzip the rear of the tent and step quietly away, I was powerless to do anything but stare.

The rear of the tent remained fully zipped. But suddenly, and with its head about to enter the front of the tent, the elephant came to an abrupt halt. Swaying and agitated, it began to back away. It was obviously concerned. Its huge ears were spread wide. It had reversed like this for some five metres when Martin finally appeared, not from the rear of the tent but from the front, walking slowly towards the elephant and gently tapping together two wooden spoons. I was amazed, even a little embarrassed, as Martin and the elephant kept going as one, the gentle rhythmic clicking of the spoons working a magic unknown to my battered jerry can. They were 20 metres from the tent when the elephant lowered its ears and wheeled to its left. Then it headed back to the river with the deceptive, ambling gait of an elephant in a hurry.

Martin lowered his wooden spoons and walked back to the tent to resume his work, but not before he had glanced across at me. Smiling, he said, 'Metal is too recent an invention to hold any deep fear for elephants. Wood is different. A long time ago, people hunted elephants armed only with spears and, as they closed in for the kill, they tapped their spears together, causing a distraction among the elephants that allowed one or two of the hunters to run in and hit their target. Even today, the elephant is born with a fear of wood knocking on wood and will run away when it hears the sound.' And with that simple explanation, he disappeared into the tent to fire up the fridges. The elephants never bothered us again.

A few days after the arrival of our intrepid Americans, we decided to make a long trek through the park to find its rare sable antelope. It would mean being on the road early.

The day began innocently enough. We were up before dawn, packing two Land Rovers with provisions and people. There were ten of us in all and for more than an hour we drove in sleepy silence, waking with the day. We had made good progress so we decided to stop under an enormous baobab, Africa's famous 'upside down' tree. I clearly remember George looking up the tree and declaring that we could proceed with our breakfast plans. We advised the tourists to stay within calling distance while we prepared bacon and coffee on a little stove and cut bread to toast over a small, contained fire.

The smoke curled up the swollen trunk of the baobab, exploring its joints and outstretched limbs. It was then, with more than half the loaf still needing to be sliced, that the first bees attacked. Only a few and I swatted them away. But there were more, persistent, clinging to the hair on the back of my neck, forcing their way in, stinging. Then I realised that the others were fighting off their own bees. 'There's a nest in the tree,' shouted George, realising he had made a mistake. 'Put the fires out and get out quick.'

We shouted and waved at the tourists, imploring them to keep away, to get up onto the track as far from the tree as possible. There was no time to see how well they had understood our frantic gestures. The four of us under the tree were in desperate trouble, scrabbling at bees, stamping on fires and, for some reason, tidying up cutlery and plates. We did not see that the tourists, thinking simply that their breakfast was ready, were already halfway back to the tree.

The first trickle of bees – which had been bad enough – became a flood and then a deluge. We dropped everything and ran, still shouting at the tourists who were now close enough to see, and even hear, that something was wrong. Everybody scattered, and I was running back along the track on my own. No thought of elephant, buffalo, lion or rhinoceros, just bees. After several minutes, I stopped running, my head pumping with pain and noise. The bees had given up the chase, their instinct to send me far from their precious nest exhausted.

We regrouped slowly, calling for each other through the intense heat of the morning. George, diabetic and having problems with his sugar levels, told us to make a fire out of dry elephant dung and to push people's heads in, to smoke off the persistent bees. One woman was a mess: she had been stung at least 300 times, but it was only after we had put her through the smouldering fire and settled her out on the ground that we realised we were one person short. Marjorie. Where was Marjorie? Nobody could recall seeing her and it was now more than an hour since the attack had begun.

At that precise moment – and given the size and isolation of Ruaha National Park it was an amazing bit of luck – a Land Rover came round the corner and stopped in front of us. The driver, whose name I never knew, let me into the passenger seat and we set off down the track. We decided to drive round the tree in ever-increasing circles until we found our missing Marjorie.

The tree stood in a battlefield. The stove was out but the little fire was still smouldering, no doubt aggravating the bees all the more. Cups, saucers, bowls, spoons, knives and forks lay broken and scattered where we had trampled through them in our panic to leave. The two Land Rovers, windows open, were crammed full of bees, buzzing angrily against the windscreens. They did not attack as I closed the windows, found and emptied a can of insect spray into each vehicle and then slammed their doors shut. We grabbed George's bag which contained his insulin pack, heaped earth over the fire and grabbed a jerry can of water.

It took three slow circuits out from the tree to find Marjorie. A hundred metres away she appeared through the bush like an apparition, standing, arms outstretched, gently swaying and emitting a low dirge-like gurgle. Her entire head and shoulders were covered with bees, a gigantic swarm held in place as each one fought its way to the centre to leave its sting in her flesh. Blood glistened in the late morning sun, oozing stickily and slowly through the bees piled over her head like some outrageous guardsman's busby hat.

In almost one movement, we jumped out, heaved her onto the spare wheel lying flat on the bonnet of the Land Rover and scrambled back before the suicidal bees erupted from her head. I was leaning out and hanging onto Marjorie's legs as we bumped our way back to the elephant dung fire. Pitching the semi-conscious Marjorie headlong into the smoke, we clawed the dead, dying and angry bees from her head, face and neck.

Her features were already grotesquely swollen. The gut-trailing barbs torn from the bodies of the bees formed a dense covering of animated 'fur' as they twitched to empty every last drop of venom into her weakened body. I reached for my camera and then stopped, knowing that I would never be able to take the photograph.

The Land Rover driver, whose timely arrival may well have saved Marjorie's life, took his leave, reassured that we would be able to manage without him and his vehicle. We sat or lay on the ground in a state of exhaustion. The heat and the smell – a pungent mix of sun, fire and fear – were unforgettable, as was the growing awareness that the worst was by no means over. African honeybees are well known for their poison and here we were, in the middle of Ruaha, with seven people in their late sixties who had been stung not just badly but, in at least one case, catastrophically.

Leaving Marjorie with her friends and George with his insulin, Don and I walked back to the tree. The bees had virtually stopped flying, so we opened up the Land Rovers and scooped cupfuls of dead and dying bodies from the floor and seats. Then we packed everything away, checked that the fire really was out, and drove back to join the others by the smouldering elephant dung. Don decided that I should squeeze everybody into my Land Rover while he raced ahead to radio for help from the Park's headquarters – assuming, of course, that the new warden had arrived that morning as scheduled.

It took us three hours to get back to camp. Marjorie, who somehow roused herself at the mention of antihistamine to say she was allergic to it, was put straight to bed. I set off for the Park headquarters where I hoped to find Don and the new warden sorting out some kind of emergency help. But Don's Land Rover wasn't there. I turned towards the house and knocked on its sheltered door, expecting no reply. But then, from within, footsteps.

The door creaked open. Into the shade of the overhanging porch a tall African emerged from an inner room, stooping to clear the frame. He moved into the light, straightened and broke into a broad grin. 'I know you,' he said. It was Alfred Labongo, alive and larger than life. Before I could respond, he hugged me from the sheer pleasure and relief of seeing someone who had been there, someone who had witnessed, even fleetingly, the havoc wreaked on his homeland by Idi Amin.

Within five minutes of finding Alfred, Don arrived with his own story. He had made good progress ahead of us until, rounding a tight bend, his way was blocked by a herd of elephants. They panicked at his dramatic approach and charged, forcing him deep into the bush where he lost his bearings trying to get back to the track. Using the sun, by remembering roughly where it had been on our outward journey, he finally made it back to camp. He was surprised and delighted to see Alfred, whom he knew much better than I did.

We sent a radio message to the Brooke Bond Tea Estate at Mufindi, 130 km to the south of Ruaha. To get there, the message had to be relayed over 1,600 km, going north to Arusha, back down south to Dar es Salaam and then west to Mufindi. As often happens, it became slightly garbled. So while we waited for the Brooke Bond plane to land on our airstrip, as requested, they waited for our Park plane to land on theirs. But we didn't have a plane. The previous warden did, but he and his aircraft were now somewhere else, so we needed them to come to us. We waited until it was too dark to land on an unlit bush strip and prepared for an all-night vigil with the ailing Marjorie. It was to be a long night of cold flannels, chatter to keep her alert and the awful realisation that she could, in fact, be dying. Even before first light, with Marjorie still alive but misshapen, pale and fluttering, the unanimous decision was that we should forget about the plane, put her in a Land Rover and drive across country to Mufindi, come what may.

It took six hours to complete the journey and Marjorie, who managed not to die on the way, was rushed into the small, well-equipped hospital, dosed with cortisone and put to sleep. The following morning, Don and I approached her room, fearful of what we might find. What we found was Marjorie, still horribly swollen but sitting up in bed, joking with the doctor in charge. She was just telling him how embarrassed she had been yesterday when she had somehow asked us to stop and help her have a pee. The doctor told us she had the constitution of an ox, had been stung more than a thousand times and that it was fortunate we had not managed to put her in an aeroplane. In the intensity of her ordeal she had perforated both eardrums – that blood glistening in the sun as it trickled through the bees – and a non-pressurised flight would have been disastrous, perhaps even fatal.

Assured that Marjorie was going to survive, that all she needed was medication and rest, Don and I returned to Ruaha, taking the longer, more comfortable route on the road. After our two-day absence everyone else was making a good recovery. News of Marjorie's improbable progress lifted their spirits even higher. George Dove, whose usual home was in the Serengeti, 1,500 km away to the north, reckoned he had not allowed for a seasonal shift in behaviour when he first examined the tree for any signs of bee activity. At Ndutu, he said, the bees would still be flying in November, not massed inside their chosen tree. Nobody thought to blame him in any way for what had happened.

Later that evening we went back to Alfred's house to drink beer and barbecue steak on an open fire. Hypnotic flames dancing in the dark were perfect for his remarkable story.

A few weeks after our evening with him in the tourist lodge on the Nile, Alfred had, indeed, been visited by officers from Idi Amin's Bureau of State Research. They informed him that there were irregularities with his employment papers that required his immediate departure for Chobe Police Station 40 km away on the road to Sudan. Protesting that seven o'clock was too late to be undertaking such a journey, Alfred promised to be there first thing in the morning. This would not do and, when he saw that the officer confronting him was accompanied by armed guards, Alfred decided not to argue the point any further. As he sat in the back of the military Land Rover, he suspected he was being driven to his death. The young guards were fidgety, avoiding eye contact and not talking. When they came to a fork in the road and took the turning that led away from Chobe, deep into the bush, Alfred knew for certain what was coming. He was primed, he told us, his brain wide awake, his senses alert to every detail of the moment.

Within half an hour, the driver stopped under the pretext of having a puncture. Alfred braced himself. 'They made a simple mistake. All the guards but one got out of the back and walked round the Land Rover on the road side. This meant they had to pass through the headlights at the front before taking up a position to shoot from the edge of the road. As they entered the bright beam, I pushed the remaining guard out of the back and dived straight into the bush. Bullets from the blinded soldiers crashed all around me as I scrambled away into the night, but I was not hit. They had killed so many people, they weren't too worried and, of course, they would not dare report that they had failed to carry out their order. They knew I would disappear after this anyway, so for them it wasn't a big problem that I was not dead.'

Alfred walked by night and hid by day until he reached friends in Kampala. Fed and rested, he set out again, soon crossing into Tanzania where he had a 1,500 km walk southeast to Dar es Salaam. He had to be careful at all times because the influence of the Ugandan Death Squads spread far over the border. In Dar es Salaam, he contacted the National Parks Authorities who told him that Ruaha would be ideal. It was isolated, a long way from Uganda and he could have the top job there because the existing warden was preparing to leave. Alfred could stay there until it was time for him to go back home. Even so, it would be a further seven years before Tanzania's invasion of Uganda forced Idi Amin into exile and cleared the way for Alfred to return to his old job in Kabalega.

The desire to continue their safari had deserted the Americans, so we left Ruaha a few days earlier than planned. A week later, Marjorie was driven back to Nairobi for a final check on her ears before being allowed to fly home to America. The only wobble in her recovery came ten days later when she blew her nose and expelled the remains of a bee which had forced itself deep inside her head. It was the trigger for an outburst of mental agony that needed to happen.

We kept in touch. Five years later, Becky and I met her in London and the three of us drove to Sissinghurst Castle in Kent. As we sat drinking tea in the garden, there were bees all around us, collecting nectar and pollen for their hives. I watched cautiously as Marjorie's eyes followed them from flower to flower.

'It wasn't their fault,' she suddenly volunteered as the little bodies filled their pollen sacs among the colourful blooms. As she had been the one to broach the subject, I found myself asking the question that had been on my mind for so long.

'Why,' I ventured, 'why were you just standing there, doing nothing with all those bees on you? What were you thinking?'

'Well, to be honest, I assumed that everybody was the same, that we were all going to die down there in Africa on one of our special safaris. And, do you know, once I decided that was what was going to happen, it wasn't such a bad feeling.'

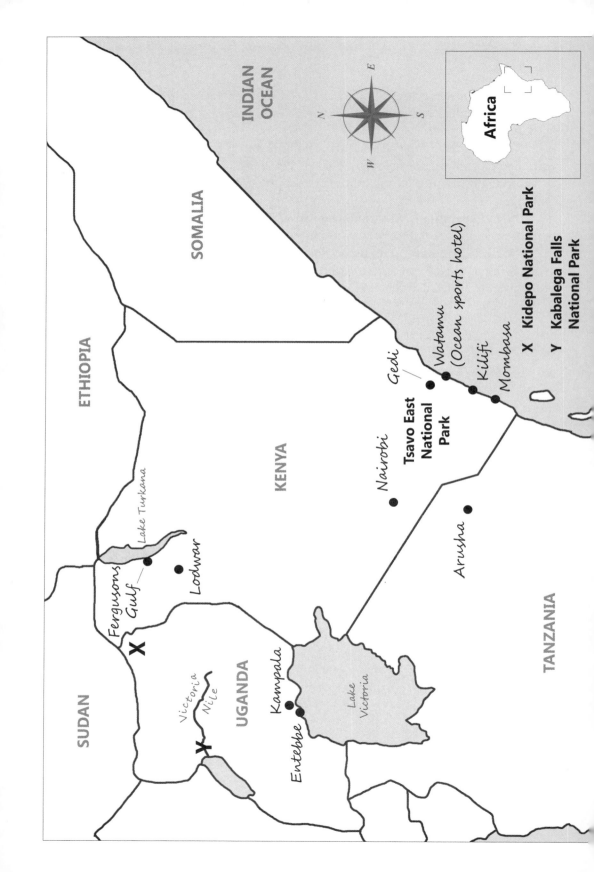

6

ISOBEL BENNETT

The museum's collection of parrots was spread before me when Don Turner's head appeared round the Bird Room door.

'I know you've just come back from Uganda, Dan, but I've got a bit of a problem. Is there any chance you could do another trip for me at really short notice?' It was, he added with a twinkle in his eye, two national parks and a week on the coast with an Australian marine biologist. All she really wanted was to sunbathe, snorkel and potter along the beach at one of Kenya's marine parks. Blonde, 28, single. Of course, if I was too busy he would understand completely.

Forty eight hours after the parrots had been hastily returned to their cabinet, and slightly nervous of what I might be letting myself in for, I was driving out of Nairobi and heading south for the border with Tanzania at Namanga. It hadn't been difficult to get further leave of absence from the museum. I was in Kenya on a grant to help reorganise the bird collection and part of the deal was that I would join some of Don's birdwatching safaris to help with the country's tourist industry.

Beyond Namanga it was a further 100 km to the New Arusha Hotel. Entering the reception area with a clear image of Isobel Bennett in my mind, I glanced round the tourists who were checking in and out. She wasn't among them. I wandered through to the dining room and she wasn't there either, so

I thought a quick tannoy announcement was required. I was halfway back to reception when she beat me to it. My name, loud and clear. And there she was, waiting, knowing it was me, her hand outstretched, the friendliest of smiles. I returned the smile and shook her hand, holding the contact with a slight feeling of relief.

Over coffee, the 62-year-old Isobel hoped, almost apologetically, that I wouldn't find our time together too boring. She desperately needed a break from Australia and wanted to do as little as possible. I assured her that I had brought my cameras and would be happy to spend my time photographing wildlife whenever she wanted to be on her own.

The deal struck, we headed away from Arusha and into Amboseli National Park. The snow-capped Kilimanjaro dominates, a wonderful backdrop to photographs of elephants and giraffes ambling between acacia trees festooned with the nests of weavers. Weavers are finch-like birds, often bright yellow, which raise their young in large, noisy colonies. The park, even then, was becoming a bit of a dust bowl, a worry for conservationists fearful of erosion from the high number of tourist vehicles charging around in search of wildlife. It was a problem for the Kenya government as well, because tourism was already a vital source of revenue, to be encouraged rather than contained. Step one was to ban tourists from driving anywhere but on designated tracks in the parks. Less fun, of course, but a necessary precaution to protect the land and its vital plants and animals.

Isobel and I drove north along the coast from Mombasa to the ferry crossing at Kilifi. There was a little albino boy who lived close to the road, whose mother smiled sweetly if you put a few coins in his tin. And then on to Gedi, where we turned right past the ruins of a 14th-century Arab town, mostly reclaimed by trees and scrub since its last human occupation 300 years ago. Gedi is loved by archaeologists, who have been unravelling its overgrown past since the early 20th century, and by naturalists who hope to catch sight of its elusive golden rumped elephant shrews. Ten kilometres beyond Gedi, we arrived at Watamu. If ever there was a tropical paradise, this was it. Five kilometres of white sand bathed in sunshine and lapped by the most beautiful azure sea. The hotel – hardly the right word for Ocean Sports – was hidden among palms that fringed the sand, cooled by a refreshing onshore breeze. The human welcome was so warm that within a few minutes we felt totally at home.

True to her word, Isobel just lazed away the days, pottering along the beach and scouring the rock pools for marine creatures to compare with those she knew from Australia. While her mind was closed to anything else, I ventured further afield with my cameras, particularly to the mangrove-lined mudflats of Mida Creek. Among the storks, egrets and kingfishers feeding in the muddy shallows, overwintering crab plovers were the special prize.

Ocean Sports was run by Chris and Mary Nichols. They engaged their visitors in a laid-back friendship that masked the efficient running of a hotel in one of the world's most glamorous locations. Mary was mostly on the move between buildings hidden among the palms, while Chris pottered around the beach and the bar, sorting equipment for trips above and below the waves. I don't think we saw him wearing anything but a pair of shorts, which may seem a strange observation given the intense tropical weather just three degrees south of the Equator. There was, though, something different about Chris's shorts. Below the top button that anchored them to his ample waist they were usually undone, a state of exposure that, from above, he had problems noticing. Those of us around him did not.

Late one morning, while we drank coffee at the bar and Chris was busy in his over-familiar attire, I turned to Isobel and said, quite spontaneously, 'From now on I think we should call this place Open Shorts.'

Later, back in Nairobi, Don laughed when I told him of our new name for Ocean Sports and from there it spread like wildfire through the tourist community, the perfectly inappropriate epithet to sell holidays to such a sexy place. Now, more than 30 years on, it is still referred to as 'Open Shorts' in *The Rough Guide to Kenya*. Of course, any claim to the coining of a new name or phrase is difficult to prove, but I wonder if anybody can positively pre-date March 1972 for this one?

It was soon time to leave Watamu. After a few days in Tsavo East National Park, Isobel and I went on to meet up with Don and a birding group in Uganda, a slight change from the original plan. Don had contacted me in Tsavo to say he had cleared things with the museum and that I could be away for another two weeks. We were soon deep into Idi Amin territory, a trouble torn Uganda ruled by a paranoid tyrant. Don made things quite clear. There were numerous road checks outside Kampala and Entebbe. I was to keep everyone's passports in the front of the Land Rover and, if we were stopped, I was to hand them out of the side window. In no circumstances were any of us in the vehicle to make eye contact with the soldiers who stopped us. After making our way cautiously past several groups of soldiers who took no notice of us, we finally came to a barrier stretched right across the road. Slowing down and taking hold of the five passports, I advised everyone to keep their eyes on a roadside tree up ahead. I fixed my own eyes on a scratch on the windscreen, slowed to a halt and slid the driver's window to one side. The next thing I was aware of was the tip of a rifle barrel sliding towards my right cheek. Maintaining my stare on the glass, I held the passports out of the open window. Nothing happened. I wanted to glance sideways but was beaten to the temptation by the slurred advice that five passports for five people was enough. The rifle withdrew and the door of the Land Rover received a thump, which I took to be a signal for us to leave. I

A male golden palm weaver and the first ring of his suspended nest, Kenya coast (Author's collection)

pulled away slowly, gathering speed and closing the window only when certain that I had done the right thing. Relieved but feeling a little shaken, I commented on the fact that he hadn't even taken the passports to check them, and got the explanation from someone who had sneaked a quick look: he couldn't, as his other hand had been busy with a near-empty whisky bottle.

Two weeks later, Isobel and I went our separate ways, I to a further nine months in East Africa and she back home to Sydney and the coastal creatures she

52

knew so well. We kept in touch, usually around Christmas, occasionally missing a year or two for no particular reason. She wrote with renewed enthusiasm in the early '90s to say she had met and become friends with Cliff and Dawn Frith and was delighted to learn that Cliff and I had worked together on the Natural History Museum Expedition to the Kimberleys and Arnhem Land in 1968.

Cliff had returned to Australia a few years after the expedition and done incredibly well, writing, photographing and publishing with Dawn while they became a world authority on birds of paradise and bowerbirds, the fieldwork alone for their wonderful books keeping them busy in Australia and New Guinea for many years. They now live in isolated splendour on their rainforest property in the heart of Queensland.

When I first met Isobel Bennett, I hadn't known, of course – having been comprehensively set up by Don – that she was 36 years older than me. Her holiday in Africa was a retirement present to herself after 40 years at Sydney University. Not that people like her ever retire. I had no idea how well-known she was, not just in Australia but in the whole world of marine biology. Some of the truth unfolded over the days we spent together. Her success in a male-dominated scientific world, where she began with no more than a passion for her subject and a determination to succeed, remains an object lesson to anyone starting out on the road to fulfilment with only their dreams to guide them.

Isobel died in January 2008, 18 months short of her hundredth birthday. She published prolifically, was honoured with medals and degrees and is an Officer of the Order of Australia (AO). She was included in the inaugural edition of *Who's Who of Australian Women*, published in 2006, yet she remained, in her letters, a genuine down-to-earth person who felt no more than incredibly lucky to have made a career out of her first love.

> *'You have no idea how much it means to a feeble old 96er, to be always remembered.'*

is how she began her last letter to me in December 2005, her indomitable spirit at last showing signs of frailty. And continuing, in her legible but spidery hand,

> *'In one way, it seems a lifetime ago since we were in Africa but it is still all so clear in my mind. Only able to get about with a frame, I find my days are filled with re-living the past…'*

Born in Brisbane in 1909, she moved to Sydney with her family in 1928. After four years with the Royal Schools of Music, she managed to join the staff of Sydney University's Zoology Department. She had 'arrived' but she was also entering, without formal qualifications, a man's world. The odds of gaining success and the freedom to develop her own work were heavily against her. But by consolidating her usefulness around Professor W. J. Dakin, becoming his secretary, librarian and demonstrator, she finally became his research assistant. One of his interests was the plant and animal life of the rocks in Australian river estuaries. Isobel helped him compile this data, and he was generous enough to give her a shared authorship of the book *Australian Sea Shores*. After Professor Dakin's death, the by now indispensable Isobel remained as research assistant to Professor Murray and it was here that she finally found her freedom. She took parties of students to Heron and Lizard Islands on the Great Barrier Reef and conducted her own research along the coasts of Victoria and Tasmania. Realising the importance of temperature in the distribution and abundance of coastal life, Isobel decided that a spell on the much colder Macquarie Island would be the perfect opportunity for a comparative study. There was just one problem. Women did not go there. The Antarctic was yet another man's world.

Undaunted by convention, Isobel leaned heavily on the board of the Australian National Antarctic Research Expeditions (ANARE). In her own words:

> 'After quite a lot of trouble we managed to get permission and in December 1959 we made our first trip on the ANARE supply vessel, a Danish polar ship, to the island. There were four of us. Hope McPherson, who had helped me along the Victorian coast, was my assistant on this occasion. Two other women also came because there were four-berth cabins and four women were allowed to go... The men who were going for the year had quite a lot of general direction and training. We had a general briefing and then the four women were called into the Director's office and he (Phil Law) spoke to us again and he finally said that on our behaviour depended the future of women in the Antarctic. We were rather unhappy about that because we thought it was quite an unnecessary remark.'

The length of their granted stay in this 'Boys' Own' world of seabirds and seals was between five and 15 days, the period within which the supply ship was expected to complete its work and turn around. Isobel knew that even the full 15 days would not be enough for them to complete their research, but they did as much as they could and left when instructed. Their behaviour, however, had been of the highest order, and the future of women in Australian Antarctica was secured. Isobel went back three times during the 1960s before publishing *The Shores of Macquarie Island* in 1971. During the same year she also published *The Great Barrier Reef*, for which she is probably best known today.

Isobel retired from the University in 1971. She thought that a holiday on the Kenya coast would show her some of the marine life on the other side of the Indian Ocean. Away from Australia she would be able to unwind from the gruelling procedure of publishing two major works in such a short space of time. So she booked herself into Don Turner's East African Ornithological Safaris for a tailor-made trip of her own.

While Isobel was on the first stage of the safari that would take her to our meeting in the New Arusha Hotel in Tanzania, I had returned to the museum after a spectacular birding trip in Uganda with Don and several American tourists. Driving north from Kampala, we met Alfred Labongo, the Acholi warden of Murchison (now Kabalega) Falls. We drove east to Kidepo National Park through hostile Karamoja country where nails deliberately scattered on the road gave us too many punctures for comfort. We stood on the Sudanese border, marked only by white rings painted round the trunks of trees, knowing that we were being watched along unseen rifle sights in the dry woodland ahead. We drove on to Ferguson's Gulf on Lake Rudolf (now Lake Turkana) where, 200,000 years previously, the first people fished for their supper, experienced the same sunsets and feared the same crocodiles. And then, my mind almost full, we drove back to Nairobi via Lodwar, famous for its swallow-tailed kites and the imprisonment of Kenya's future first President, Jomo Kenyatta.

Settling back into my new life in Nairobi, where I had arrived from England only a few weeks earlier, it was a relief to know that I wasn't going anywhere for a while. I could now get on with some of the bird work planned for me by Alec Forbes-Watson, particularly those parrots. But then, quite unexpectedly, Don's head popped round the door.

Although Isobel's last handwritten letter to me was dated December 2005, we were kept in touch by her devoted sister Phyllis, who was immensely proud of her big sister's achievements. Phyll, it should be noted, was all of three years younger, yet she drove every day to the nursing home that supported Isobel in the last two years of her life, and she missed her company at home dreadfully. She read letters written to Isobel by overseas friends and replied to them on her sister's behalf. Phyll and I continue with our Christmas cards and letters, as though carrying a torch for Isobel.

Just occasionally since our short African journey in 1972, I have wondered whether, without Don's 'blonde, 28 and single' ploy to lure me from those museum parrots, I would ever have left Nairobi to make friends with one of life's special people.

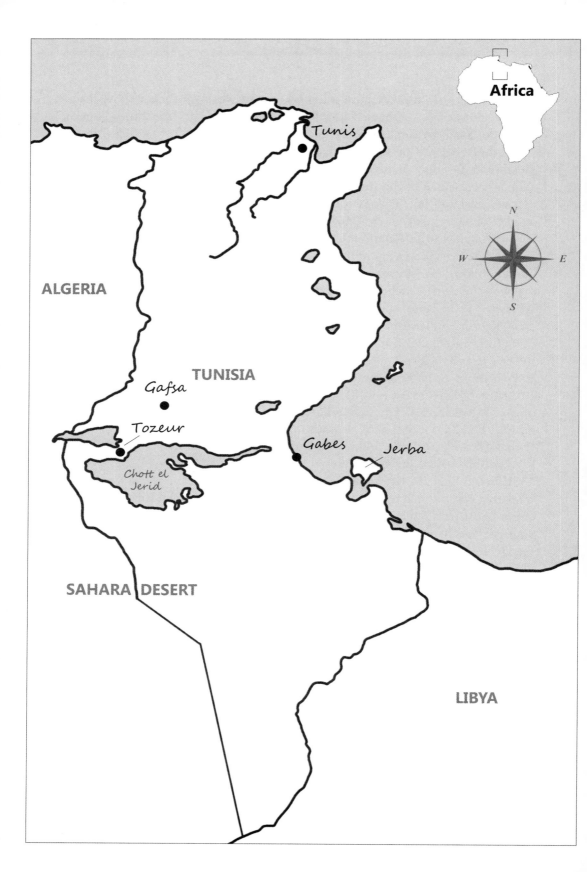

7

A CHRISTMAS CAMEL

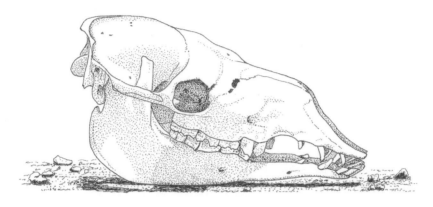

From out of nowhere, a shadow fell over my hand. My eyes followed a
dark red line up navy trousers and there, towering above the shiny black
shoes that had stopped a few inches from my nose, was a Tunisian *gendarme*.
The policeman tilted his head, raised an eyebrow and enquired rhetorically,
'Anglais?' And then, with a brisk 'Follow me, please', he turned on his heels and
set off towards a distant row of huts.

Thinking this might be the last straw in a somewhat eventful week, I gathered
up my camel's head and trudged after him. There were people watching from
the huts and I assumed I would have a bit of explaining to do. But the policeman,
instead of taking me inside, led me to the back of the first hut. With a flourish
of his hand and a triumphant 'Voilà!', he pointed to a tap. And added, with the
faintest hint of a smile, 'And not so many people to see what you are doing!'

When I had cleaned the skull well enough for it not to smell too badly on the
plane, I wrapped it in thick plastic and set off to find John. I had left him filling
the car with petrol and parking it as far from the Hertz office as possible. 'Let's
hope they don't find it until after we've taken off', we had agreed as we drove
into the airport. We were insured for the damage to the car so that wasn't the
problem. The problem was our embarrassment.

We hurried through to the check-in desk. Thirty minutes later, at the front
of the queue, our sharp exit from Tunis came to an abrupt halt. We had managed

57

to miss our flight by the small matter of 24 hours. They were sorry but today's flight was full. They would try and get us on one tomorrow. We were tired, disappointed and the last of our money had gone into filling the car with petrol. Air France, however, was not going to abandon its English customers in their hour of need. We were given enough local dinars to cover a taxi into town (and back), a basic hotel and even a meal. Two hours later we were sitting in a little restaurant, showered and wearing the freshly-laundered clothes from the night before. On the table in front of us were two bowls of couscous, each with its own little pile of cabbage and mutton perched on the top. We felt good. It was hard to believe we had been away from England for no more than a week...

Three weeks previously, early in December 1973, I had been sitting at my desk in the Portland Place office of Time-Life Books in London, a staff writer on their series *The World's Wild Places*. One of the books in production was *The Sahara* and I was writing a picture essay and captions to accompany the main text by Jeremy Swift. The essay was about animals living in desert conditions but I needed more pictures to go with the words.

The following day, John A. Burton – whose middle initial sets him apart from other John Burtons working with nature – came into Time-Life to do a bit more work on a similar book on Europe. We went out for lunch. John and I were good friends from Natural History Museum days and, since we both took photographs, he had an idea. Two phone calls later, we were booking our flight to Tunisia, funded by picture agent Bruce Coleman who trusted us – correctly as it turned out – to recoup his £300 investment from the future sale of our pictures. To minimise disruption to Time-Life's schedule, John and I decided to be away over the Christmas holiday.

Two weeks later, we drove out of Tunis airport in torrential rain, any notion that we were heading for the sun-baked Sahara the stuff of dreams. At one petrol station we were told to proceed with caution, particularly if driving at night. Bridges had been swept away and people had died. But after an uneventful day and overnight stop it was, miraculously, away with the umbrella and out with the sun cream. In just a few hours we had made the transition from the Mediterranean to the Sahara, from green to yellow. We were still in Tunisia but now there was no rain and the sun shone at a pleasant 20°C. The contrast between north and south is such that there are birds that can spend the summer in Europe and the winter in Africa but never leave Tunisia. With the Mediterranean Sea twinkling on our left, we took a right turn opposite the tourist island of Jerba and headed inland for Tozeur, a small oasis town in central west Tunisia, 200 km along the road to Algeria.

The Hotel Splendid could offer us only the bridal suite, which we took on condition that another bed was added. Next day, with our passable French and the help of the hotel receptionist, we found Ali, a Paris University student home for the Christmas vacation. Bored with desert life after only a few days away from France's swinging capital, the 20-year-old jumped at the offer of a badly-paid week as our guide and interpreter. Each morning we took food and water and were led by Ali through deep gorges, rolling sand dunes and oases surrounded by date palms. Tozeur, with its hundreds of bubbling springs where water escapes from the great artesian basin beneath the Sahara Desert, is a fertile place. It has an irrigation system, based on a 13th-century design, that enables dates to be grown on a massive scale.

We photographed desert scenery, wild palms growing in rocky nooks and crannies and little insects – mostly ants and beetles – as they scuttled over the hot sand. There were special desert birds to photograph, like Moussier's redstart (did we take its first-ever picture in colour?), hoopoe larks, mourning wheatears and what we thought were pale crag martins. The unexpected amphibian was a large toad burrowing into damp sand by an irrigated palm grove. But of reptiles and mammals, important players in my Time-Life essay, we saw absolutely nothing, however many stones we turned, holes we investigated or footprints we followed.

One evening, Ali followed us to our honeymoon penthouse and consumed a third of a litre of duty-free whisky in one go. Before the potent liquid had a chance to register on his senses, he was violently sick into our bidet. To avoid embarrassment with the staff, let alone blame for the horrible mess, we spent at least an hour poking the bits down into the pipe through the immovable grille and flushing them away. Ali made his own way home in disgrace.

The following morning at breakfast, the hotel manager approached for a quiet word. Our instant guilty reflex was that he had found out that Ali, a Muslim, had been drinking in the hotel and we were responsible. But no, it was just a slightly ironic request to see if we would like to order special French champagne to accompany our Christmas Day meal. The bill for this alcoholic bonanza, which came, on the day, with an equally intoxicating display of belly-dancing, would be half the weekly cost of the bridal-suite-with-two-beds.

Ali was still suffering later that day when we set out along a sun-baked dirt road towards a pink line shimmering on a distant lake. They were greater flamingos on Chott el Jerid and, while not vital to my desert story, photographing them would make a nice change from the days we had spent on our hands and knees with insects and plants.

We were halfway there and battling slightly with the deep ruts appearing on the track when something large and furry caught my eye, 20 metres away in the sand. 'I've got to have it,' I screamed above the noise of the engine. Swinging the car round through almost 360°, John brought the car to a halt by a pile of stinking, rotting flesh. Ali fell out of the back of the car, clamped a handkerchief over his face, ran off upwind and collapsed in the sand. John, more understanding of the whims – and whiffs – of wildlife fanatics (being a bit of one himself), looked on benevolently and didn't say a word. But I still felt the need to explain.

My father was a painter and, to satisfy his anatomical phase, I was under orders to return from each of my trips abroad with a skull for him to incorporate into a painting. The phase began simply enough with sheep, cows and horses found at home in England but, as my visits to exotic places increased, so did the possibilities for more exciting creatures. He already had kangaroo, crocodile and impala but had done nothing with them. Surely the skull from the 'ship of the desert' would spur him into action? Hacking off the putrid head and wrapping it loosely in plastic, we resumed our journey to the flamingos on the shimmering desert lake.

There had, in fact, been rain here quite recently. The lake had a covering of shallow, salty water – hence the feeding flamingos – and we were grateful for the 50-km causeway that guaranteed road access to Tozeur from the east. To approach the flamingos we had to turn onto a side track, its rutted and crumbling surface so thin in places that it seemed it might collapse at any moment. It took just 200 crunching metres to convince us that the safest option for our little car was to turn round and leave the distant birds to their brine shrimp lunch. We managed a few long shots of them rippling in the haze but they were wary, always moving away, never coming closer. My real prize, though, was in the boot behind us and it was already obvious that I would have to deal with it soon. We drove back to town in stinking silence, Ali's head hanging out of one of the open rear windows, his eyes rolled to the heavens, the handkerchief still clamped to his nose.

In Tozeur we were always followed by hordes of little children. Happy, laughing kids who, like Ali, had found in us something to relieve the monotony of their desert lives. They took a particular fancy to our hire car, which was where its troubles began. Despite the horrible stench wafting from its locked boot, the kids crawled all over it, bending its windscreen wipers and removing its wing mirrors and wheel trims. We even caught one of them, his mind set on the prize of a shiny bulb, attempting to lever out a headlight glass with a length of rusty metal. Luckily, our lights remained intact and the wipers could be bent back to a more-or-less usable shape without breaking. We were still able to see most of what was going on in the sandy swirls – they were hardly storms – that occasionally blew our way.

Ali lived with his sister in the town, so we got to know this side of his family quite well. They invited us into their house for mint tea when we dropped Ali off in the evenings and quizzed us about life in England and whether we were married with children and, if not – the case for both of us – why not? One of these conversations – there were times when they felt more like grillings –

concluded with our being invited to a Bedouin (Muslim) desert festival in a nearby gorge. It was to take place towards the end of December, at night, just two days before we were due to return home.

We drove out of town in the Ali family truck at the end of another day in the desert. From every direction, people were converging on the distant gorge. Many were on foot but others were crammed into or onto donkeys, bicycles, cars, lorries and buses, hooting, honking and shouting. Above it all was the bleating of goats as these little animals were dragged and carried, terrified, perhaps, by the inevitability of their last journey. During the light of day, out in the open, it was, despite the fate of the goats, all good fun. But then the sun crept down behind the sandstone ridge, taking her light and warmth with her. The concentration of human bodies increased alarmingly as they converged on the dark entrance to the gorge. As we entered on foot, both John and I, the only Europeans visible, and probably the only ones there, felt a certain trepidation. Too many eyes were glancing our way. Too many heads were gesturing. The children who loved us in the town were keeping their distance.

It was a narrow gorge with steep sides clearly outlined against the evening sky. From top to bottom its rocky faces were dotted with fires whose brilliance grew with the darkening night. Above dancing shadows and excited voices came the desperate bleating of the goats being sacrificed to the prophets. Skinned, dismembered, roasted and consumed. Gunshots echoed around the walls of rock whose night-clad peaks merged into the stars with a towering menace. As the people of the night danced hypnotically to the rhythm of their drugs, we felt our first real pangs of fear.

Now firecrackers were being hurled across the gorge, too often and too close to our little party perched on its narrow ledge. People seemed to be remonstrating with Ali's family. Was our presence too much for them? A musket was fired over our heads. Children screamed and clung to adult legs. The barrel of another musket exploded as it was fired, causing panic on a nearby ledge as its owner slumped into a crowd of excited onlookers. What on earth was happening? Ali turned and told us to get ready to leave. 'It is too dangerous for the children to stay here with these madmen.' We thought we knew the real reason for our sudden departure. We should never have been there in the first place. Christians may not have been forbidden but our presence was clearly an error of judgement.

Through the pungent smells of cordite, burning flesh and the illuminated smoke from fire, gun and thunderbolt, we crept along the narrow ledge. The flickering fires were not enough to light the way safely, so we were feeling in semi-darkness, pushing against the flow of human bodies where they were still scaling the gorge. Rocky footholds were wet and slimy with the blood, shit and entrails of dead or terrified goats. Somehow we reached the bottom of the gorge intact. Lost in the dark and the milling throng of people too preoccupied to notice, we slipped quietly away to the family truck and the safety of the open desert.

John and I fantasised over breakfast that the gorge was full of scavenging crows, eagles, vultures and jackals, and what a wonderful photo-opportunity such a gathering would present. But when Ali came to the hotel we could hardly ask him to take us back. The party might not yet be over. Ali was apologetic and embarrassed, refusing to make eye contact, betraying, we now knew for certain, the real reason for our hasty retreat from the gorge. So we said nothing and sat in silence until he spoke. He managed a faint smile as he reminded me that the car was still stinking and would I *please* do something about it.

But that wasn't going to be possible. This was our very last day and we desperately needed to be out there finding a few more things to photograph. So off we went for the last time, taking our smell and a reluctant Ali with us. Giving the gorge a wide berth, we spent ten hours chasing distant gazelles across the sand, but all we ended up with was a few more shots of the same old things – small birds, insects, plants and scenery. That night we packed our bags, hoping that we had at least got enough material for my essay and for Bruce to recoup his generous and slightly risky outlay.

Cramming our gear into the boot with the plastic-wrapped camel, we set out on the first 200 km that would take us back to Gabes near the coast. We had given ourselves plenty of time to reach Tunis for the flight the following afternoon so, when we saw a signpost pointing to Gafsa Zoo, a northwesterly round-trip detour of about 300 km, we didn't hesitate. If there was just one new animal there for us to photograph, it would be worth it. We could always make up the lost time by driving through the night.

The Director of the Zoo was delighted to receive foreign visitors. His eyes lit up, though, when he heard that John and I had both worked in London's Natural History Museum, which he had himself visited many years previously. Sensing we were onto a winner, we asked if we might, perhaps in return for a £10 donation to the zoo, be allowed to photograph some of his animals.

'For £10, my friends, I will let them all out of their cages for you. I have the perfect place, a small oasis here in the desert with a fence.' So out they came. Gazelles, fennec foxes, hedgehogs, vipers, sand boas, lizards and scorpions. In two frenzied hours of creatures being spilled out onto the sand, we used up our final half-dozen rolls of film. With most of our subjects safely returned to their cages, there was just time for a refreshing cup of mint tea before we were back on the road to Gabes. Our spirits were high. We had turned our trip round on a final day of fortune.

Backtracking and then driving north from Gabes late in the afternoon, we slipped out of Africa and back into Europe. The cool, wet Mediterranean climate had not relented. At a petrol station we were again told to proceed with caution. Apparently, bridges were still being swept away, roads were impassable and 900 people had died during the week. Grateful that our headlights and wipers were still intact, we vowed to stop only when we reached the airport at Tunis.

The dark night sky was heavy with cloud. John was driving while I navigated through the torrential rain. I must have missed a road sign because, all of a sudden and without any warning, the car was aquaplaning and then lifting off into black space. Before either of us could move a muscle or take in the possibility of a 100-metre drop into a raging torrent, we landed with a resounding splash. A moorhen cried out in alarm.

Our headlights settled on open water. The car felt as if it was bobbing gently. If there was a current, we could be carried from our only exit point from the river. Stepping out into no more than half a metre of water, we swung the floating car round so its lights illuminated the bank behind us and guided it slowly back. We were fortunate to have driven into what must have been no more than a long, shallow depression that rarely filled with water and needed neither bridge nor causeway for a safe crossing when dry. With a final heave as the wheels came up against the muddy bank, we pushed ourselves clear.

With its electrics thoroughly drenched, the car had no intention of starting, so we pushed it 30 metres back to the sign – yes, there was one, though it was more of a warning than a detour (we must have left the main road at a right angle bend and carried on up a little sidetrack) – and climbed in to wait for it to dry out. Impatience got the better of us and within half an hour we had flattened the battery. There was now nothing to do but wait, in the hope it might recharge sufficiently to fire up again later. Wet to the knee, we polished off the remains of Ali's duty-free whisky, wrapped up against the cold and prepared for a long, uncomfortable night. It was still only eight o'clock.

An hour later, headlights appeared on the road and a pick-up truck skidded to a halt. Its driver sloped towards us, peering inquisitively. He seemed disappointed to find people inside the car which he perhaps had thought might be his for the taking. He grinned widely when we explained our predicament. 'No problem,' he shrugged. 'I pull you.'

He looped one end of the rope over his towbar and then lashed the other end round and round our front bumper. When he had finished, the two vehicles were less than two metres apart. John and I exchanged knowing, fearful glances, acknowledging that we were powerless to intervene. We tossed an imaginary coin and it was John who got to stay behind the wheel. With our car in third gear and the clutch fully engaged, we were hauled off into the night at breakneck speed. The short journey was as terrifying as it was unreal. My feet were pressed hard against the floor of the car, my hands braced against the dashboard. The back of his truck was right there, its bright red lights mostly lost in the swirl of mud, sand and smoke that our wobbly windscreen wipers could barely push to one side. At times, we were being hauled along at 60 k.p.h. by a man who seemed to have forgotten that we were right behind him, firmly attached to his towbar and with every chance of smashing into the back of his truck if he suddenly hit his brakes.

The white-knuckle ride ended after ten fearful minutes when the engine of our car suddenly roared into life. With third gear engaged, John had to act quickly, slamming his feet down onto both brake and clutch to stop us ramming the back of the pick-up and to prevent our engine from stalling. Successful, he hooted madly, the brake lights in front of us flared and we ground to a halt with John somehow managing to avoid the skid that would have been equally disastrous. Our saviour emerged from the cab of his pulsating truck, still grinning. While I was cutting the rope off the hideously-buckled front bumper that was now missing its number plate, the haggling began.

We had little money, no food and no cigarettes. We didn't mention the cameras and I was definitely not going to offer my camel skull. He shrugged, still grinning, leaving the ball very much in our court. In an attempt to engage his sympathy, I explained that all we had were the clothes we were wearing. That was a mistake, because the clothes would do 'just good'. We protested. The Cheshire cat smile dropped from his hardening face. And, then, right on cue, the engine of our car cut out.

Perhaps sensing that there was nothing here to detain him any longer, or perhaps desperate to get away from the dreadful smell that might even have been coming from our coveted clothes, our friend grinned another hideous grin, threw us a smug 'Au revoir', climbed back into his cab and roared off into the night. However serious our predicament – the car was dead and we had no idea where we were – our chief emotion was relief that he had left.

We crawled back into the car, vowing to refuse any further offers of help. At 11 o'clock, as a final gesture before an uncomfortable night, I turned the key in the ignition. The engine sprang back into life. We crossed over a bridge that had withstood the gales and the floods and by midnight were back on a road to Tunis. Wet, tired and drained of energy, we were determined to keep going until we reached the airport.

It was long past midnight when we drove into a little town that was mostly dead to the world. But there, on the right, were the lights of what looked like a little hotel. As we pulled alongside the tables on the pavement, an elderly couple appeared, and we asked if they could sell us any food, despite the ridiculous hour. They could do better than that. We were offered rooms, hot baths and a wonderful vegetable stew in exchange for whatever money we could afford. They even insisted on washing and drying our clothes. In the morning, as though reflecting the generosity of this little family business, the rain clouds had gone and the sun was shining brilliantly. The car started first time. We were back on the road with renewed warmth, the horror of the night before a distant memory.

Later that day, in the wreck of a car which shamed us both and with our getaway tickets at the ready, we drove into Tunis airport. The camel skull was in a desperate state and, although we didn't have much time before our flight, something had to be done about it. Looking around the open tarmac while John dealt with the car, all I could find was a small puddle twinkling in the sunshine, so I sat down and began scraping, washing and picking.

And then, from out of the blue, the *gendarme*'s shadow fell across my hand.

POSTSCRIPT

However impressed my father was with the camel's skull, it never found its way into any of his paintings. Perhaps this was because, by then, his brushes were mostly worn and his palette was beginning to dry. After his death in 1988, my mother took the skull to the art class she ran in Bembridge on the Isle of Wight and it immediately took centre stage in the still-life sessions. The skull, minus one tooth, has finally returned to me and it now sits alongside the bottled spider that gave me my own James Bond experience in Australia in 1968.

John A. Burton has dedicated his post-Tunisia life to conservation. After formative years with the Fauna Preservation Society (now Fauna and Flora International) in London, he created and runs the World Land Trust (www. worldlandtrust.org), a charity which uses public donations to purchase wild places around the world in conjunction with local partners. At the helm are the patron saints of wildlife and cricket, the two Davids, Attenborough and Gower, and the World Land Trust has been a spectacular success. It still has a long way to go.

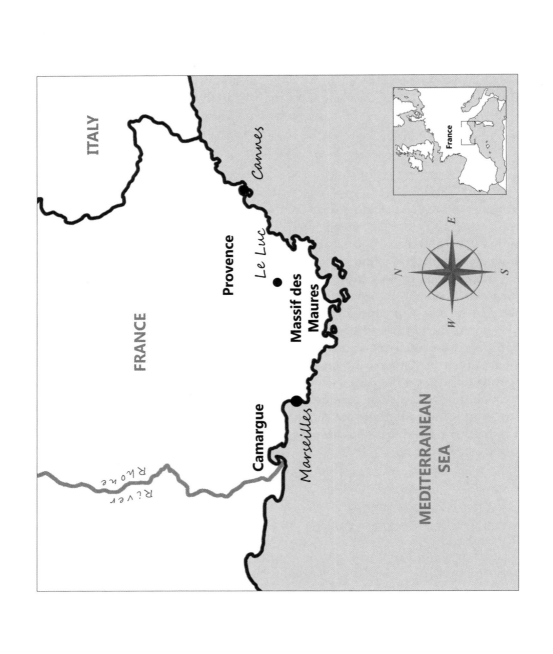

8

TONY AND THE LAST TORTOISE

With one final effort, she hauled herself out of the olive grove. She lay quietly in the long yellow grass, weakened by the heated rituals of the past few days. Despite her own late arrival, she had managed to find an equally late partner, and had now buried their fertilised eggs. With luck, the hungry badgers and beech martens that came sniffing in the night would not detect them below the surface. Beyond burying her eggs, there was nothing more she could do to protect them. It was time to go. Ahead lay the long, arduous journey back to the forest, where the canopy of leaves would shield her from the desiccating summer sun. The open, man-made olive grove might be perfect for her developing embryos but it was no good for her. Shade is a tortoise's ally and she would seek it out and rest wherever she could.

Five thousand metres above the olive grove, Tony Allen and I were descending on our flight from London to Marseilles. At the airport we picked up a car, stowed our filming gear and headed east towards the little Provençal town of Le Luc, a leisurely two-hour drive into the mountains of the Massif des Maures, overlooking the blue Mediterranean Sea. It was hot, even for the end of June, the relentless music of cicadas filling the air.

Close to Le Luc, we followed the instructions on our schedule to the front gates of a hideous tourist hotel. We had been booked into this piped-music monster by a Production Assistant who knew exactly how much we had to spend: our daily BBC rate for the south of France. But one look was enough.

We cancelled our reservation and drove down into Le Luc. A *gendarme* on traffic duty, slightly agitated by our disrupting his work with questions about places to sleep, dispatched us to a narrow side street where, not seeing anything even remotely hotel-like, we assumed we had drawn a blank. But then a door opened and a middle-aged man stepped out onto the pavement. We made eye contact, held, perhaps, a little longer than necessary, as though each of us was expecting a response to explain the other's presence. Finally, I asked if he knew of a hotel nearby. He smiled. He *was* the hotel, the door behind him its front entrance. Were we, by any chance, looking for somewhere to stay?

I explained that we were here for a week and that we would, if possible, like to stay in his hotel until next Sunday when we would be flying back to London. He stroked his chin thoughtfully. Yes, but there was one little problem. His hotel was for travelling salesmen, so it was open on only four nights of the week, Monday to Thursday.

Every Friday he returned to his family in Paris for the weekend. Today being Monday, he had just arrived from the north to begin another short week's work. June was always a busy month for him but he could fit us in for four nights if we didn't mind sharing a room - with each other, of course. Our first thought was that we should take up his offer and look around for somewhere else to stay for the weekend. We could always go back to the hideous hotel if we were desperate.

But he beat us to it. We were, he now beamed as though he had just solved the riddle of the universe, welcome to stay here until Sunday morning. He would leave us a front door key to post through the letter box when we left. He would charge us about £15 a night for the room and we could eat all our meals at the little bistro not 50 metres away. But then he went even further. He would also leave us the key to his wine cellar and we were to put the money for whatever we consumed into the box by the cash register. There were no fixed prices, just an adequate amount would do. No problem!

By early evening we had unpacked and were confronting our favourite French meal: steak, *frites*, salad, bread and red wine. Over coffee, we studied the map carefully. The next morning we were to meet the on-the-spot tortoise expert, Bernard Devaux, in the square of a little village high up in the sun-bleached hills of Provence.

Bernard was puzzled. He wanted to know why we had come looking for mating tortoises in the middle of June. The peak time for this activity was the slightly cooler month of May. There might still be stragglers, of course, but we had missed the real action by weeks.

This was embarrassing. I had chosen this particular week because the expert in England had told me that the tortoise-mating would continue until at least the end of June. The olive grove Bernard was to show us was the official study

Tony Allen filmed nesting black-headed gulls in England before setting off to find tortoises in France, 1986 (Author's collection)

site and we had been assured of finding enough tortoises to construct our sequences in just a few days.

Early in 1986, in the BBC canteen in Bristol, I had bumped into Ian Swingland, for whom, ten years previously, as an undergraduate, I had carried out a project on rooks around Oxford. Over coffee, I had told him that I was making a film about animal decision-making and he immediately offered his own research for consideration. Ian was working with Hermann's tortoises in the South of France. As with other reptiles, the eggs of these tortoises hatch as males or females according to the temperature at which they have been incubated. But if the females buried and abandoned their eggs to be kept warm by the sun, how could the sex ratio of the whole population be maintained at 50:50?

The key question for Ian was whether each clutch produced an equal number of males and females or whether half the clutches produced all females and the other half all males. Either way, the sex ratio would remain the same. To help answer his question, Ian needed to know if individual females were genetically programmed to choose sunny or shady places or how, if this was not the case, there was sufficient temperature variation within each nest to account for the different sexes. It sounded like an intriguing piece of animal decision-making and I was happy to include it in the film. We had nothing else planned for the middle and end of June, so the date was set.

Tony and I would film tortoises mating, digging their shallow nests and then burying and abandoning their half-dozen or so eggs. Back in England, we would film Ian talking through the questions he was trying to answer and we would illustrate the key behaviour with our short, edited clips. It would make an interesting few minutes on a once-popular but little-understood household pet.

Had Tony and I managed to find just a few of the mating tortoises we had been promised, we would have covered the sequences well enough. But, as it turned out, we spent five days scouring the olive grove to no avail. There was virtually no chance that our quarry had avoided detection. Tortoises are just a little slower, a little more conspicuous and a little less bothered than other creatures we have filmed. Had they been there, we would have found them.

We lunched on *baguette,* cheese, chocolate and water in the olive grove where we hunted high and low for sight and sound of tortoise activity. But Tuesday became Wednesday, became Thursday and then Friday. And still nothing. The heat and the endless rattling of the cicadas were intoxicating, dream-like, blurring the edges as we wandered to and fro with mantids, shrikes, lizards and snakes all around. But no tortoises. Bernard had been right.

Late on the Saturday afternoon, we gave up and made our way back to the car. I sat behind the steering wheel of the Mercedes parked at a precarious angle 40 metres from the track leading into the olive grove. The sun was still shining.

Tony packed his last camera case into the boot, closed the door and gave the all-clear. I started the engine, put the car into reverse and encountered a slight resistance, followed by a sudden, sharp bang. A puncture on this uneven ground, with the equipment stowed, the shadows beginning to lengthen and both of us feeling depressed about the sequence that didn't exist.

I switched off the engine, opened the door, and went round to join Tony. He was already kneeling by the nearside rear wheel, holding his head in his hands. There was no puncture. Just the exploded shell of what may have been the very last tortoise to bury its eggs in the olive grove that year. The same one, perhaps, who hauled herself away not long before our plane came down from the clouds a few days earlier. But now her eyes were closed and her head lay gently on a small rock to one side of the tyre that trapped her broken body.

Her job done, her buried eggs destined to hatch as both males and females, she had been resting nearby when we parked early that morning and created another bit of shade to help her journey back to the forest.

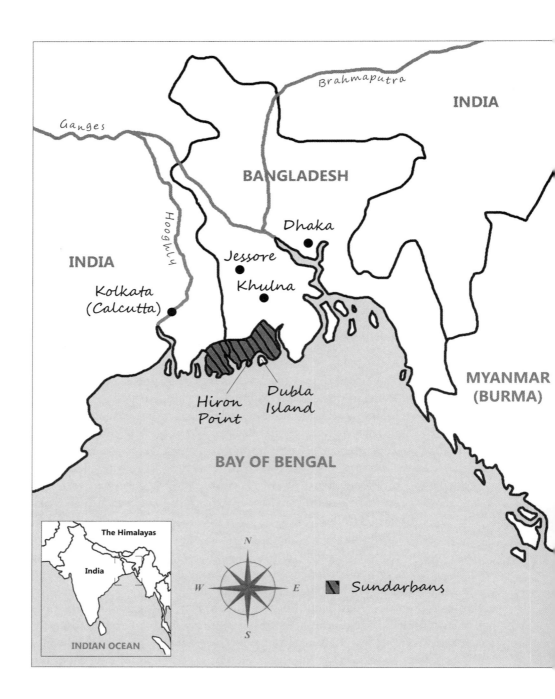

9

MANGROVES AND MAN-EATERS

The sun was lifting from the sea like an enormous egg yolk when we set out from Hiron Point in our double-engined boat. Hugging the coast for the first kilometre, past the graceful silhouettes of Bengali fishermen setting their nets, we turned into the mangroves and chugged up a narrow creek, an open highway through a tangled maze of roots and branches. We were nervous and I remembered the warning on the *Rocket* paddle steamer as we travelled south from Dhaka to Khulna on our way to the Bangladesh Sundarbans.

'But why,' the stranger had enquired, 'why would you ever want to go looking for man-eating tigers? They eat so many people who are doing their best to avoid them, you are certain to be killed.' *Rocket* logic, perhaps, but not science.

Tessa and I were nervous that morning because we had left Alan 18 hours earlier in what we hoped was a tiger-proof hide. The hide was 30 metres from the carcass of a deer which we had staked out above the high-tide line. If a tiger came after sunset, Alan would not be able to film it in the dark. His only chance lay with it staying on the deer until the sun rose, giving enough light for a decent exposure. It was a long shot, not without danger. The forest's beauty is overshadowed by its man-eating tigers. Accepting the risk of being eaten is a condition of entry.

The people of Bengal have lived with this occupational hazard for at least 4,000 years. But as their population has increased, particularly during the past century, more and more of them have needed to enter the forest to make a living; maybe 50,000 people each year, from both India and Bangladesh, because the Sundarbans is divided between these two countries. The forest offers food and firewood, but to collect such vital resources means confronting the tiger. And tigers must kill regularly: spotted deer, wild boar, rhesus monkey and people. All are vulnerable to attack.

The forest stands on a maze of islands where two great rivers, the Ganges and the Brahmaputra, merge after their long and separate journeys round the Himalayas. Where they empty their freshwater into the tidal Bay of Bengal, they form a vast delta, their shifting, muddy deposits so fertile that nearly 10,000 sq km of mangrove trees, criss-crossed by innumerable rivers and creeks, clothe the entire region. The Sundarbans takes its name from its most common mangrove tree, the sundari, a Bengali word for beautiful. Among the trees is a wealth of wildlife, and presiding over all others is the guardian of the Sundarbans, the Royal Bengal tiger.

The sun had climbed high above the trees when we secured the boat and set out on foot with our mandatory armed guard. His cumbersome .303 rifle, accurate over several kilometres, would, we all knew, be of little use in a hurried close encounter with a tiger in the forest. The mud sucked at our feet, slowing our progress through the stubby mangrove roots exposed by the low tide. We were still nervous, and not just for Alan. We had heard stories of how tigers take their human prey, of two men standing near each other, talking, one turning away and then turning back to find he is alone, muddy pugmarks on the tree. The tiger, unseen, had been watching, waiting for averted eyes before springing, ghost-like, hitting the tree and pushing off, whisking the unfortunate man away in its vice-like jaws. All, we were told, in a split second and with barely a sound.

We caught sight of the hide through the trees and stopped to search for the carcass with our binoculars. We were hardly expecting to see tigers, although some less sensitive scavenger like a monitor lizard might have been tearing at the decaying meat. We could see nothing that Alan might have been filming and moved slowly forward. There was no response to our arrival from within the hide. A jungle fowl, the first chicken, called nearby; and then a monkey. Had they seen us, or a tiger? Tessa called again, quietly.

'Thank God you're here,' came the strained reply which, despite its anxiety, filled us with relief. But then Alan continued in a loud whisper, 'Have a look round the back of the hide and tell me if there are any pugmarks.' We found a single line of them leading to the tower supporting the hide. They stopped at the wooden ladder we had wrapped in barbed wire the day before.

Alan emerged from his sleepless ordeal. In the middle of the night he had heard more than one tiger on the kill, a two-hour crunching of flesh and bone.

And then silence. Complete and utter silence until he heard a slight sound immediately behind him. Somehow, he thought, one of the tigers had left the kill, circled the hide and approached slowly from behind. How slowly is impossible to tell, but the tiger, if indeed one had arrived, would have taken its time, gently transferring its weight from one paw to the other as it moved silently and purposefully through the thick, cloying mud.

Alan had hardly dared breathe. His arms were folded on top of his camera, cushioning his head while he listened for sounds from the kill in front of him. That nagging silence, he had been thinking, could have been explained because the tigers had made short work of the remains of the deer. But were they now sleeping nearby, or had they already gone? His best thought, before the slight sound, was that they were taking a break and would resume feeding closer to daylight when filming might be possible.

In his heart of hearts, though, Alan was not convinced that all was well. The feeding had stopped too abruptly, as though the tigers had been disturbed. What if they had detected his presence and were indeed coming to investigate? If they were man-eaters and the deer had not been enough, could they, at that very moment, be weighing up their chances of another meal? And then, to give shape to his fear, there was that first slight sound immediately behind him.

It was dark and Alan was alone. Perhaps he was imagining things and should try to relax. But then there was something else: a gentle sniffing. It really was a tiger, and it really was less than five metres from his back. Despite the urgency of his brain, his muscles froze and his hand refused to reach out for the machete that stood beside him. At that moment, it felt as though the barbed wire counted for nothing. Defenceless, he waited for whatever might happen next.

There was no attack, no audible retreat, and yet, within half an hour, Alan knew he was alone. The tension had lifted, the tiger had gone. As the colourful wash of dawn embraced the night-soaked trees, Alan's muscles re-engaged with his brain. A cold mist rose from the muddied floor, thinned and drifted away. Birds took to their song posts and a jungle fowl approached the scant remains of the deer and began picking at its bones.

There was no knowing where the tigers were when Tessa and I returned to the hide in broad daylight. We drank black coffee with Alan while our guard kept watch with his rifle, trying to reassure us that we were not actually in that much danger. Had he missed something? There were, he continued, very few tigers in the Sundarbans. It was just that they are so persistent that they will follow boats all day long under the dense cover of the forest, biding their time. That was why local people thought there were so many man-eaters. Whenever they stopped, a tiger would always appear and attack them.

There are probably at least 300 tigers in the Sundarbans, more than in any other stretch of habitat in the big cat's entire range. Some of them – more, it would seem, than can be explained by the accepted causes of old age and injury – have taken to feeding on human flesh. It is a real problem for the people of an impoverished region who must enter the forested delta throughout the year. Filming the relationship between people and tigers in the Sundarbans was the task we had set ourselves. Alan McGregor, Tessa Woodthorpe-Browne (soon to become Tessa McGregor) and I were here in 1984 to make a BBC *Natural World*.

Nobody knows how many people are killed each year by tigers in the Sundarbans. Twenty? Forty? Two hundred? A lot of people enter the forest illegally to avoid paying their dues and their deaths are not recorded. Officials are reluctant to publish figures that might give the mangroves a bad press. The truth might deter those people who enter the forest legally and pay permit fees and taxes on their collections of wood, fish and honey. We heard stories about where some of this money might be going but the Minister for Forests in Dhaka later denied that corruption was at work. He did not know how much we knew about him, that his first job when taking up office had been to tour the Sundarbans to negotiate a good percentage of the money raised in the forest for himself. And he was not, in our own limited experience, the only minister on the make.

One evening at our filming base at Hiron Point, the pilot rest house where the mangroves face the open sea, we were invaded by a workforce, each person armed with a mop and bucket. They had been despatched from Khulna on the morning boat and now, after a ten-hour journey, they began spring cleaning every single room. Our minder from the Bangladesh Film Corporation, Saiful Haq, told us that the Minister for Telecommunications was coming to stay for a few days and that he would decide when he arrived whose rooms he would commandeer. Saifal advised us to be half-packed and ready to move. To our relief, because we had the best rooms, the Minister stayed on board his boat and even after two days we had not seen anything of him. At nine o'clock on his third evening, however, we saw his brightly-lit boat slip its mooring and head for the open sea, suggesting that he was not returning upriver to Khulna, at least not via the shortest route. Curious, we went down to the jetty and asked the guard on duty if the Minister would be returning. With the straightest of faces, he told us that he would be back in a few hours when he had finished hunting, and hopefully with a tiger.

We grabbed lifejackets and a spotlight, jumped into our boat – all 120-horsepower of rigid-inflatable inshore rescue craft shipped out from England – and, with Alan at the helm, sped off into the night. Cutting the engine every few minutes, we soon located the much larger craft and its noisy, excitable crew. The resounding report of a 12-bore shotgun echoed through the forest. The Minister, seated at the prow of his boat, and with the bank illuminated by hand-held lights, was firing randomly into herds of spotted deer, half-blinded before they could flee back into the forest. Amid excited shouts loud enough to attract the curiosity of any number of distant tigers, minions were being sent overboard to retrieve the deer that had been killed or wounded. Apart from the obvious danger to life and limb, it was mayhem, a real disgrace. It was time to do our bit for Greenpeace.

Under the challenging glare of our own spotlight, and rather taken aback by our sudden appearance, the Minister composed himself quickly. He was not impressed by my advice that animals in a Nature Reserve are to be conserved, not killed. 'They are conserved for me to shoot,' he laughed back, before adding threateningly, 'and I know who you are, you are the BBC and I can have you thrown out of my country any time I like.' We decided at this juncture – he was leaning over the rail above us with a loaded gun – to retreat and face him in the morning at Hiron Point. But by the time we splashed water on our faces at first light, his boat had already gone, and of his threat we heard no more.

Tigers need small amounts of fresh water every day or so. The Sundarbans is tidal, inundated twice a day by saltwater sweeping in from the Bay of Bengal to contaminate the fresh water carried down by the rivers from the north. So here in the mangroves, many animals must get their 'sweet' water from their food. Some eat succulent leaves, particularly in the morning when these are also wet with dew. Tigers are carnivores. They might chew on wet grass at times but, if they can't drink, their water requirement can be metabolised, at least in part, from the meat of the animals they kill. Deer and wild boars are special tiger food but they are not easy to catch unless ambushed at close quarters. Monkeys, another favourite, will stay up in the trees unless they have to cross open spaces or they feel the need to beachcomb for small creatures in the mud exposed by the falling tide. By dropping leaves from the treetops they attract the deer to come and feed beneath them. Now, in a reciprocal arrangement and with the deer providing the low-level tiger-spotting eyes, the monkeys can descend from the trees with more confidence. But once down they must now be on the lookout for giant pythons slithering along shallow tide-drained creeks, and crocodiles floating close to the shore, mostly submerged but alert and ready to lunge.

People are easy prey for tigers in the Sundarbans. They are clumsy, their senses inadequate and when they try to escape on foot they blunder, stumble and cry out, their hearts pounding with fear. Their only advantage is technology – weapons, fire on demand, boats. Without these, they are hopelessly vulnerable. On occasion, we ventured forth without our armed guard, feeling that we were probably safer without him than with him. We found ourselves one day, just the three of us, gliding along narrow creeks, chatting away and looking for things to film while the tide was low. Limpets on the exposed trunks of golpata palms seemed to say 'tidal' in a strong visual way, so we stopped. Alan and Tessa set up their tripod and camera. Curiosity led me out of the boat, up the slimy 2-metre-high bank and into the thin covering of trees. I could see ahead for at least 30 metres and, given that we had neither seen nor filmed a tiger, was not at all concerned. Alan called up for me to be careful, not to wander too far out of their sight or out of their minds. Five metres ahead of me, a shallow gully was reduced to just a trickle as the water seeped in from the surrounding mud. A line of indentations caught my eye and I went to have a closer look. Pugmarks, and fresh enough to have been made quite recently. Closer still and I could see how sharp the edge of each print was in the soft mud. And then I saw that they were slowly filling with water. The pugmarks to my right were already full, losing their sharp outline, while those to my left, leading towards a clump of bushes 40 metres away, were still empty, fresh and crisp. Had a tiger been following us? Had it been alarmed by my climbing the bank? Had it already made for the little bit of cover where it could ambush us as we passed by in the boat? Had stopping to film the limpets saved us from an attack? Keeping my eyes firmly on the bushes – it is generally accepted that tigers prefer to attack from the rear – I backed off through the mud and slid down the bank. Alan and Tessa had completed their filming, so we packed up, turned the boat round and slipped quietly away.

Boats travelled down from Khulna to Hiron Point quite frequently. From one of these stepped a man with a cow which, he said, was for us to put out as bait. He had heard of our repeated failure to film a tiger and thought this might be a way to help us succeed. Not wishing to reject his concern and his generosity, we agreed to take the cow for a few days, on the understanding that if it was killed we would pay for it, and if it survived we would hand it back with a small payment as rent.

The cow was tied up in a grassy clearing a day or two before the full moon, just beyond a large tree where Alan could sit safely in a hide mounted on a platform. He spent more than 50 hours in that hide, staying overnight so he was ready to film for a few hours after dawn. At the end of three tigerless sessions, we were feeling so sorry for the cow which, it seemed, mooed more from loneliness than fear, that we decided she should go back to her owner unscathed. Setting out from Hiron Point with two guards, I left one in the boat and walked with the other through a narrow corridor that had been cut through the mangroves to let people in and out. We then crossed the clearing to meet Alan at his tree. We quickly dismantled the hide, untied the cow and made our way back across the open glade.

Halfway down the little corridor, not ten metres from the boat, Alan realised he had left his binoculars hanging in the tree. On the spur of the moment we left the cow – it could not turn round and the guard on the river was just ahead of it – and retraced our steps. We were back within five minutes. The cow had gone. Up ahead, the guard in the boat had seen and heard nothing.

If the cow had managed to double back, we would have seen her, or at least her tracks. The two guards examined the evidence, what little there was, and declared that a tiger had followed us from the grassy clearing, keeping low under the tangled mangrove roots close to the corridor. When we turned to retrieve the binoculars we had given the tiger a choice – us or the cow. It had chosen the cow, they continued, either because there were three of us or because it was not a man-eater. Having crept up to the unfortunate cow and killed her with a dislocating bite to the neck, the tiger had picked her up by the small of the back and lifted her clear of the roots that hemmed her in. Then, carrying the cow across a raised lattice-work of roots – the guard now pointed to some mud as his evidence for this conclusion – the tiger had dropped down some distance from where we were standing. She would, even now, be crouched over her victim, her eyes fixed firmly on us. If the guards had painted an accurate picture of what had taken place in those few minutes, it was more than just spine-chilling. It demonstrated an incredible feat of strength, as though the cow had just vanished into thin air. We felt sad, guilty and vulnerable, and returned to the boat in silence.

When a human has been killed by a tiger, the warning is a stick flying a fragment of cloth, often red, placed in a prominent creek-side position. People are reluctant to go ashore. They tie up in mid-river, anchored to a pole driven deep into the mud. But tigers are not averse to swimming out to moored boats to secure their human victims, even in the dead of night. Isolated in the mangroves, people are both vulnerable and available, 24 hours a day. Sundarbans tigers know this and they have acquired a reputation for being more nocturnal than tigers elsewhere. They come and go as they please, like phantoms of the night.

The lack of fresh water in the forest is a greater problem for people than it is for tigers. Beyond drinking every day, people need water for cooking and washing. They must carry it with them on their boats. In some places it has been possible for the authorities to dig large pools to catch rainwater or to fill slowly from below ground where the fresh water table is high enough. People, deer, monkeys and wild boars focus their lives on these vital places. Tigers are also attracted to them, and not just to drink.

In New Delhi, before travelling down to Kolkata (Calcutta) to have a look at the Indian side of the Sundarbans with Alan and Tessa, I talked to the staff of Project Tiger about the problem of people being attacked so frequently. The difficulty, they said, was that while you can control people entering a normal tiger reserve from the fixed point of, say, a village, and you can monitor the whereabouts of the small number of tigers as well, the Sundarbans are very different. People can enter the forest from all directions. They slip in by boat undetected and once inside the vast labyrinth of islands and tidal creeks, they are untraceable. It is impossible, I was told, to keep determined people out of a place they depend on for their livelihood. At whatever risk to their lives. And matters are hardly improved if you don't know how many tigers there are and where they might be from one day to the next. All you do know is that too many of them kill and eat people, and when it happens, of course, it is too late, not just for that person but also for their family. We talked at length about the man-eaters and why there should be so many of them. One suggestion was that tigers are more likely to attack people who are bending over to cut wood, adopting the profile of a deer, a bit like those surfers who are said to be attacked by great white sharks because their profile, from below, is similar to that of a seal.

We touched on the idea that the sea could contribute to man-eating. Tigers need small but regular amounts of fresh water to drink. Because, in the absence of standing water, some of this can be derived from the meat of their prey, they must kill frequently, even if they consume only a small portion of meat from each of their victims. In other words, they may be killing frequently to satisfy their thirst as well as their hunger. The incoming tidal flow may displace or contaminate the meat, providing food for crocodiles, crabs and monitor lizards. So, instead of returning to a carcass, the tiger simply looks for a fresh replacement. And deer, monkeys and wild boars, the tigers more natural prey, are much better at saving their own lives than we are at saving ours. With so many people out in boats or on foot during the year it is little wonder that tigers turn on them as often as they do.

If 'killing to drink' does contribute to man-eating in the Sundarbans, it would be reasonable to expect all its tigers to be eating people, but they obviously are not. If they were, human deaths would run into many thousands a year. People would stay away and the tigers would either perish or their numbers would be reduced just to those who could make a living out of the forest's other animals. It has been suggested that, because the salinity of the Sundarbans varies throughout the forest, the need to kill people varies from place to place and that this would explain why not all tigers are man-eaters. Having lived and worked there for weeks on end, I can't help coming back to the idea that it is also a question of efficiency. In conditions more favourable than mangrove swamps, tigers manage to kill their natural prey just once in every ten to 15 attempts. If that figure is doubled or trebled in the Sundarbans, where thousands of people flounder around, detached from the nuclear safety of a village, it is not difficult to see why human flesh might be added even to a healthy tiger's diet.

The conversation turned to people wearing masks in the forest. The masks are worn on the back of the head, not the front. The logic of this is that tigers prefer to attack from the rear, waiting for a person to pass before pouncing. The surprise to the tiger of seeing another face walking backwards, and now staring straight at it, has no doubt been a lifesaver.

Around the freshwater pits, another experiment has been a full-sized human model wired to a battery delivering a nasty 240 volts to an attacking tiger. This, too, has reduced the number of human kills. We agreed, though, that what was required was variety, that tigers would soon learn about shocking dummies and two-faced men. A succession of other devices would be needed to keep these crafty cats guessing, delaying their attacks long enough for another life to be spared.

From meeting the Project Tiger people in Delhi, I flew to Kolkata to meet Alan and Tessa at the Fairlawn Hotel in Sudder Street. They had somehow managed to squeeze past a long waiting list and even reserve a room for me. The Fairlawn is an integral part of Kolkata's history, a colonial outpost perfectly adapted to a modern self-ruling India. Owned and run by Ted and Vi Smith, they busied themselves round their guests, making them feel they were the most important people in the world. Vi sold me material for curtains to take back home and when I had bought almost the entire stock of hand-embroidered sheets, cushions and shawls from the man from Nepal who ran a little stall in the hotel grounds, it was Vi who sent me off to New Market with one of her employees to have a suitcase made for the flight back home. Intrigued by the hustle and bustle of the covered market, I went back on my own a few days later to take some photographs. In a nearby street I came across a sight that made me wonder just how much we had progressed as a species. Tons of rotting vegetables lay knee-deep in a yard by a back entrance to the market. Four young boys were sifting through the putrid pile on their hands and knees, searching for anything that might still be edible. Dogs and crows were paying close attention. They, too, were hungry. The closer the dogs and the crows could get to the foraging humans, the better their chances of success. So while the boldest birds perched on a boy's head, the most daring dogs followed at his shoulder. The moment his hand lifted something for inspection, the crows and the dogs piled in, flapping and fighting. The smell and the mess were dreadful, but they were nothing compared to the human degradation. My camera remained in its case.

Alan, Tessa and I soon realised that the heart of the Sundarbans was a 'man's world', no place for women and children. It was the men who shouldered the yearly tasks of collecting wood, palm leaves, fish and honey to be taken home or sold in local markets bordering the mangroves. The women stayed at home in villages outside the forest, where they were occasionally at risk from tigers, but they did not venture into the forest to work. Around Hiron Point, whose pilot rest house was our home in Bangladesh, fishing was big business. Freezer ships did the rounds, collecting fresh hauls and packing them off to international markets, particularly Japan, where fortunes were paid for the large Bay of Bengal prawns. We guaranteed our own supply of these delicious shellfish by paying the fishermen more than they were being paid by the racketeers who exploited their labour and left them exposed to the threat of tigers. The heavily-skewed sex ratio of workers in the mangroves means, of course, that virtually all the people killed by tigers are men, leaving a growing number of husbandless wives, fatherless children and brotherless sisters. Beyond the Indian Sundarbans, a special 'Widows' Village' is home to the women and children who have been deprived of their men by man-eating tigers.

There is, however, one time of the year when women and children are actively encouraged to enter the Sundarbans. It is the November full moon, the cyclical beginning of a four-month fishing season that will see a large number of men killed and eaten by tigers. It is a time for the whole family to pray to the Mother of the forest, Bonobibi – the only permanent woman in a man's world – for the safety of its breadwinner over the coming months. We decided to film this festival, sensing it could make a strong opening sequence to the programme.

The day before the festival we went across to Dubla Island, intending to plan the filming and sleep over, ready for a pre-dawn start the following morning. Our overnight accommodation turned out to be a ramshackle wooden shed next to a small freshwater pool. While we considered the pros and cons of putting up with a night's discomfort before a long filming day, news came through that someone had been killed by a tiger within the last week. To spend the night in an unprotected shed close to a supply of drinking water that could be visited by a man-eating tiger would, as our guard put it, be a bit risky. Heeding his advice, we decided to head back to Hiron Point and return early in the morning.

It was cold and dark when we walked down to the jetty from the rest house. The sky reverberated with a brilliance of stars never seen in polluted parts of the world. The usual pre-dawn silence, such a delight before the day wakes, had gone. Boats were everywhere as families and friends emerged from the mangrove creeks and set out across the open sea. Dimly-lit lanterns swayed from the bows and masts of their little boats as they bobbed their noisy, excited way towards Dubla Island.

With an embarrassing, peace-shattering roar of our engines, we left the flotilla far behind. In no time at all we were unloading our gear onto the white sand above the high-tide line, hauling our empty boat ashore and setting up camera and tape recorder. We had to be ready for the rising sun that was such a crucial part of the festival. Alan decided on a little burst of film to check his camera, a trustworthy Arriflex that had never let him down. Nothing. He tried again, still no response. Was this the moment we would rue that deliberate decision, for the sake of mobility, to leave the back-up camera behind – just this once? Now the flotilla of boats we had passed on the sea was arriving, people were spilling

Land Rover problems in the north Australian bush, 1968 (Courtesy of Cliff Frith)

From left, Tony Hiller, Brian Booth and Ralfe Whistler, with Harry Butler and Dan Freeman skinning a wild pig, Kimberleys, Western Australia, 1968 (Courtesy of Cliff Frith)

Elephants destroy baobab trees for the high water content of their spongy fibres, Tanzania (Author's collection)

A sleeping leopard is easily missed – the tail is the giveaway! Serengeti, Tanzania (Author's collection)

Clockwise from left: Sunset with white-browed sparrow-weaver nests, Transvaal, South Africa (Author's collection)

Black rhinoceros, Zimbabwe (Author's collection)

Black-maned lion, Ngorongoro Crater, Tanzania (Author's collection)

A metre-long grey monitor lizard from the Sahara Desert in Tunisia (Author's collection)

John A Burton suitably camouflaged in the Saharan sandstone of Tunisia, 1973 (Author's collection)

Moussier's redstart in a Tunisian oasis (Author's collection)

Above: Alan and Tessa McGregor film travelling musicians on Dubla Island, Sundarbans, Bangladesh, 1985 (Author's collection)

Below left: Alan and Tessa McGregor filming nipa palms in muddy mangroves, Sundarbans, Bangladesh, 1985 (Author's collection)

Below right: People draw fresh water from specially dug pits among coastal mangroves, Sundarbans, Bangladesh, 1985 (Author's collection)

Above: Mangroves grow further out to sea as their roots trap silt carried down by rivers, Sundarbans, Bangladesh (Author's collection)

Below: Hauling nets at low tide, coastal Sundarbans, Bangladesh (Author's collection)

Above: Drying fish for export is a major industry of the Sundarbans mangrove forest, Bangladesh (Author's collection)

Below left: An isolated boat is vulnerable to attack from man-eating tigers, particularly at low tide, Sundarbans, Bangladesh (Author's collection)

Below right: A floating sheep is a meal for scavenging jungle crows, Khulna, Bangladesh (Author's collection)

Above: Low tide inside a mangrove forest exposes a spikey carpet of aerial roots, Sundarbans, Bangladesh (Author's collection)

Below: Mangrove forest from the air, Sundarbans, Bangladesh (Author's collection)

Above: The lifelike skeleton of a dead crab lying in samphire grass, Camargue, France (Author's collection)

*Below: Boats are the only form of transport through and around mangrove
forests, Sundarbans, Bangladesh (Author's collection)*

Waterbird breeding colony (pale area on trees) on the Orinoco/Apure flood plain, Venezuela (Author's collection)

Above: Young red-bellied piranhas, filmed in captivity for a 1997 BBC Wildlife on One *(Courtesy of Tony Allen)*

Left: Tony Allen braves the piranha tank to carry out vital repairs, Hato El Cedral, Venezuela, 1994 (Author's collection)

*Above: Chris Catton (left) recording a red-bellied piranha feeding frenzy
induced by Tony Allen, Venezuela, 1994 (Author's collection)*

Below left: Chris Catton, still smiling after a long day filming piranhas, Venezuela, 1994 (Author's collection)

Below right: James Gray filming on the rough Atlantic coast of Dominica, West Indies, 1991 (Author's collection)

Dan Freeman photographing young male elephants, Zimbabwe, 1998 (Courtesy of Tony Allen)

Nigel Tucker sound recording young male elephants, Zimbabwe, 1998 (Author's collection)

*Nzou the elephant, matriarch of a herd of domestic buffaloes,
Imire Game Reserve, Zimbabwe (Author's collection)*

Left: Production assistant Jim and local girl selling hand-embroidered bags, Kunming, China, 1987 (Author's collection)

Right: Michael Richards filming black-headed weaver nesting colony, Mtito Andei, Kenya, 1980 (Author's collection)

Below: Mike Potts filming in reedbeds, Lake Neusiedl, Austria, 1980 (Author's collection)

Clockwise from left: Male black-headed weaver and the early stage of his nest, Kenya (Author's collection)

Male black-headed weaver displaying to attract females to his completed nest, Kenya (Author's collection)

Michael Richards filming on Cousin Island, with Praslin Island in the background, Seychelles, 1980 (Author's collection)

Clockwise from left: White-tailed tropic bird at nest entrance in casuarina tree, Cousin Island, Seychelles (Author's collection)

A young fairy tern grows up where it hatched on a bare branch, Cousin Island, Seychelles (Author's collection)

Wright's skink on drying out coconut, Cousin Island, Seychelles (Author's collection)

Above: Tony Soper talks to camera (Ron Bloomfield directing for the BBC's Discovering Birds*), Farne Islands, England, 1982 (Author's collection)*

Below: Gannet breeding colony, Bass Rock, Scotland (Author's collection)

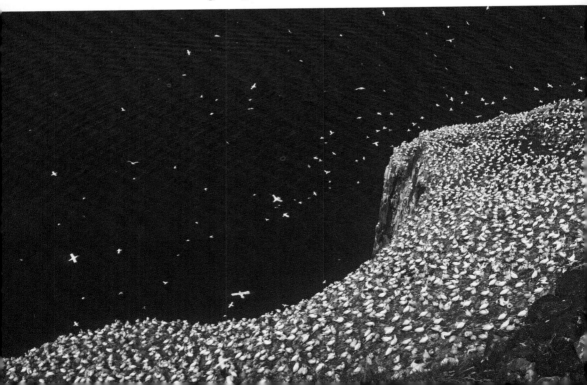

Above: Incubating gannets are just out of reach of each other's sharp beaks, Bass Rock, Scotland (Author's collection)

Below left: Florida Everglades with alligator in the foreground, USA (Author's collection)

Below right: Tony Soper (right) and Sonny Bass rehearse a piece to camera for BBC Birdwatch, Robin Prytherch directs, Florida Everglades, USA, 1986 (Author's collection)

Emperor moth on yellow iris, Camargue, France (Author's collection)

Above: Female Barbary macaque with baby, Rock of Gibralter (Author's collection)

Below: February sunset inside the Arctic Circle, Bodø, Norway (Author's collection)

out, shouting and laughing, and children were running and playing in the sand. Now they were coming along the beach from left and right in their hundreds, with less need for torches because the sky was brightening fast. We were ten minutes from sunrise and the camera wasn't working. Alan's last resort was to expose the circuit board, an array of fuse-like connections hidden on a plate beneath the camera body where its various filming speeds are controlled. He took out several pieces, cleaned them and put them back. The camera burst into life. Not a moment to lose. The sun was peeping over the skyline, the band was playing and the powerful, rhythmic dance was underway. Alan discarded his tripod and went in among them, hand-holding his camera. Centre stage belonged to the drummer who glared defiantly at the lens as he pounded out the rhythm that honoured Dakshin Rai, Bonobibi's Muslim consort, who ruled over all the Sundarbans.

It was potent stuff. Around the band, people lay in the surf offering prayers, fruit and rice to keep their men safe from tigers. Garlands of flowers floated in the gently-swelling waters that broke over the sand, littering the beach with a colourful sway of petals and leaves. And to Bonobibi herself, went the plea:

O Mother
Thou who lives in the forest,
Thou, the very incarnation of the forest,
I am the meanest son of yours.
I am totally ignorant.
Mother, do not leave me.
Mother, you kept me safe inside your womb
For ten months and ten days.
Mother, replace me there again,
O Mother, pay heed to my words.

And then it was over. The band broke up and its half-stoned members trudged off to the boat that would take them away from Dubla for another year. We were back at Hiron Point by early evening after a little detour to film a large saltwater crocodile, another man-eater, which had been seen floating offshore just round the coast. We heard nothing more of the Dubla man-eating tiger and decided, perhaps a little cynically, that our guard had invented the story because he hadn't liked the idea of spending the night in that dilapidated shed. But then neither had we, and the comfort and safety of Hiron Point was, after all, only three-quarters of an hour away.

Khulna was always our starting point to reach Hiron Point by boat, the only way down through the mangroves. Getting to Khulna from Dhaka was the more interesting stage of the journey. There were three ways and we experienced them all. We travelled first class on the *Rocket* paddle steamer, an entertaining 24-hour experience that appears to be known to no-one in the world except Michael Palin, who ended his memorable television journey along the Himalayas on board, and our neighbours Barbara and John Dalton, who live less than 100 metres from us in Bristol and who undertook the journey long before we met them. The quick way is to fly from Dhaka to Jessore and complete the journey to Khulna with a 40-km taxi ride, an efficient system for business people with nothing on their minds but to arrive, work and depart as quickly as possible. The most interesting journey was the one that involved being driven all the way from Dhaka by minibus, taking the whole day, crossing rivers on crowded ferries and getting a wonderful feel for the countryside and its people. We did this journey at different times of the year, on well-constructed, all-weather roads raised above the surrounding land to protect them from flooding.

In the pouring rain of the wet season, the land on either side of the road was under a metre of water. Beyond the lush green of roadside vegetation, irrigated fish farming was in full swing. Egrets, cormorants, herons and kingfishers took advantage of this concentrated supply of food. They perched on picturesque and ingeniously-constructed bamboo frames operated by one or two men walking along a central bamboo. As the men moved away from the centre, their weight tipping the bamboo downwards, a net was lifted clear of the water level on the opposite side. Near-naked children and their colourfully-dressed mothers scooped out the fish, competing with the birds that jumped in and grabbed what they could before the laughing, stick-wielding children ran in from the sides. When the fish had been removed, the frame-walkers transferred their weight back across the turning point and the net was again lowered beneath the sparkling surface. A gentle flow of water was maintained through the net by the opening of sluice gates that allowed fish to pass from one section to another as they were ready to be caught and eaten.

Other sections of land were kept deliberately shallow so that paddy fields could be maintained during an extended growing season. We would eat the most wonderful meals of rice and fish by the side of the road, washed down with cool, refreshing coconut milk that we drank straight from the punctured nut. There was something reassuring about drinking this juice, knowing that it was pure and safe and would not lead to tummy upsets later in the day.

To stand at the same roadside spot towards the end of the dry season, when there had been no rain for months, was incredible. There were no nets, no children having the time of their lives collecting fish and beating off avian competitors with big sticks. Just hard-baked, dusty ground strewn with the withered yellow grass that once stood so lush and green. There were no people visibly occupied in the desolate heat of the day. Instead, intriguing columns of smoke rose from the parched plains at neatly spaced intervals. Bricks were being baked in underground kilns. In this labour-intensive, unmechanised world, perfectly-formed clay bricks would be brought to the surface and transported to the roadside. Here, armies of women smashed them with hammers, reducing them to hardcore to repair old roads and lay the foundations of new ones. Nearby, whole bricks were stacked high. They were ready to build the walls of homes that would be overlaid with the leaves of the golpata palm, cut under the threat of man-eating tigers and brought up from the Sundarbans on the few rivers that flowed at this time of year.

There was one other way to get down to Hiron Point, and that was by helicopter. There was a helipad on the flat-roofed pilot rest house, keeping it accessible throughout the year in case of emergency. We would need to do some aerial filming at some point and, in Dhaka at the end of one filming trip, I approached the Minister for Foreign Affairs, who was my main Government contact. He thought it might be possible, though the Sundarbans was a politically and militarily sensitive area. He explained that the border with India ran down through the mangroves on the western side and that the border with Burma was not too far away in the east. Added to this, within the Sundarbans themselves there were still outlawed freedom fighters, hiding after the war of independence with Pakistan in 1971. A low-flying, hovering, back-tracking army helicopter might not send the most tactful signal to them. And besides, they were heavily armed.

We had, in fact, already met one of these outlawed people, a Major Zed who knew all about us from his network of contacts in Dhaka long before we got anywhere near the mangroves. He tracked us for weeks before making an approach, inviting us to lunch with him and his band of followers. On a little island deep in the Sundarbans, he was happy with our explanation of why we were there for so long. He had smiled when we told him that we might one day be filming from a government helicopter. After lunch with Major Zed, I left Alan and Tessa to their filming and set out on the long journey back to England. I had to get Alan's film processed, printed and logged. Then I would spend several days looking at it and wondering, in the increasing likelihood of us never filming a tiger, what sort of programme we might end up with.

When I returned to Bangladesh, clearance had been granted for our aerial filming, though it would have to be carried out from Dhaka, with us on board from the outset, rather than with the helicopter coming down to meet us at Hiron Point. I somehow felt that this was a ploy to guarantee that we would pay for the entire time that the helicopter was away from its army base in Dhaka, rather than paying just for the filming time once it arrived to pick us up after a 200 km flight. Filming from helicopters is always expensive, wherever you are in the world, so I had built a figure into the budget that would allow for a four-hour return flight and three hours' filming. A seven-hour day that would cost £5,000. The money had to be paid in cash, in advance, and I was whisked off to a bank in an official car and then driven under armed escort to the military base to hand over the bundles of notes. When the formalities were complete, we arranged a date when Alan, Tessa and I would return to Dhaka, load our gear onto the helicopter and fly back down to Hiron Point.

Three weeks later we were back in the Dhaka Sheraton, welcomed with a broad wink by the doorman, no doubt one of Major Zed's contacts in the city. We were waiting for the weather to be just right for a long day in the air. It was, when it happened, a memorable flight, though lacking the intimacy of the people and the rhythms of their lives on the ground. A bonus was a helicopter pilot trained for war, one able to duck and weave, hover, fly backwards, sideways and almost upside down as Alan filmed from every conceivable angle. From the safety and comfort of the skies, despite the thought of Major Zed's sharpshooters, the mangroves looked far less intimidating than we knew them to be down at ground level.

One thing we did not know in the mid-1980s was that the rising tide of climate change was preparing to engulf the low-lying Sundarbans. Environmental changes in the Bay of Bengal will be complex, brought about by such things as increased Himalayan meltwater and the expansion of the sea as it warms up. Changes in coastal salinity will affect the distribution and behaviour of both animals and plants. Ten thousand square kilometres of mangrove forest will be severely threatened. The lives of millions of people will be disrupted. Several islands have already been washed away and more are set to follow, perhaps 20 in the next ten years. The pressure on people and wildlife will build for the next half century, after which satellite images may simply show the Bay of Bengal reaching 100 km further inland than it does today. In the absence of stabilising mangroves and their cargo of wildlife, including those 300 precious tigers, erosion will continue apace. Tigers will be forced deep into human territory. Despite winning a few bloody battles, they will eventually lose their war against people. But people will also lose. Villages and towns, no longer protected by the forest, will be exposed to cyclones driving up from the south with devastating effect. Even by 2020, a huge number of people will have been driven away from their drowning homes and fields. They will seek refuge in cities where they will live in vast slums. To combat poverty and infant mortality, they will continue to have large families. Many of their children will be forced into hard labour. They will work 12 hours a day, six days a week, to add a few meagre taka to the family pot.

Despite the prediction of our fellow traveller on the *Rocket* paddle steamer from Dhaka, we came through our two-year mangrove experience physically unscathed. But other people did lose their lives to tigers while we were there, despite their devotion to Bonobibi and Dakshin Rai. Exactly how many people died is impossible to know. During our longest filming trip, a local figure of 40 deaths was double the official figure entered into a Forestry record book in Khulna.

A discrepancy of 20 deaths in just three months gives an idea of how difficult it is to keep track of what is really happening in this wild and beautiful forest: a forest that remains wild and beautiful, for the time being at least, because it is the forces of tiger and tide, not people and profit, that are the true masters of the Sundarbans.

Sondrio

Alps

②

Y

Aosta
Milan

①
Venice

Turin
River Po

Tuscany
Archipelago
Florence

ITALY

Pescasseroli

Ceresole
Reale

Civitella
Alfedena

Gargano
Peninsula

CORSICA

Pontine Marshes

④

③

SARDINIA

Rome

⑤a

⑤b

MEDITERRANEAN
SEA

⑤c

National Parks (1992)

① *Gran Paradiso*

② *Stelvio*

③ *Circeo*

④ *Abruzzo*

⑤ *Calabria:*

    ⓐ *Sila Grande*

    ⓑ *Sila Piccola*

    ⓒ *Aspromonte*

Gambarie

X

SICILY

X Mount Etna

Y Dolomites

🔺 **Apennine Mountains**

N

W   E

S

10
THE ITALIAN JOB

We sat at a long wooden table in a small stone house. A refreshing breeze entered through half-shuttered windows and the outside door creaked on its ancient metal hinge. Three generations of women, each dressed in black, moved along the sunken path connecting the cool of our open room to the kitchen next door. Their short journeys yielded carefully cured or preserved delights from the Italian countryside. Fungi, artichokes, gherkins, olives, oil, butter, cheese, meat and wine, all gathered or handmade by the household and stored for the coming winter.

I had asked our guide Renato to book a restaurant for ten people as my final gesture of thanks to the Wescam aerial film crew who had worked so hard during the year. Their contribution to the series of films I was producing on Italy's National Parks had been vital. But autumn in Calabria is not the best time to find restaurants open to the public. On the afternoon of our arrival in the south, Renato had asked where the helicopter would land after our aerial filming the following morning. Armed with a local map and the fervent hope that the weather would not change our schedule, he had set out in secret to find a family that would open its home and its larder to visitors from the north.

Our filming over Aspromonte – the most southerly of the three sections of Calabria National Park – went better than expected. After a hesitant start, the sun broke through to brighten the beech trees hugging the rocky slopes, their late

October leaves dancing red, yellow and gold as we swept overhead. The helicopter landed at 11 o'clock high in the mountains, exactly where Renato was waiting for us with flasks of sweet, black coffee. We decided that things had gone so well that we would do more aerials in the afternoon only if the light was exceptional. With several hours to fill and with that special lunch in mind, I turned to Renato, expecting a long drive down into the valley. Instead, he smiled, turned on his heel and led the way to a little outcrop of mountain homes.

Claudio, the pilot, drank nothing, the rest of us a little more. There were no restrictions on what we could eat so we worked our way through the simple and beautifully presented local food. During lunch, clouds bubbled up from the mountains to blacken the afternoon sky. At two o'clock, we cancelled any thought of more filming, and those of us without responsibility for the safety of others were free to sample the local wines a little more vigorously. But finally, after four wonderful hours at the long wooden table in the small stone house, it was time to go. The family was reluctant to be paid for its hospitality, but Renato managed to persuade them to accept enough money to at least replenish their shelves. It was the perfect compromise. As we walked back to the helicopter with a warmth fuelled by generosity, alcohol and fresh mountain air, I could hardly believe that I had turned the job down when it was first offered to me.

It had been during the autumn of 1990 while I was working on film sequences about animal sex that the call came through from Milan. Richard Ronan of Filmgo, an Italian production company, had been given my name by BBC producer Pelham Aldrich-Blake. Would I, Richard wondered, consider making a series on Italian national parks for home video and Italian television?

Why would I say no to an unsolicited offer like that? For one, I was far too busy working on this sex series. Beyond that, I was committed to making a film in the Caribbean. But if Richard needed any advice on how to tackle a wildlife series, something his company had never before attempted, he could ring me any time for an informal chat. I gave him my home telephone number.

Christmas came and went. The sex series was seriously dragging its feet and would not now be delivered until well after its scheduled date. This time, I was not able to renew my already extended contract and I began working in Dominica for Television New Zealand's Natural History Unit. But the first stint on a beautiful Caribbean island came to nothing. TVNZ underwent radical changes and independent productions were the first to suffer as the jobs of permanent staff were protected. It was June 1991, when the second filming trip to Dominica was about to be cancelled and I had no other work in the pipeline, that I appreciated the folly of closing the door on the offer from Milan.

But then, quite unexpectedly, came another phone call from Richard Ronan, apologetic and wanting to ask a bit of a favour. Filmgo had its '£1 million-plus' sponsor for the film series but the money was likely to be withdrawn if, at their final presentation in two weeks' time, they could not show they had a proper wildlife production team lined up to do the job. Could they, Richard wondered a little cautiously, put my name down as series producer and, if I agreed to this, could I give him a list of people I would employ – camera and sound crews, directors, editors, etc – to complete the task? I agreed and got to work drawing up a list of people who confirmed that they, too, would not mind having their names put forward in such a way: Tony Allen, James Gray, Chris Catton, Robin Prytherch, Ramon Burrows, David Shale, Neil Bromhall, Sean O'Driscoll, Rogier Fredericks and Ross Couper-Johnston.

Richard's next request was more demanding. He needed an outline script for each of the parks, showing what could be filmed in the quickest, cheapest and most guaranteed way. His sponsors, he explained, needed to see what was going to be done in the time and money available before committing themselves to a contract. This was a difficult one. Given the whims of wildlife and weather, very little can actually be guaranteed, and Richard wanted the two half-hours on each of the five parks to cover all four seasons. I asked him to give me a week to come up with something that might convince his sponsors.

I sat down with Ross to make sense of what I had agreed to do. There were no internet search engines to help us in those days. We would have to visit bookshops and libraries and wade through the animal, plant and general literature for anything we could find on Italy. Pelham had made a BBC *Natural World* on Gran Paradiso National Park in the Alps, so that would be a useful start. But we were soon wondering how long it might take to come up with something even half decent. But then a breakthrough.

I was waiting to have my hair cut in Bristol when I noticed in the pile of magazines on the table in front of me an *Al Italia* airline supplement on, I could hardly believe it, the national parks of Italy. There it was, just lying in front of me, an impressively-horned male ibex adorning its bright blue cover. It must have been a good omen. Each of the five parks – Gran Paradiso, Stelvio, Circeo, Abruzzo and Calabria – was covered in exactly the detail we required, particularly their geology and climate. These would be the guaranteed components of each programme, with plants and animals fitted in once camera crews were on location. The one bold commitment we made was to include Italy's special mammals – bear, wolf, ibex, boar and chamois – which would be worth a bit of extra time and money to give each park a 'flagship' species.

By the end of the week, I had faxed Richard the information he required, including a summary that extolled the scenic and wildlife virtues of Italy's national parks. James, Tony and Chris agreed that Filmgo could cite their own Green Films Productions as the UK company that would be contracted to do the work and I sent videos of films we had made for major broadcasters around the world.

Ten days after the deadline, Richard telephoned to say their sponsors were delighted with the proposal and had agreed to release the money for us to get to work. Richard thanked me for helping win this reprieve. We were now free to become 'unavailable' while he looked around for someone else to do the job. But there was no need for him to look any further. I was by now so interested in the project that I came clean and said, to Richard's delight, that we would be able to make the series ourselves.

The following week, Becky and I flew to Milan. The Filmgo staff who would be working directly with us – Richard, production manager Marina Clerici, and Renato Marchiano, our senior driver/guide/fixer – were instantly warm and friendly. We would all be answerable to the somewhat irascible leadership of Giorgio Oldani. The main objective of these two days was to work out how to handle the money and the logistics. I was a bit uneasy about arranging things from England and then reporting to Filmgo before they would release the money to pay for it. We finally agreed that Filmgo should have this responsibility, with me telling Marina in advance how many people I wanted to be where, and for how long. Our only expense from England would be the excess baggage, for which we would be reimbursed.

I was really pleased with this division of labour. It would pay us less but it would allow us to concentrate on the filming, the most important thing. For their part, Filmgo saw the benefit of retaining financial control of the whole project. We also agreed that the post-production, which we had originally costed to be done in England by Ramon Burrows, would be undertaken in Milan by Filmgo, with me supervising.

With a total of 310 camera days, we would be able to film in each park for about two weeks at a time, four times in the year. This would be tight but Filmgo agreed that the 62 days per park would include neither our travel days nor the aerial filming, which we hoped might provide as much as five minutes for each programme.

The aerial filming promised to be special. Richard and Giorgio proudly announced that Filmgo owned its own Wescam unit, a gyroscopically-mounted camera housed in a perspex bubble outside a helicopter and worked remotely from controls inside the craft. It was so stable that the operator, the highly-skilled Lucas Grigioli, could zoom the lens between wide angle and tight without any vibration from the helicopter interfering with the stability of the picture. This meant that the film could be run at normal speed (24 frames per second) rather than at 40 or 50 fps which would otherwise be needed to iron out any wobbles in a slow motion effect.

The Wescam system isn't cheap, which is why by the early 1990s it was still being used only in advertising and feature films. It was out of range to normally-budgeted UK wildlife films until the BBC's *Planet Earth* pushed the boundaries of budgeting way beyond the norm a decade or so after us, using their own 'Heligimbal' system. With a following ground crew for maintenance and refuelling, our own Wescam outfit worked out at around £20,000 a day – expensive, but an exhilarating first for a wildlife television series.

We filmed our national park aerials in three sessions during 1992. The most exhilarating of these was a ten-day flight in early summer taking Lucas and me from the Alps in the north to the coast overlooking Sicily in the south. My final flight, the occasion of the farewell lunch with the crew, was in the autumn, when the helicopter was transported south by road to film Calabria's most southerly section, Aspromonte.

We began our ground-level filming in the autumn of 1991, a gentle journey into the delights of Abruzzo to film the extensive beech forests turning from summer green to the brilliant yellows and reds of autumn. On our first day in Pescasseroli, a small over-commercialised town high in the Apennines east of Rome, where a liberal sprinkling of brown bear logos offered visitors an unspoken promise that could never be kept, we attended a meeting of park directors. This was a fortuitous gathering, we assumed, of all the important people we would need to know and work with during the coming year. At one point I stopped to have a cup of coffee with Renato to discuss our progress. On the table next to us, two park directors, one from Stelvio in the north and the other from Calabria in the south, were sitting drinking and chatting like long-lost friends. Renato saw me looking at them and he leaned forward and whispered, 'Dan, these two people, to you they are two Italians talking together, yes?' I nodded. 'Well, you know, in Italy, some people, from the north, will see a German talking to an African!'

During the conference we spoke to four of the five directors and they were delighted at the prospect of our filming in their parks. Strangely, the one person we didn't meet was the Director of Abruzzo, Franco Tassi. However hard we tried, he always seemed to be just beyond our reach and we excused him for this on the grounds of the conference being on his patch. We assumed, despite a slight feeling that things weren't quite right, that we would meet him in a day or two when his busy schedule had calmed down. Dr Tassi was an expert on bears and we would need his help if we were to film them in his park. Before making another appointment, Renato managed to arrange for one of his rangers to take us into the public sector of the park.

'*Orso!*' came the confident cry before we had even climbed out of the vehicle to look around. '*Orso!*'

The ranger was pointing to the mountainside rising out of the valley mist, trying to convince us that he had seen a bear within a few minutes of our passing through the park boundary. Even with binoculars I could see nothing but bushes, trees and grass, although there was a single dark rock a few hundred

Sean O'Driscoll at a wood ant nesting colony, Abruzzo National Park,
Italy, 1991 (Author's collection)

metres away. It was certainly big enough but it definitely wasn't a bear. For a start it didn't move a muscle during the 20 minutes we scanned the hillside around it. And yet the ranger seemed to be pointing straight at it. Perhaps his bear was now hiding *behind* the rock? But there was also the slight suspicion that he was trying a little too hard to deliver a bear at the first opportunity. Renato was already convinced that Tassi did not want to help us. Could the ranger have been under orders to encourage us to think we could film bears right here at the very edge of the park, so we would be happy to stay here until they came strolling along?

Even from the very beginning there had been something different about Abruzzo National Park. Unlike the other parks, it seemed to make its own rules, implying that it was able to operate outside official directives. Permission for us to film there had been accepted a little reluctantly and Marina had already sensed that co-operation was far from guaranteed. And, strangely, Franco Tassi would never meet us. Hence Renato's suspicions. Following the park directors' conference, Tassi broke another appointment, leaving his assistant to apologise for an 'unexpected' departure for Rome, 80 km away, never mind that we had flown from England and driven 500 km from Milan. It happened again a week later after we had settled into a nearby hotel with all our equipment. Tassi, once again, was too busy to see us.

Renato was convinced that while Tassi might be obstructive he would never actually stop us going anywhere. We had our centrally-issued permit to film in Italy's national parks, so all we had to do was take the initiative. This, he suggested, would work because Abruzzo could not afford to be omitted from such a high-profile wildlife series. Its director, it emerged, had a reputation for being fiercely protective of his bears and was unwilling to co-operate with anyone who wanted to film them. That meant, above all, trying to keep people like us out of the central, core area of the park where the highest concentration of its 80 bears was to be found. Taking the initiative, Renato had left a message in Tassi's office that we would be ready at seven o'clock the following morning to look at areas of geological interest. We would need a ranger to accompany us. It worked. But it might also explain why, with no personal axe to grind, the same ranger magically conjured up a bear the moment we stepped inside the park boundary.

One piece of good fortune was that Hans Roth, a Swiss scientist, was working on bears in Abruzzo while we were there. He gave us as much information as he probably dared, even marking a little map with the most recent sightings. He warned us that he had radio-collared some of the bears and that, if we happened to find one of these individuals, their bulky collars might spoil our filming. He also suggested that late summer of the following year, with the bears fattening up on berries for their short hibernation, would be the best time to see them in broad daylight. But berries are so widespread, allowing the bears to forage way beyond the core area of the park, that finding them would still need Tassi's co-operation.

'Yes,' a young ranger volunteered after two hours at our hotel bar, 'the bears walk a long way in the summer, but they come back to the middle of the park for winter because we put out carcasses for them. It means they can get fat for their winter sleep in a safe place and it stops them going into hunting areas or making a nuisance of themselves with people living at the edge of the park.'

This was how we should try to film Abruzzo's bears, and now that we understood the rules of engagement with Franco Tassi, it was possible to make plans to sit in a hide over one of these carcasses. We broadcast our intention to every member of staff and, before long, it was accepted that this was what we would do the following year. We never applied for a core area permit and we left clear instructions that Marina was to be telephoned in Filmgo's Milan office when the meat was put out for the bears the following autumn. Like Renato said, 'It's up to him now.'

James was filming otters in the north of Italy when the surprise but welcome call came from Abruzzo in October 1992. A red deer carcass had been laid out in the core area, above the treeline, and they were confident that it would be visited by at least one bear during the coming week. The timing was perfect. James was about to leave Milan for southern Italy to film wild boars feeding on beech mast and he could easily spend a night or two in Abruzzo on his way to and, if necessary, from the private estate where the boars lived.

But on both occasions, the bears came to feed in the dead of night and James could not film them. His reward for braving those freezing hours of darkness, though, was to film a wolf on the carcass at dawn, with foxes and hooded crows wary of its presence. Against the wonderful backdrop of the Apennine Mountains with their first dusting of winter snow, the sequence was well worth James's efforts.

We never did see any bears in Abruzzo, although we were able to use some footage of a bear and a wolf at the same carcass, shot by one of the park rangers a few years previously. It became a standing joke between us to work out which was the more elusive, the bears or Franco Tassi (whom we also never saw).

Our gentle journey into Abruzzo in the autumn of 1991 preceded two difficult winter sequences. Filmgo, we knew, would be looking for perfect results to convince their sponsors that we had been a good choice to make the series. An expensive filming trip that failed to deliver, for whatever reason, would not be acceptable to people who did not understand the fickle natures of wildlife and weather. The next two species on the flagship list were the chamois and the ibex. Both live in the Alps and both conduct their courtship in the high snow and ice of November and December. We needed them for the winter programmes on Stelvio and Gran Paradiso. We would never normally have chosen animals as difficult as these so early in the filming. But we had to go for them and they had to work. Filmgo and their sponsors would want to see an early return on their money. The scenics of Abruzzo, however well received, were not enough. They wanted animals doing dramatic things.

Robin went off to Stelvio with Renato, Tony and Sean in November. They huffed and puffed their way to the top of the mountain where the park rangers knew the chamois were living. They were rewarded with a wonderful piece of luck. Stelvio's theme was its melting glaciers, with fast-running streams pouring out from the ice at their leading edges. Having found, followed and filmed their chamois courtship, they were on hand when a massive avalanche thundered down the mountainside opposite. Tony was able to film the falling snow with the chamois scampering away to safety in the foreground. With this unlikely key shot secured, he was then able to build the melting sequence even further, highlighting the perils of being a mountaintop mammal juggling sex and survival in a challenging winter world.

If our own survival depended on the success of these two sequences, we now had the comfort of one of them being safely in the can. We had no idea, though, what would happen with the ibex in Gran Paradiso. Of higher profile than the chamois, this was the sequence Filmgo was watching intently.

I had allocated four precious filming days, but all we knew was that the ibex rut would happen sometime during December. Our departure from England had to be precise. Marina telephoned the park rangers every day from the end of November until, two weeks before Christmas, one of them reported signs of increased activity. Two days later, we met Renato at Milan's Linate airport and off we went through Turin to the deserted pre-Christmas town of Ceresole Reale. Marina had found a little hotel and persuaded its owner to open up and give us rooms for the few days we would be there. No food, just rooms. Renato came equipped with the Italian equivalent of the *Good Food Guide* which we used to fuel our gastronomic ramble through Italy for the next 12 months. Here, in the freezing Alps, we took to the mountain delights of gnocci, polenta, grolle pots and grappa-laced espressos.

Each morning we woke at five to the dark, perishing cold of an alpine winter. Struggling out of bed and pulling and strapping on layers of winter warmth, boots and gaiters, we loaded the filming gear into Renato's wheel-chained and de-iced Renault Espace and sat for ten minutes while its ageing diesel engine warmed to the task of battling through sub-zero temperatures. And then a quick stop at a local bar to meet two park rangers huddled round a blazing fire. Here, we encountered the Renato ritual of cigarettes and double espressos before he was pumped up and ready to go. We followed the rangers to the end of the road at the base of Gran Paradiso Mountain, loaded ourselves with equipment and set off through the night's fresh fall of snow.

The three-hour walk was invigorating. It was hard work hanging on to Renato's shirt tails as he scampered ahead, jet-propelled by caffeine and nicotine, carrying what appeared to be no more than a bag of feathers on his back. But then he and the equally energetic Sean were still in their twenties, their birthdays on the same day in February. Tony was just a few months away from forty, a milestone I had passed six years previously. The two rangers, so at home in the hills, were ageless.

By mid-morning we reached a recently-built refuge 2000 metres above sea-level, a small pine cabin which soon had a fire blazing in its open hearth. Shedding clothes, we made ourselves at home and tucked into the packed lunches that Renato had ordered from the bar the night before. Although it was nearly midday, the sun was only just beginning to light the top of the far wall of the valley opposite the cabin. Tony set up his camera and filmed the alpine choughs that gathered round for scraps of our food.

One of the rangers suddenly announced that he could hear something – a resounding but distant echo as testosterone-charged male ibex squared up, walked side by side and reared up to come crashing down into each other's horns. For the next three hours we listened and waited. It sounded a little like sporadic bursts of thunder echoing all round us. A few small groups came within 100 metres or so, hidden against the dirty brown rocks where the snow had not covered their grassy food, but the big males were not with them. Even so, Tony was able to begin constructing the sequence. He got some good establishing shots in the sun: the groups of females and young males set the scene and established the reason for the aggressive behaviour of the dominant males. We left the refuge at half past three with two rolls of useful material. We knew the competing males were in the next valley, so perhaps tomorrow they would come a bit closer.

It had been an exhilarating first day: fresh air and strenuous exercise are the ingredients for an intense feeling of well-being. In the evening, Renato fed us mountain food in one of his chosen restaurants and we were back in our ghostly hotel and asleep by ten o'clock. In no time at all, our boots were back on and we were witnessing Renato's early morning ritual in the bar. The bond already forged, we greeted the rangers like long-lost friends and set out on our long trek to the refuge.

The females and the young males were just where we had left them the previous afternoon. And so, unfortunately, were the big males. We waited hopefully, but they would not come any closer. We climbed a bit higher and a bit further in their direction, but to no avail. They were at least a kilometre away, hidden in the next valley. If we went to look for them, we could easily push them even further away. We decided to return to the refuge and wait. We still had two days in hand. There was no need to panic. We again filmed the females and young males, some of whom engaged in mock fights, giving us more useful build-up material to the real thing. But we also knew that this footage would be worthless *without* the real thing, so it was a slight risk to be exposing expensive film stock with no guarantee of being able to complete the sequence. Tony and I talked endlessly about this, as we have on every filming trip over the years. Given the exorbitant cost of film stock, together with its processing and printing, the amount we put through the camera was a serious consideration. It does not apply to modern video cameras which can record, be viewed and then wiped for re-use on location if the desired behaviour is not there. Our decision was to get whatever footage we could from the refuge and just hope the fighting males would come into our valley by the next afternoon. If they still kept their distance, we would go and look for them on our final day.

We were now halfway through and Renato had already reported our lack of success to Filmgo during his obligatory evening phone call to Marina. We were beginning to feel the slight frustration of putting in a lot of effort for little reward. But a hallmark of my work with Tony, which spanned 20 rewarding years on five continents, was that neither of us ever panicked when the odds appeared to be stacked against us. We also knew that it was such an important sequence for the Gran Paradiso film that we could make a case for coming back next year while we were still editing in Milan. But the key thing this year was to impress Filmgo that we could actually come up with the goods.

Day three began just like the other two. The night sky was clear and the icy walk to the refuge, though still hard work, was already shortening through familiarity. We offloaded, stretched our legs and began cleaning and laying the fire while the rangers went to investigate the valley. Ten minutes later, one of them tapped on the window, beckoning and banging his gloved fists together. We knew exactly what he meant. Abandoning the fire, we assembled the camera and tripod and crept out into the snow.

We heard them before we saw them, the noise so loud that it seemed they must be right on top of us. They were. Overnight they had come out of the far valley to join the females we had already filmed. Now they had come even closer. What's more, there were even more females and youngsters, giving us 50 or 60 ibex of all ages within 300 metres of the refuge. There were ten very large males, each as impressively horned as the one I had seen on the *Al Italia* supplement back in Bristol. Above us, a cloudless sky, not yet criss-crossed with the vapour trails of planes descending on Milan and Turin. The omens were good.

As the sun highlighted the sheen of their thick winter coats, the big males warmed to their task. The light was good enough for Tony to begin filming on his 150–600 mm zoom lens using a 2x converter. It's very difficult to eliminate shake at what becomes the 1200 mm end of the cine camera, even with the hefty tripod we had lugged up the mountain, and yet that was where Tony had to begin. We had no idea how long the ibex would stay or what they would do. Incredibly, they came close enough for Tony to remove the converter to get even steadier and sharper images of the sparring males. To be honest, though, the fights were not as impressive as we had hoped they would be. There were no headlong charges, no sustained battering rams, just these sudden clashes between males who sized each other up while walking side by side for a few metres. With neither backing down, one would then rear up on its hind legs and brings its horns crashing down into its opponent, who turned inwards to face the challenge. Within the hour we knew we had our sequence, that tomorrow we would be polishing and refining, hoping for a single dramatic shot that might make all the difference.

Renato could see how relieved we were. His nightly phone call to Milan would be that much easier, and not just because we had our promised sequence. He was beginning to understand and appreciate the challenges of wildlife filming. For the following 12 months he would be a staunch supporter of the way we worked.

The following morning, the clouds were down below the level of our hotel and we crept to the bar in almost zero visibility. The rangers were not there. They had already telephoned to say our filming was over. The freezing fog that had descended during the night was not expected to lift for several days. We flew home from Turin knowing that we had even managed to save one of our precious filming days. The two high-risk winter sequences on which so much depended were declared a resounding success. We breathed a collective sigh of relief, and prepared ourselves for the hectic year ahead.

'Of course,' I was assured by the Director of Circeo National Park on the coast 50 km southwest of Rome. 'We have an enclosure with three big pigs near the visitor centre. You can go in and film them.' It was a start but we had been here before with other animals. There were those two captive wolves in Calabria University who lived on a tennis court-sized patch where they were being studied, leaving us wondering what they were actually being studied for. On another occasion we were offered captive wolves to film in Abruzzo National Park and set out for the remote village of Civitella Alfedena with our usual 'Well, it's worth a quick look' cynicism. What we found at Civitella, though, was the most perfect setting imaginable. A long wall overlooked natural woodland and a high fence ran down from either side of the wall, through the woods and out of sight behind a central hill. The seven wolves were given meat every day or two and we could easily film from the top of the wall which rose ten metres from their side but, on our side, was only a metre or so above the level of the road. It could not have been better.

The wolves were cautious and would slink through the thick undergrowth, offering wonderfully furtive shots as they edged towards their meal, often squabbling among themselves. It was the ideal place for filming, summer and winter, and the wide shots of the pack on top of the distant hill had a spectacular backdrop of snow-covered mountains.

The chances of Circeo's wild boars offering anything comparable to the wolves of Civitella were remote. We arrived at the park headquarters and were taken the short distance to the enclosure. My heart sank. A high wire-netting fence ran round a dense stand of fir trees. The fenced area was about 20 metres by 20 metres and the netting was visible all the way round. Beneath the trees, in almost total shade under the thick evergreen canopy, two large boars lounged in the dust. The third lay up against the wire netting close to where we were standing.

If we could manoeuvre them into a bit of sunlight, they might be useful for big close-ups of their faces, but nothing more. And no, they could not be released because they would run away and that would be the end of the big tourist attraction. Aware of the need not to offend anyone, I said it was good to know they were here and that we would come back and film them later. Turning round to go, I looked straight into the faces of three wild boars standing barely five metres away, right out in the open.

'Oh, those,' said the keeper of the 'prize' specimens still snoring and snuffling in the enclosure. 'They come from the forest every day and we give them food as well. It's a mistake. They follow us everywhere until we feed them. They are a nuisance because they dig up flowers and can hurt people. We think we have to take them away from here, somewhere safe for them and us.'

'Would they follow us along the road and into the flooded forest?' I asked hopefully.

'Why not? If they are hungry and you have food, they will follow you anywhere.'

Circeo National Park protects the last vestiges of the famous Pontine Marshes that stretched along Italy's west coast between Rome and Florence. The string of freshwater lakes, marshes and forest, isolated from the sea by sand dunes and once riddled with malaria, were a hazard to humans who have tried for centuries to drain them and use them for farmland. The last serious attempt to do this was by Mussolini in the 1930s. Fortunately, he did not succeed completely and the national park now protects what he left behind: four lakes (Sabaudia, Monaci, Caprolace and Fogliano) set in reclaimed farmland and forest, overlooked from the south by Monte Circeo. One section of forest is flooded annually by rain and meltwater coming down from the Apennine Mountains and it was in this forest that we had hoped to find and film wild boars. But they were just too wary and, though we saw plenty of evidence to suggest they were there in good numbers, not one would come within reach of our cameras.

And so, ten days later, a funny little procession set out from the park headquarters and along the busy main road southwest of Rome. Chris and James were loaded with filming and sound recording gear and one of the park rangers followed them with a large box of fruit. Trotting obediently behind them were the three wild boars, happily following the promise of another free meal. Inside the forest, the food was scattered and buried, and the three pigs spent the rest of the day rooting around to find it. They fought obligingly over their trophies, running and squealing excitedly from one to the other as James filmed from every conceivable angle. No one was happier with our success than the park director who understood that we had no further need of the boars in his little enclosure.

Further south, in Calabria, the park director was less happy, even apologetic, when we again raised the subject of filming wild boars.

'No,' he said, 'we do not have any captive wild boars near here and there are so few in the park itself you will be lucky to see even one.'

He was telling us this in a little roadside café in the northern section of the park, Sila Grande, where we drank hot chocolate in front of a roaring wood fire. But soon after the director left, we were approached by a complete stranger. He had overheard our conversation and wondered, rather strangely, if he could help. He spoke in Italian to Renato, handed him a card, and an appointment was made for ten o'clock the following morning. As we drove back to our hotel, I asked Renato who the man was.

'Well, you know, we are in the south of Italy,' came the uneasy reply, and I thought of nearby Sicily whose snow-capped Mount Etna was visible from Aspromonte, the southern section of Calabria National Park, and of an Italy still reeling from the recent car-bomb assassinations of the judges Giovanni Falcone and Paolo Borsellino.

'But we don't have to go if you don't want to.'

'What do you think?' I threw back at him.

'Well, should be OK, I think.'

I trusted Renato totally. 'OK, then let's go.' I wondered whether he would tell Marina during his nightly phone call to Milan.

The next day we left the hotel early, giving us plenty of time to lose our way and still keep our appointment. Turning off the main road, and again off the secondary road, backtracking once, we followed a rough track that ran alongside a high fence. It came to an abrupt halt in front of an impressive set of gates. Renato pressed the bell and announced himself to the disembodied voice. The gates opened slowly but closed the moment we had driven through. Ten metres in, we stopped in front of a small wooden building.

The man we had met the day before appeared on the doorstep, smiling a welcome that put us at ease. He offered coffee and cakes set out on an untidy table and chatted casually to Renato. But I had begun to feel uncomfortable. How could this man help with wild boars that didn't exist? What did he really want from us? Where on earth were we? Were we safe? I remembered how quickly the big front gates had closed behind us.

After two double espressos, Renato's elixir of life, the subject of wild boars actually began to take shape. He spoke in English, his questions now directed at me.

'How many do you want?'

'How many have you got?'

'Plenty for you, my friend. You want to see?' I nodded. He made a quick phone call.

So off we went, on foot, through the house and out towards the forest. The forest, he told us, was on the edge of the national park and my first thought was of the director who had denied any knowledge of captive wild boars. We were met by the animal keeper whom we followed through a maze of empty enclosures to a large muddy field that rose in front of us and fell away to the forest behind. The area between the crest of the hill and the forest was hidden. The animal keeper whistled loudly, banged a trough and suddenly, from nowhere, they came, over the brow of the hill and straight towards us. We ducked behind the gate as a screaming hoard of at least 200 wild boars skidded and squealed to a halt, clambering over the troughs and each other with a disregard driven by the promise of food.

It was an impossible situation for filming but Renato and I were now curious to know what was going on. If we were on the edge of the park, why didn't the director know about this? Or perhaps he did and he didn't want us to know that he knew? So why, then, should we have been approached by a complete stranger who must have known exactly who we were talking to in the little café with the roaring wood fire? All our man was prepared to tell us was that he was allowed to breed wild boars here on the condition that a certain number were released annually into the park. The remainder would be transported all over Italy and released on private estates where they would be hunted. This, clearly, was where money was changing hands, but it was still not easy to see how or where the park director fitted into the picture, unless, of course... But Renato assured me that there were things going on down here that we did not need to know about. Before we left, though, I had one final thought. 'Do you get wolves here, attracted to the wild boar?'

'Yes,' came the interpreted reply. 'This is why we have such a big fence all round, but even this is not enough in the winter when the wolves are very hungry. They dig. They climb.'

'What do you do then?'

'We have to shoot them.'

'You shoot wolves? From the national park?'

'We *have* to shoot them. You know, a hungry wolf is like a man at a discotheque. He cannot help himself!'

Six months later, I again travelled south to Calabria, this time to join James and Chris for a few days in Aspromonte. It was late afternoon when I arrived at their little hotel outside Gambarie, and James suggested we sit on his balcony to go over the details of what they had been able to film. At precisely six o'clock, James held up his hand and told me to listen. I could hear approaching vehicles, nothing more. But these were not ordinary vehicles. They were army vehicles and each one – we counted 19 – contained at least four gun-toting soldiers. Alighting from their jeeps, which were then neatly parked, the soldiers filed into the rear of the hotel and went to the kitchen, part of which had been sectioned off as an armoury. Here, James and Chris had established, they accounted for their bullets and handed over their empty rifles before disappearing to their rooms. We would see them later in the dining room and bar.

'So the Italian army is in town. What's so special about that?' was all I could think to ask.

'Well, they're on honey buzzard patrol, that's what's so special.'

'Honey buzzard patrol? You can't be serious? There must be nearly 100 of them, all armed to the teeth!'

But James and Chris *were* serious. They had done their homework in the hotel bar. The soldiers were here following a request by the German government. Their job was to prevent honey buzzards from being shot on their long migration from Africa to Europe. Apparently, they were being slaughtered every year in southern Italy and their breeding success further north in Germany was declining. On the face of it, the story had credibility, but if you considered that honey buzzards are among the most abundant of European birds of prey, are highly secretive and that they cross from Africa to Europe over the whole length of the Mediterranean, not just over Italy, then it began to look a bit questionable.

That evening in the bar we made another breakthrough. As we negotiated the price of a lump of marzipan for me to take home to Becky, a local farmer informed us that if a Calabrian man does not kill a honey buzzard on its northerly migration, his wife will be unfaithful to him before the end of the year. So each spring, when the first of these birds appears in Italy, the loaded shotguns are waiting for them.

But ahead of the raptors' arrival, the army positions itself at strategic places around Calabria, and our hotel at Gambarie happened to be one of their bases. Their orders were to apprehend and even shoot anyone who attempted to kill one of these insect-eating birds of prey. One person, we heard, had died this year already.

'Of course,' Chris chipped in with his usual cynicism, 'the army being in Calabria for a few weeks has got absolutely nothing to do with honey buzzards! It's just an excuse for a seasonal show of strength to the local Mafia, to remind them that they don't totally rule the roost down here. If someone dies while this point is being made, well, the soldiers are just doing their job protecting birds of prey, and there's not much anyone can find wrong with that!'

By the end of 1992 we had shot nearly 500 rolls of film and my remaining task was to put the programmes together with Filmgo's editors. For three months I commuted weekly between Bristol and Milan. As each film was edited, I wrote the commentary and faxed or carried it to Milan to be translated into Italian.

A final gesture from Filmgo remains with me. Halfway through our contract I returned to the Milan office after a quick trip to the Po delta. We were working in the delta, an important wetland, because part of our commission was to make four extra short films on wildlife areas outside the national parks. The other three were the Dolomites, the Tuscany Archipelago and the Gargano peninsula. Back in the Filmgo office I told Richard how Becky was quite happy for me to return home with stories of wading around the delta near Venice, but that it would be more than my life was worth to say that I had actually gone into the city without her. Richard found it hard to believe we had never been.

When we were close to the end of the project, Richard handed me an envelope. Inside were two vouchers for a long weekend in Venice's Hotel Saturnia. Leaving our three boys with Becky's parents, we flew to Milan for the final narration check on a Thursday and took the train to Venice the following day.

It was a lovely and much appreciated ending to a two-year experience that was not only successful but which was also one of my most enjoyable. The two, of course, go hand in hand; people who are treated well work well.

But it wasn't quite the end. A few months later the series reached the finals of the Jackson Hole Wildlife Film Festival in America, and the Calabria summer film won the Invitational Prize at the Sondrio Festival in Italy. The following year, 1994, I was invited back as a member of the international jury at Sondrio. With the air miles I had accumulated on more than 20 journeys between England and Italy, I was able to take Becky, Michael, William and Edward with me. The high spot was a day out in nearby Stelvio National Park in the Alps where the boys, like the old ranger in Abruzzo, tried to convince me they had heard, and even seen, a bear within minutes of our crossing the park boundary.

CARIBBEAN SEA

Caracas

Barinas

Apure River

Merida

San Fernando

Orinoco

Llanos

Hato
el Frio

VENEZUELA

San Cristobal

Hato el
Cedral

COLUMBIA

Orinoco
River

BRAZIL

N

W   E

S

AMAZON BASIN

11

RED-BELLIED PIRANHAS

The chattering and the squabbling that had been faintly discernible when we cut the engine grew louder and louder. And then, as we drifted even closer, we were hit by the ammonia-rich stench of baby birds, hundreds of them, even thousands, the youngest confined to their nests, the oldest out exploring, crossing boundaries and fighting on flimsy branches. Adults were circling, landing, emptying fishy food into grasping beaks and setting off once more to hunt in the open, shallow water.

After half an hour in our Everglades-style airboat we had arrived at bird city, an outcrop of trees on the floodplain of Venezuela's Orinoco River. The place was bursting with energy, an organised mass of bird homes from the lowest branches touching the water to the top of the tallest tree. Cormorants, darters, herons and storks were packed together, like a seabird colony on a coastal cliff. And for the same reason, too: thousands of square kilometres of prime feeding but with very few places to build a nest and raise a family.

As we floated between bushes, two egret chicks began fighting, spearing at each other with their wickedly pointed beaks, jabbing this way and that until one of them lost its balance and went tumbling into the crystal-clear water. Dark shapes formed, flashes of silver and orange. As the young egret struggled towards an overhanging branch, the shapes and the flashes intensified, converging on the little body, tearing, distorting and dragging it beneath a mat of floating leaves.

Piranhas. Red-bellied piranhas. I stared at the fading ripples, shocked and disbelieving.

For years I had associated piranhas only with the Amazon River meandering through the deepest rainforest of central Brazil. I was not prepared, on my first visit to South America, for them to be here, 1,000 km further north, in the vast openness of the Orinoco floodplain, the *llanos*. And yet here they were. The water that had looked so fresh and inviting became sinister and forbidding.

Between May and November each year, thousands of waterbirds gather to raise their young in these isolated colonies across the *llanos*. Their plentiful droppings fuel a rich food chain in the water below their nests. Fish come to fatten and breed, among them the red-bellied piranha, one of the many different fish released by the rains from the dry season waterways that confine them for the remainder of the year. As their own breeding gets under way, the piranhas turn almost black and females are courted by the smaller males. Eggs are released and fertilised, and the males guard them for two days until they hatch. It is these breeding piranhas that are the real danger to other animals. A protective and territorial impulse infiltrates their ranks and they patrol the shadows of the deep, primed to attack anything that moves. A stick falling from a nest has hardly stopped bobbing on the surface before it is hit powerfully as piranhas shoot up from below. Some of these 'sticks' turn out to be baby birds, particularly egrets which climb out of their nests before they can fly, invade each other's space and are beaten back. Once down in the water, they stand little chance of escape. If the piranhas don't get them, caimen or anacondas will.

As the breeding season recedes so, it seemed to us, does the red-bellied piranha's aggression. Later, in the dry season, when they are crowded in evaporating lagoons, they appear to have lost that instinct to attack anything that moves, despite a universal assumption that this is when they are desperate for food and are at their most dangerous. The fittest of them, in fact, have as much food on hand as they want. As well as being killers and scavengers, red-bellied piranhas are also cannibals.

I never forgot the visual and emotional impact of the piranhas demolishing that young egret. Eight years later, in 1991, and working with Tony Allen, James Gray and Chris Catton of Green Films Productions, they surfaced in my mind while I was thinking up ideas to present to programme commissioners around the world. I knew that piranhas eating egrets would make a strong sequence but also that this would not, in itself, be enough for a whole film. The sequence had to be part of a story that would hold its audience for at least half an hour.

It was one of those dramatic dry season/wet season African wildlife films that provided the answer. Graphic shots of waterbirds feasting on stranded fish in a drying-up waterhole were the key. What if my piranhas ate the young waterbirds in the wet season and then, in the dry season, the tables were turned and the waterbirds ate the piranhas stranded in the dwindling lagoons?

What followed were six visits to two ranches in Venezuela, El Cedral and El Frio. Between 1993 and 1996 Green Films Productions and I made two films, the first for National Geographic (*Explorer*) and the second for the BBC (*Wildlife on One*). Being based in Caracas enabled me to renew a friendship with Mike and Juliet Wood, who had left London to work and raise their family in South America. Apart from being good company, they were invaluable in getting things under way.

The Venezuelan Audubon Society in Caracas put me in touch with Donald Forbes, a local tour operator who turned out to be the single most important reason why we succeeded in our filming. It was Don who organised the import and export of all our equipment, booked hotels, organised inland flights and who negotiated with the ranch authorities over how long we could stay and what it would cost us. It was Don who took control of the day-to-day difficulties that beset filming trips and sorted them with a willingness that went beyond the call of duty and into friendship. He had a cheerful optimism that said you did the best you could and trusted the system to take care of the rest. When a glass window – two metres by one metre, 2 cm thick and costing US$800 – arrived in Caracas from Florida on its way to the aquarium we were building at El Cedral, our immediate problem was how to get it the next 600 km to the ranch without suffering so much as the tiniest of scratches. Don just bundled Tony and Chris onto their waiting plane with a 'Don't give it another thought.'

While Tony and Chris flew with the gear to San Fernando, where they would be met by a ranch bus for the three-hour drive to El Cedral, I stayed behind with Don. We had a little business to sort out, namely the finding and buying of a suitable vehicle. Don had already struck a deal with the ranch owners. In return for our board and lodging on at least two filming trips – it turned out to be three – we would buy a brand-new, open-backed, four-wheel drive Toyota that we would leave with the ranch when we finished. They would use it and service it between our trips but it would be ours exclusively while we were there. It was a perfect arrangement, both for the ranch, because our keep would actually cost them very little, and for us, because we would be guaranteed a reliable vehicle without any of the hassles of hiring.

Don beavered away on the phone and finally came up with the only new Toyota for sale in the country. It was hundreds of miles away, over in the Andes, a two-hour drive south from Merida, close to the Colombian border. We could collect it the following day. 'And this,' smiled Don, 'is where the fun begins. We have to have the money, all of it, credited to their bank account by this evening. And that means cash.'

The taxi dropped us off close to Don's bank in Chacao, leaving us to walk the final 200 metres. The manager politely confirmed that he would not accept my American Express travellers' cheques, though why not, with his own country's economy tied so strongly to the US dollar and with the Venezuelan bolivar devaluing daily, seemed a bit puzzling. Don suggested he pay them into his own account and withdraw the appropriate amount for me in local currency. In what became a total cloak-and-dagger operation, in a bank teeming with people who seemed to have no good reason for being there, we carried out the transaction. In went my signed travellers' cheques and out came the best part of 3,000,000 bolivars. Don, slightly out of character, was sweating a bit as we stuffed wad after wad of grubby notes into my old, battered briefcase and then pushed our way to the door. Outside, we had those same 200 metres to negotiate before the taxi stand. Don set off at a canter making things, I thought, a bit obvious. He just told me to hug the briefcase and keep going but when I protested a second time he turned and hissed, 'You don't understand. We could be mugged just for the briefcase. If they open it later and find all that money inside, that's a bonus!'

From the relative safety of the gas-guzzling taxi, weaving and hooting through the Caracas traffic, Don told me more about the people milling around inside the bank.

'They can be paid to watch out for people like us, giving descriptions on their cell phones to people outside. Or they may bump into you and leave a chalk mark on your shoulder so you can be followed. I won't say we were lucky but these things do happen.'

One hour later, with US$13,000 fed into an anonymous bank account on the strength of nothing more than a verbal agreement over the phone, Don was beginning to relax, confident that everything was going to plan. He had worked in Caracas for many years and had learnt when to trust his instincts. 'Oh well, you're insured,' he chuckled from the comfort of a third beer in the Hotel Avila where I was staying. It didn't bear thinking about.

The massive, snow-capped Pico Bolivar and Pico Espejo came and went as we dipped down through the clouds early next morning to land at Merida, *el techo* – the roof – of Venezuela.

The air was crisp and clear, invigorating, a welcome change from the stifling mugginess of Caracas in its exhaust-filled valley. And then the 200 km drive south to the Colombian border, downhill and increasingly hot. We found the garage near San Cristobal and were given tea and a sandwich before the receipt of the money was gratefully acknowledged. We were ushered through to the workshop and there, having its final polish, was the bright red Toyota we had been promised. To protect our investment in the vehicle, Don became its legal owner and when the formalities of insurance and licensing were complete, we set out to drive the 200 km northeast to Barinas where we would part company.

It was a wonderful drive up through the cooler mountains, stopping at tumbling rivers to watch torrent ducks brave white-water rapids and humming birds flit between exotic Andean flowers. We climbed even higher and the road took us out into the *paramos* moorland above the treeline, a desolate place clothed in mist, permanently wet and cold. A little roadside shop provided us with hot drinks and biscuits. I rummaged through the local produce for sale, settling on a number of colourful jumpers to take back home to my family. In Barinas, a noisy industrial town at the foot of the mountains, we were back with the heat and the humidity. I checked into a hotel and Don, after a farewell beer, took a taxi to the airport and his flight back to Caracas.

I still had 300 km to drive the following day to meet Tony and Chris at El Cedral. Don had taken care of my scant dictionary-led Spanish, which could have posed a problem at any of the roadblocks, by arranging for a taxi to drive the whole distance in front of me. It would cost around £20 and would protect me from getting lost and from having to explain to the police someone else's brand-new vehicle. The taxi driver was to say that we were delivering the vehicle to the ranch, which was true, and I was to smile cheerfully and say nothing.

It was a lovely drive across the open flatness of the *llanos*, with its islands of trees and abundance of waterbirds, the heat a welcome change from the damp cold of the previous day. I was back with Tony and Chris in time for lunch and then the three of us took off for a late afternoon drive. The raised dirt roads of cattle country are laid out in grids which confine wet season water to deep lagoons. As the dry season progresses, water is moved between the grids via sluice gates, promoting the growth of fresh grass for grazing. This would provide a useful fall-back for us: when we returned at the height of the dry season, it would still be possible to film any important wet-season detail we might have missed.

Three days later, while we were considering how to convert a caiman breeding pen into an underwater river scene for piranhas, the ranch manager, Heriberto, came over and announced that a taxi for National Geographic had arrived at the lodge on the main road 7 km away and was now on its way down to the ranch. Intrigued, we went to meet it. After a long wait, a seriously old and battered car slithered through the mud and slewed to a halt at the entrance to the tourist lodge where we were staying. There were four people inside: the driver, his wife and two children, all crammed onto the front seats. Pride of place, the entire back seat, was taken by our piece of glass for the tank, as bubble- and blanket-wrapped as possible and in perfect condition. The driver, used regularly by Don's company in Caracas, was as proud as punch to have delivered it successfully, and his family had never had such an exciting holiday. For two days they had crept along the worst roads in the country, perhaps even in South America, sometimes in torrential rain, never leaving the car unattended and even sleeping in it at night. We gave them a hefty tip and, refusing all offers of hospitality, they turned round and set out on their 600 km journey back to Caracas.

The little Cessna aircraft came buzzing down over the trees and settled onto the grass runway behind the tourist lodge. It was Yolanda Carbonell who, like Don, was to become a good and reliable friend, ready to help at a moment's notice. On this occasion, she had come to the ranch with a photographer but agreed to help with our aerial filming at a later date. She knew Don and also lived in Caracas, and it would be easy to organise the dates. She, in fact, flew us on several occasions over the next three years and, although filming from a light aircraft with its door removed is not ideal, it gave us the viewpoint we needed and was much cheaper than a helicopter.

Yolanda was one of those people whose personality gets them into and out of difficult situations, particularly in a male-dominated society like Venezuela's. She procured permits from nowhere and challenged, charmed and disarmed bureaucrats to get her clients and her friends through the most resistant of red-tape barriers. She was a passionate conservationist and would never work for or with anyone who did not, in her opinion, have Venezuela's best environmental interest at heart.

Yolanda become the country's first woman pilot. She flew tourists and film crews while slowly financing her own tourist lodge, Las Nievas, on the edge of the more southerly rainforest. Her husband had suffered a fatal heart attack the day they took their first group to their new lodge and she told us how she flew back to Caracas with his body and their two small children in the little plane, reflecting that perhaps she should just keep going over the coast, and the sea, until they ran out of fuel.

It was a few months after our final visit to Venezuela in 1996 that Don telephoned me in England to say that Yolanda was dead. Her little Cessna, the one we knew so well, had hit a power line during an impossible landing in a storm. She will be missed.

Before our first visit to Venezuela, Chris had had to explain to his wife why 50 non-lubricated condoms had made their way into his luggage for the nine-week filming trip. He managed to convince her that they were for professional purposes only. Chris would be doing the sound recording and, for piranha feeding frenzies, he needed to be able to record underwater. This would mean dipping his microphone below the surface where it would need the protection of a membrane that would not interfere with the quality of the sound. Satisfied that non-lubricated condoms would keep his tackle dry, we flew out of London's Gatwick Airport confident that they would at least partially cover a soft microphone-housing 40 cm long and 10 cm wide.

We sat in the boat under the enormous waterbird breeding colony, lost in the raw energy of young birds fighting among themselves and clamouring to be fed. Tony was fitting the 40 cm-long periscope lens to his camera, I was easing a 5 kg lump of beef onto a butcher's hook and Chris was stretching a gossamer-thin sheath over the microphone held firmly between his legs. Half a dozen split condoms later we climbed out of the boat and into the mid-thigh deep water. We were a bit nervous. Nobody had ever tried to film piranhas like this. I lowered the meat and Tony positioned the front element of his periscope lens up against it, the camera body just above the surface. It wasn't going to be easy. Visibility in the muddy water was down to just a few centimetres and the gear and the meat were heavy. Chris submerged his partially-covered microphone listening for any telltale sounds of piranhas coming our way.

'I don't think there are any piranhas here,' I suggested after we had taken the strain for several minutes and nothing had happened.

'No,' replied Chris without so much as a smile, 'that caiman probably ate them long ago.'

'Caiman? What caiman?'

'The one about three metres from your backside.'

I was crouched over the water, facing Chris, just able to peer back between my legs. Even upside down, the unmistakable head of a South American alligator was moving slowly towards us, lured, no doubt, by the disturbance and the meat. Chris calmly leant into the boat, cut off a sizeable lump of beef and tossed it over my back towards the sinister snout. With an explosive snap, it vanished.

We took up our positions again and this time Chris, inspired by his success with the caiman, threw small pieces of steak two metres in front of us, just where the shallow swamp we were standing in joined the deeper river that ran round the back of the bird colony. The water swirled and the meat shuddered on its hook. Tony whispered urgently, 'Here they come,' and flicked the switch on his camera. Seconds later, they were all over us, thrashing and shearing off lumps of flesh with their powerful, razor-sharp teeth. Some were leaping out of the water, others were colliding with our bare legs and all were interested in just one thing, the lump of bloody meat on the end of my hook. And then, just as it had begun, it was all over. The hook was hanging free, the piranhas had gone and we still had flesh on our legs.

I was ecstatic. This was our first attempt to film a piranha feeding frenzy underwater, in the wild, and it had worked. Tony, typically, wouldn't commit himself on the all-important things like focus, visibility, steadiness and composition. But he was prepared to agree that it had been a good beginning, that at least we now knew one way of getting the piranhas to come to us that was only mildly life-threatening. Chris was the least enthralled. He thought he had got some interesting sound but he was sure it had been ruined by Tony shouting at me to keep the meat still or pressed against the periscope lens while he filmed.

Neither of us remembered any of this, so we listened to the recording. Chris was right.

Satisfied, though, that the piranhas could be attracted quite easily, there was no further need to do the sound recording and the filming at the same time. We improved on both techniques and were so busy that, after a couple of weeks, the inside of our metal boat was a complete rubbish tip.

'One day, when it's raining and we can't film,' I vowed, 'I'm going to come down here and give it a thorough spring clean.'

It didn't happen. A week later, while the weather was still too good for domestic chores, I was up early and the first person I met was the manager, Heriberto.

'Dan!' he said. 'Today I need your boat. It is the biggest one we have and there are eight people coming for a trip on the river. I will give you the smaller boat for today and you don't have to worry about this because my men have already gone to change them over for you.'

It did occur to me, fleetingly, that they would find the large boat a bit of a mess and that the alternative boat was, in fact, too small, but that was all. And then I bumped into Chris, lured out of his room by the irresistible smells of bacon and coffee.

'But the condoms, they'll find the condoms!' Chris interrupted while I was trying to explain why our filming plans had to be changed for the day. It was the condoms, not the filming, that concerned him most. He wanted to use them again, but now they were almost certain to be thrown away.

'The boat,' he continued over breakfast, 'is littered not just with any old used condoms, but with horribly stretched and torn ones. Never mind that I want to keep the intact ones, what on earth will they think we've been up to?'

It was too late to worry about that now. The men were back and we were reassured by Heriberto that the little boat was waiting for us and that it would be replaced by the larger one the following day. We spent the whole day filming on land, once even waving to the tourists as they sped along the river in our large boat. But had they gestured and waved back a little too enthusiastically? After a slightly uncomfortable evening meal, convinced that we were being targeted by telltale looks and sly comments from all sides, we set out the following morning to continue filming in the colony. The large boat was just where we had left it. It was remarkably clean, and empty, except for a local cigarette packet tucked carefully under one of the seats, between a supporting metal strut and the side of the boat. Curiosity got the better of Chris. Inside were seven neatly rolled up condoms, all intact. The men, perhaps in keeping with local tradition, must have understood that if they were still in good condition and we had not thrown them away, then we might want to use them again. We did, but not on the piranhas.

The most obvious wild animal in the *llanos* is the capybara, the world's largest rodent. Once hunted almost to extinction in Venezuela, they thrive today on the ranches that have long had an interest in conservation and which have now set themselves up to attract wildlife tourists. Capybara are known locally as *chiguiri*.

'Mrs Chiguiri holds her breath and goes below the surface,' explained Mauricio, the manager of El Frio ranch, 150 km east of El Cedral. 'All you see is Mr Chiguiri jigging on top of the water with a big smile on his face. It is finished when she runs out of air and has to come back up. But if he is a very lucky Mr Chiguiri, she will take another big breath and go down for him again.'

For eight years Mauricio had been the manager of El Frio, a biological research station that recruited its academic and domestic staff from Spain. Such was his isolation, it was not surprising that he knew the mating habits of the wildlife around him better than he knew his own, try as he might to impress the female tourists who came his way for just one or two days at a time.

Mauricio had obviously studied capybaras down to the last detail, which wasn't too difficult. They are large, docile creatures, easily approached. They spend their days in water, feeding on vegetation, especially water hyacinth, or lolling around on dry land, preferring the warm mud-baked roads where it is difficult to move them out of the way. Hawk-like caracaras and tyrant flycatchers perch on their backs, watching for insects on and around their bodies.

What's interesting about capybaras is that they have, over millions of years, evolved to live as semi-aquatic mammals in piranha-infested waters. They would not be living like this today if their ancestors had been eaten whenever they dipped their toes into the rivers and lagoons. The first piranha attack I ever saw on television, probably around 1980, was on a capybara setting out to swim across a narrow river channel, only to be demolished in minutes by ravenous piranhas. In our experience, though, piranhas do not attack living capybaras and the local ranch people confirmed this. We were told that their fur is too thick and prickly, that their skin is too tough or that they give off chemicals that piranhas don't like. On several occasions we saw male capybaras, some even with open fight wounds, swimming in deep water where there were lots of piranhas. Just once, though, a large male, leaking fresh blood from a gash on his shoulder, re-fused to dive into the water when we approached. The rest of his group had gone in, alerted by the barking alarm of one of them, diving in head first with a series of clumsy splashes. But not this big wounded warrior. He hobbled along the riverbank all on his own, refusing to enter the water even when we approached him on foot to see if he would jump. For some reason, he just wouldn't do it. He chose to remain with us on land rather than swim 50 metres to a small island of grass to join what we assumed, perhaps wrongly, to be the rest of his family.

If, however, you are a dead capybara in the water, things are different. Red-bellied piranhas are perfect scavengers. They are the vultures of South America's waterways, picking up the slightest scent of decaying flesh and homing in fast. A 50 kg capybara really will be demolished in minutes, hit from all sides by hundreds of piranhas, each one shearing off a mouthful of flesh, retreating, swallowing and turning to wrestle in for more. Because the piranhas come from all directions, the carcass begins to spin, and as it turns the fish can be thrown clear of the water, flashing orange and silver in a violent sunlit spray.

Having filmed living capybaras feeding and swimming untroubled by piranhas, we now wanted to film the piranha reaction to a dead one. We put out a general call for a roadkill and, a week later, one was brought in. It had been dead for two or three days and the finder had obligingly removed its bloated innards to save us from the awful smell, leaving a neatly scooped-out body cavity. This was not what we wanted because it would now sink like a stone and we would not be able to film the piranha response to it at the surface. It needed buoyancy. We had an idea.

Chris had now finished his underwater sound recording but he still had a supply of used and unused condoms. The following day, Tony and I sat on the river bank under a blazing South American sun blowing up condoms like balloons at a birthday party and tucking them carefully into the capybara's empty body. When it was crammed full, with a few added stones to keep it tummy-down in the water, we stitched it up and there it was, larger and lumpier than life but all set for its maiden voyage.

When Tony was ready with his camera, 30 metres downriver from me to allow for drift, I waded into the shallows and sent the rotting rodent on its way. As a precaution, I had tied a length of nylon fishing line to one of its back legs. The capybara settled low in the water but it didn't sink and the lumps and the stitching were well enough hidden not to interfere with the filming. Five metres into its journey, it shuddered visibly as the first wave of piranhas came up from the deep. Within a minute, its body began to wobble and spin under the force of the attack, which threw up enough spray to mask the details of our operation on its body. Thirty seconds after that, piranhas were being hurled clear of the water.

'The condoms!' yelled Tony. 'I can see the condoms!'

The piranhas had bitten clean through the stitching, and our birthday balloons were now bobbing to the surface and rocking off downriver. Deprived of its buoyancy, and with the added weight of its remaining stones, the capybara was sinking fast. It took an enormous effort to haul it back to the shore on the nylon line, a real live tug-of-war with piranhas determined to have the final say. Fortunately, they not only lost but they had also eaten no more than the capybara's belly. It looked good enough on top for another try. The little flotilla of condoms, meanwhile, had bobbed off out of sight. If they survived the attentions of local fishermen, they would be carried down the Apure River to San Fernando where they would enter the main flow of the Orinoco 800 km from the Atlantic. We wished we had put a message in one of them.

We drove slowly along the riverbank until we found what we were looking for. Tony set up his camera while I dragged the capybara towards a dead tree that lay half in, half out of the water, hooking the body over a submerged branch as though it had just floated along and become entangled. Looking for the best picture composition and lighting, Tony decided to wade into the water and hand-hold his camera, steadying himself against the branches of the tree. He was thigh-deep and not wearing waders as I backed away and began throwing bits of meat into the deep channel just beyond the fallen tree.

It took 20 seconds for the piranhas to find the capybara. But this time there was a different problem. Because it was caught on the tree, it could not spin and the feeding frenzy was not developing. There was hardly any spray and very few fish were coming to the surface. And then, before we could act, the capybara began sliding off its perch, lower and lower in the water. Tony was also sinking but he had to keep filming because the edge of a large black cloud was creeping over the sun and there was no chance of retrieving the carcass through the branches of the tree. When the capybara finally went, he switched off his camera and tried to lift a leg out of the muddy sand. It stuck fast. He lost his balance and began toppling backwards into the branches of the tree. 'The camera!' he screamed. 'The camera!'

As I lunged forward and snatched £35,000 of equipment from his grasp Tony, all 100 kg of him, went down through the branches and straight onto a submerged anaconda that had been quietly watching from the safety of its coils. These it unleashed like a gigantic spring and three metres of seriously startled snake exploded out into the deeper part of the river. With perfect timing and characteristic good humour, Tony, checking first that his camera was safe, peered up from the water and the branches and the mud, and enquired, 'Any chance of a pay rise?'

We returned to the ranch to find that we were being joined for dinner that night by a small group of tourists. We could see that Mauricio had already singled out one of the girls by the way he kept looking at her. But before we got even close to recounting our day with the *chiguiri* on the river, Mauricio caught my eye, looked across at the girl, back at me and Tony, and then, with a big grin and to the utter amazement of everyone else, took a deep breath and disappeared beneath the table!

When we told Don in 1995 that we were coming back to make another piranha film and needed a different location, he contacted El Frio, a research station about 100 km from El Cedral. The managers there had heard about the success of the Cedral tank, which had been kept going as a tourist attraction, and were keen to have one of their own. Don faxed us their plans to fine-tune and we ended up with an amazing home for the fish. It had a large viewing window in its own comfortable dug-out area that would be ideal for filming. The pump system was perfect, as was the provision for keeping the water clear of algae. We had had difficulties with algae in Cedral where, despite shielding the surface from the sun with a black cloth and using a liberal amount of non-toxic algaecide, the 18,000 litres of fresh water still needed to be changed every few days. Once a fortnight we had to give the walls and the branches a thorough scrubbing before the tank could be refilled. We had none of these problems with the Frio tank, which held twice as much water.

We did, though, have problems at El Frio with vehicles. The arrangement that Don came to here was not a patch on that with El Cedral and the brand-new Toyota. Here, we would pay reduced board and lodging but only if we paid for new air-conditioning units to be installed in the two rooms that were being renovated especially for us. The importance of these new and expensively-equipped rooms to El Frio in the future was such that they also offered us the exclusive use of one of their Toyota station wagons. All we had to do, Don smiled again, was pay for it to be 'fixed'. We kept our part of the bargain by paying for our first stay and for the vehicle repair weeks in advance, and by then leaving the money for the air-conditioning units with Don in Caracas when we travelled down to the ranch for the first time. As soon as these units came in, he would send them down to be installed in our sparkling new accommodation.

When we arrived, the rooms were a building site and the Toyota was about the worst vehicle any of us had ever seen. The one thing it did not do was work, despite the advance payment to fit new starter motor, battery, tyres and more. Our arrival was the catalyst and, as Alexis, the ranch foreman, set to work, we added the finishing touches to the piranhas' new home. After a week of nailing down the astro turf we had brought from England, painting the walls and filling the place with logs and greenery, the set was ready and the old Toyota was grumbling back into life. The accommodation worked out as well. Having failed to provide the promised rooms, they had to put us into the relatively new and mostly empty block where there were twelve good rooms for the occasional tourists.

The Toyota was a disaster. It must have been about 30 years old, sounded like a tractor and had virtually no springs which, while not good for us, was seriously bad for the equipment. At every pothole or bump, whoever was driving had to slow down to a virtual stop and crawl forward as carefully as possible. But its greatest problems were overheating, cutting out and failing to restart. On more than one occasion, I walked several kilometres back to the ranch to get help. It just had to go.

It was with the greatest sense of relief, then, that we sped down the main highway for 20 km in the ranch's much younger Toyota which we had been lent for the day. It was a pleasure to turn off the tarmac road for a bouncy but comfortable cross-country ride to one of El Frio's two waterbird breeding colonies. It was August and the rains were in full swing, so we needed to be careful. The arrangement was that if we weren't back by 8 p.m. – it was dark by six and the colony was more than an hour from the ranch – they would come looking for us. On this particular day, we worked late in the colony and it was already dark when we set out for the main road. It had been raining heavily and there was one particular swampy section of about 50 metres that still had to be crossed. Chris slowed right down as we approached the spot, letting Tony and me out into the mud. We would walk across first, testing the ground below water for deep ruts. Chris had to memorise our precise pathway and then get through in a low four-wheel drive gear, keeping the engine revs up and slipping the clutch to prevent it from stalling. From the far side, and with the mosquitoes already out in force, Tony and I watched as he engaged second gear, dipped his lights and set out for what needed to be a short, sharp dash to firmer ground. So much for the theory. Halfway across Chris was dragged sideways, over-steered to get back in line and plunged into something even deeper, bringing the Toyota to a juddering halt with an ominous cracking sound. He restarted the engine to find that all his traction had gone. We were stuck fast. It was half past six and Alexis would not be here until at least half past nine.

We unloaded the equipment and carried it carefully through the mud to where we could just make out the tracks we had carved during the previous weeks. The mosquitoes were unbearable and it was fortunate that, along with a bit of insect repellent, I had half a bottle of whisky in my bag to accompany a packet of local cigarettes left in the Toyota. We wrapped up, huddled up, drank up and puffed smoke at mosquitoes for an hour until we felt quite ill and didn't mind being bitten any more. By nine o'clock we could see headlights coming towards us every few minutes but they turned out to be stars or fireflies preying on our minds. And then, as if by magic, they were real. It was Alexis and a lorry-load of helpers with – the loveliest of touches – a crate of cold beer. We were soon back at the ranch, showered and tucking into rather crispy lasagne and Mauricio's local rum. The car would spend the night in the swamp. In what I considered to be a magnanimous gesture, I offered to pay for half its repair as long as we could use it whenever we wanted. My trump card was a promise to say nothing more about the air-conditioned rooms that were clearly never going to happen, even though Don had already sent the paid-for units down from Caracas. We had a deal and, while the vehicle was being recovered and mended, we concentrated on filming piranhas in the tank. All 120 of them were behaving in a very relaxed way, which was a relief given the ordeal of their capture.

For this, we had needed the help of a large open-backed truck carrying six 44-gallon drums and half a dozen people. We half-filled the drums with lagoon water as close as possible to the catching site, so there would be no dramatic change in temperature for the fish. Then we set to work with nylon lines, each with a thick steel trace 30 cm long ending in a medium-sized hook with a pronounced barb. Onto this went bite-sized chunks of steak. We also had a bucket of bloody offal which was heaved into the water to bring all the fish in the neighbourhood within range of our lines.

Three or four people fishing for two hours can easily catch 100 piranhas. There is nothing difficult or skilled about it. You pay out line so that it lies in an organised mass around your feet, whirl a couple of metres of the heavy, baited end round your head like a lasso and fling it out into the water. If the offal has done its job – and quite probably even if it hasn't – a piranha will be onto your bait in a fraction of a second, taking it straight under so that you feel an immediate pull. You then haul the piranha back to the shore. It doesn't always make it. If you pull too hard, the meat and the hook may come out of its mouth, or the fish may bite through the line or even the metal trace. Occasionally, a piranha remained hooked but itself came out of the water half-eaten. A feature of the piranhas we observed in the tank was that they would always investigate an individual behaving differently, and when its behaviour was too different, it would be attacked. A piranha dragged through the water at high speed was bound to attract a certain amount of attention and could be attacked even when the water was muddy.

Most of our piranhas were landed intact. Then it was a question of getting them off the line and into the drums on the lorry. A hooked piranha is not the easiest to deal with. It is, for a start, primed to snap viciously at any object that enters its mouth. It helps if this is not a finger. We learnt, after the odd minor mishap, to use small pliers. These, once the suspended piranha had been firmly grasped behind the gills, could be pushed into its open mouth to grasp the embedded hook.

Once the fish had been bounced along the dusty or muddy road to the ranch, they were carefully emptied into the tank and concentrated in about 500 litres of their own water. We then introduced the hosepipe and filled the tank slowly from the ranch's underground supply. This water was very cold and we did not want to shock the fish. With pumps and filters running smoothly, the piranhas needed three or four days to settle before we felt they were behaving naturally enough to be filmed. On overcast or rainy days, when we couldn't film anywhere else, we would sit and watch them as they patrolled the tank in a tight squadron, turning this way and that with a togetherness that looked like pre-planning. Only once did we see them attack a different kind of fish of the many we put in with them.

While the larger El Frio tank was the better of the two for filming captive piranhas, it was the Cedral waterbird colony that gave us the best footage of piranhas in the wild. One day at Frio, when Yolanda was with us, she suggested taking us over to El Cedral – a 20-minute flight away – to say hello to the people on the ranch. As familiar landmarks and buildings came into sight, I asked if we could do a little detour over the bird colony for a final look. To our horror, it was completely empty. Ramon, the local guide who knew the wildlife of Cedral better than anyone, confirmed that the birds did not use the same colony every year. Their droppings, he explained, were so acid that they would soon kill the trees and the bushes. They move around the *llanos*, perhaps also to avoid nest parasites, using a different location every two or three years. He did not know where there was another breeding colony, so they obviously move quite a long way. We were stunned to be given this information so casually, having been reassured on our first recce – though not, admittedly, by Ramon – that the birds were there every year. We thought back a couple of years to when Tony, Sean O'Driscoll and I had come out in August at a moment's notice because we felt our colony footage needed improving. To have gone all the way back out to the *llanos* to find the colony empty would have been little short of a disaster.

In both films, there were underwater shots of young egrets being demolished by piranhas. These were filmed in the purpose-built tank with captive fish, the only way to get the dramatic viewpoint to cut with the above-water shots that had been filmed in the wild. Every day, before filming in the colony, we paddled, waded and climbed through the bushes to collect any dead egrets hanging in the branches. These were the ones that had fallen but not reached the water. We put them in a cool box to be transferred to the ranch freezer in the evening. From there, we could take them out and feed them to the fish in the tank when we were ready to film the details of their underwater frenzies.

Although we weren't the first people to film wildlife in the *llanos* of Venezuela, the high profile of our two films and their graphic feeding frenzies generated a rush of interest. Within a short while, sequences and whole films on piranhas, capybaras, anacondas and caimen were being commissioned around the world, some even with daredevil presenters who braved the water to wrestle or engage with their fishy and reptilian subjects. El Cedral and El Frio were visited by film crews looking for quick-fix encounters and dramatic footage of the 'will he or won't he be eaten?' kind to satisfy a growing trend in programming.

Mauricio remembered our lengthy stays at El Frio with affection, as though we had broken through a barrier to become part of the ranch's extended family, rather than just its paying customers. When I sent him copies of the BBC *Piranhas* film, he didn't just say that he liked it and we got the credits right. It was, he added, something extra to remember us by. As though the experience of having us under his care for all those weeks of good-humoured banter about *chiguiri* sex, air-conditioning units and a Toyota that should have been scrapped 20 years before, were a necessary part of mucking in to do a worthwhile job. Neither Tony nor I would have had it any other way.

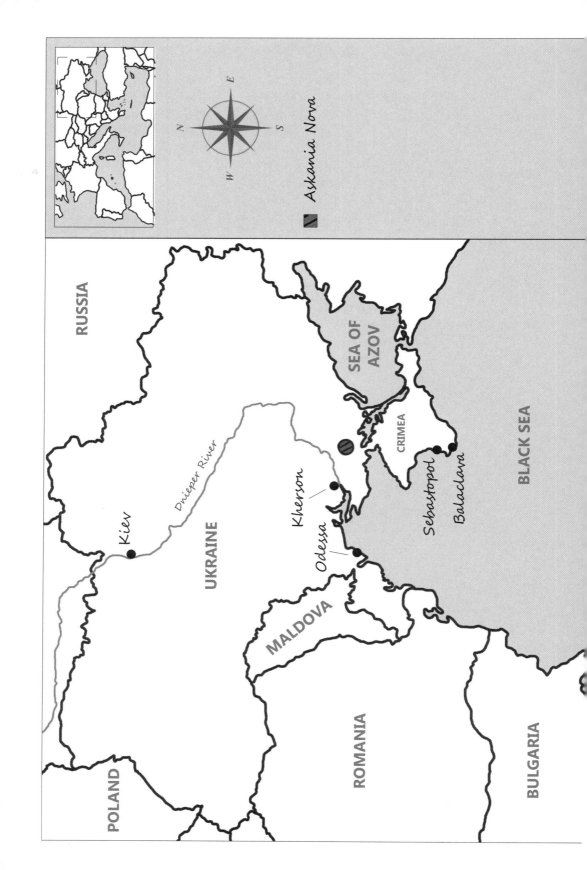

12
JAMES AND THE GIANT OAK

'Yes, have maybe seen two, or maybe three, in last week,' the Ukrainian Igor confirmed as we set out across the frosted grass. 'Not so much for you but expecting soon. It is good time, but cold.'

Reassured, we crunched even further from the vehicles on the track, our attention held by dwarf irises and tulips, islands of purple and pink in a sea of frozen green. The sun broke through the low cloud, bringing warmth to the day. Grass softened underfoot. We saw the first of our own 'not-so-much-for-you-but-expecting-soon' snakes. And five minutes after that, we saw another. Igor shrugged and pulled one of those 'This is nothing' sort of faces. But within the hour, and with the sun climbing higher and hotter, we had seen at least ten more. Surely, Igor, this was different? Couldn't this be the real thing? Igor knew that it was. Transformed from his earlier pessimism, he swelled with pride, his job done. At last, the snakes were wide awake and the penetrating heat of the late April sun was enticing them out into the open. More by luck than by judgement we had, on our first morning, walked out into the vipers' peak emergence after their long winter sleep below ground. And now they were crawling up through the grass or sitting in their coils to soak up the new heat of the year. There must have been hundreds of them, warming and watching, whichever way we turned.

One of the drivers, Alexander, shouted urgently. Had he been bitten? Side-stepping the coiled and slithering bodies, burdened with filming gear, we hurried towards him. No, he hadn't been bitten but what he had found was special – a writhing mass of at least ten males piled onto a single, larger female, each trying to make contact with her body. It was viper sex, and we had to have it.

James dumped his backpack, attached the long straightscope lens to his camera and plunged it into the middle of the heaving bodies. We stood well back and watched as he hugged the ground, oblivious to any danger from the snakes around him. We managed to keep them off but more were coming, lured to the orgy by chemicals carried on the morning breeze. They were sliding all round him, but James did not flinch. He had time for a quick lens change for a wider shot before the sex-mad serpents unravelled and went their separate ways.

James stood up, stretched his aching muscles and realised for the first time in 15 minutes that we were still there. He grinned, evidently pleased with what he had seen through his viewfinder, but added, drily, as we stepped through the dispersing reptiles to help him with his gear, that perhaps he might be given just a little bit of notice when required to lie down in a field of poisonous snakes. We smiled. Although we had known that the emergence of Orsini's viper happened around this time, it was pure luck to have stumbled on such a large number of these half-metre long snakes so early in our filming.

Tiptoeing back to the cars, serenaded by skylarks whose female partners were now sitting on eggs in snake-infested grass, it was strange to think we were less than 60 km from the Crimea. This famous peninsula jutting out into the Black Sea in southern Ukraine holds a prominent position in many a schoolchild's history textbook. It is the site of a mid-19th-century war, when the combined forces of Britain and France invaded Russia, rapping Europe's most powerful country soundly across the knuckles to curb her greedy land-grabbing. By the end of the Crimean War in 1856, the Charge of the Light Brigade, Balaclava and Florence Nightingale were among the new names etched into the annals of European history. But the reason for *our* being here went way back beyond a strategic war between European superpowers.

One hundred years before the Crimean War, Catherine the Great of Russia had the idea of inviting foreigners to colonise the grasslands north of the Black Sea. It was to this desolate windswept plain that the German army deserter Johann Fein trekked in the 1760s. His family prospered, his daughter married into another immigrant family and the Falz-Fein dynasty was born. Over the next century, it expanded its territory and imported animals from all over the world to create a special zoo on the steppe grassland. The estate became known as Askania-Nova, the name of a nearby settlement. In 1985, after 200 years of regional unrest, revolutions and world wars, the Askania-Nova estate became a government-controlled Biosphere Reserve of almost 34,000 hectares. It is today run by the Ukraine that rose from the ashes of the USSR in 1991. Now, six years later, James Gray, Nigel Tucker and I had just witnessed the mass spring emergence of the Reserve's Orsini's vipers.

We were here to make part of a film on European grasslands for Green Umbrella, a Bristol company run by Peter Jones and Nigel Ashcroft. Askania-Nova remains the only patch of ancient grassland in a continent that has largely been cleared for agriculture. Filming the Reserve's rich wildlife would be our contrast with the impoverished wheatfield 'grasslands' that are so widespread in Europe today. None of us had been here before.

After a first night in Ukraine's capital, where we dined a little predictably on caviar, chicken Kiev and vodka, we set out on the ten-hour drive south to Askania-Nova. There were six of us in two vehicles driven by Anatoly and Alexander. Our interpreter and fixer, Yuri, was a passenger, wanting to know all about life in England. We drove down through open rolling country with its mix of industry and agriculture, the ploughed fields full of women planting potatoes by hand. We crossed the Dnieper River north of Kherson in the dark, at least an hour behind schedule. This was new territory for both drivers, so it wasn't until half past eight that we entered Askania-Nova village and began looking for the Hotel Kanha. Round and round we drove, past identical blocks of flats with dimly-lit windows and not a person to ask for help. Suddenly, three women appeared out of the evening gloom and, as we wound down the window to speak, one of them stepped forward, poked her head inside and said, 'Freeman?' in such an exasperated and rhetorical way that it was obvious that not many people came this way at all. We simply could not have been anyone else. 'Good, follow please.' Within a few minutes we were unloading at the hotel where they had spent the last two hours waiting for us. We had met them just after they had decided to organise a search party. The hotel – the best and only in town, for which each of us was paying US$50 a day – was not worthy of many stars of approval. It was cold, badly lit and scruffy, and neither the showers nor the lavatories worked properly. But it was dry, there were plenty of spare blankets and the people were very welcoming, having been genuinely concerned that we might have been lost on a freezing windswept plain. They had been detailed to look after us and we were their responsibility.

Having dumped our bags in our rooms, we were told that food was ready and that we should hurry. We were almost frogmarched the 200 metres to a private house – the catering for our week's stay having been contracted out because the hotel kitchen was open only during the summer – where we were ushered into a back dining room with a beautifully-laid table full of the most appetising food. There was salad, cold meat, eggs and bread already on the table and, when we had eaten these, out came steak and chips to be washed down with the thickest, creamiest yoghurt drink, *smetana*. Over coffee, Anatoly produced a bottle of Kiev's finest vodka and we went through a little 'welcome to Askania-Nova' ceremony of two rounds of 'down in one'. The aged owners of the house, the providers of the lovely food, stood quietly by, smiling tensely as they observed a younger generation and foreigners indulging themselves and socialising in a way that would not have been possible a few years previously. On the short walk back to the hotel, Yuri assured us that we had offended no one and that it was just the older people who wished for a return to the stark and disciplined 'but at least you know where you stand' days of rule from Moscow. Younger people felt a great sense of freedom and many were busy making plans to go and live in Canada.

It was 11 o'clock when we returned to the hotel. We had been on the go for 16 hours and now, well fed and primed by Anatoly's vodka, we were ready for a good night's sleep. But the three women had other ideas. They were still waiting for us in the hotel lobby, and they were not alone. Several men had joined them while we were away eating. All, it turned out, were park employees with responsibility for different animals, plants and locations. We were now expected to go through the plans for the next ten days, beginning with the next morning when we would be taken to film flowers and steppe marmots from a hide, already erected and overlooking a den from which young would be emerging any day. The park had obviously lost none of its old-regime efficiency, and we felt immediately grateful that the species list we faxed from England several weeks earlier had been taken so seriously. We had given the vipers no more than a question mark, yet Igor had been out looking for them every day.

We soon got used to the quirky ways of the hotel. We confined our use to those lavatories that worked reasonably well, gave up any thoughts of washing properly and engaged the staff's patience head-on with our pre-dawn starts and trails of mud when we came in from the wet. The highlight was my returning one evening with a stinking red deer skull that I left 'fizzing' for three days in a bucket of bleach. Twice, I saved the skull from being thrown away – in protest, apparently, against the misuse of a hotel bucket – before scraping it clean, giving it a final soak and producing it as a gleaming white trophy to be taken home. The skull, in the end, was much admired by the hotel staff who examined it closely, hardly able to believe it was the same one that had caused them so much concern. And when Nigel pointed out that, after its enforced bleaching, even the bucket had never had it so good, everyone fell about laughing.

It was several days before we met Askania Nova's manager, Mr Havrylenko. I had already sensed that he was a distant and slightly sinister figure by the way, even after diplomatic translation by Yuri, he was referred to by his staff and the fact that he didn't bother to meet us on our first full day at the Reserve. This slightly vague impression of him was fully endorsed within seconds of our first encounter. Yuri and I had been summoned, and advised to be on time.

Mr Havrylenko was a busy man. So, at 5 p.m. precisely on our fourth day, we presented ourselves and were ushered into his office. He was dressed largely in black, had one arm, a patch over one eye and carried a gun. With his head, he gestured towards two empty chairs strategically placed in front of, and dwarfed by, his large, important desk. He wanted to know if everything was to our satisfaction, that we were being properly fed and whether I had enough money to pay the balance of Green Umbrella's account following the initial down-payment by credit transfer. He mellowed visibly when I said everything was as good as we had hoped for and that I would be happy to release the additional US$1200 the day before we were due to leave Askania-Nova.

There was, however, just one small problem I wanted to discuss with Mr Havrylenko. The previous night, at our evening meal, I had asked Yuri how the old couple managed to feed us so well on US$15 a day per head when money was so tight and shop prices so high. Yuri had told me that the money from Green Umbrella had been advanced to the caterers and that they were spending it all, making nothing for themselves. Yuri had then turned to them, because they were hovering as usual, and said that we would like to thank them for doing so well on US$15 per day per person. Their jaws dropped, they went bright red, looked hard at each other and stalked out of the room shouting at the tops of their voices. Yuri hurried after them and after several minutes of heated discussion came back with the embarrassing news that Mr Havrylenko was paying them only US$10 a day per person. Which meant that he would be pocketing US$300 at their expense. This was a serious amount of money now that funding from Moscow had dried up and people never knew from one week to the next what might happen to the value and availability of their already meagre salaries, savings and pensions.

Yuri translated my concern. Mr Havrylenko thought for a while and then assured me that the matter would be resolved. In what must have been a bit of a climbdown, he later offered the old couple an extra $2 a day per person which made a big difference to them, even though he was still able to hang onto $180. Our caterers were absolutely delighted with the extra money, but Yuri also hinted that their real delight lay with knowing that someone had stood up to Havrylenko and made him grovel.

Our ten days in the Reserve went very quickly. We left the hotel around dawn on most mornings, usually walking James into one of the hides that had survived its overnight battering from the wind, before driving round to rescue those that hadn't. A few days after the unexpected success with the snakes, we found baby steppe marmots venturing out into the open with their parents for the very first time. We found a rookery crammed into a small stand of roadside trees, its lowest nests almost touching the ground. We scoured its hundred or so nests one by one with binoculars and soon came up with a number of

A steppe marmot on lookout duty on the grasslands of Askania-Nova,
Ukraine (Courtesy of Nigel Tucker)

lesser kestrels, red-footed falcons and long-eared owls guarding empty nests or incubating eggs. There was even one long-eared owl chick that was out and about, learning to fly. Birds of prey come to the steppe grassland to feed on its small rodents. Their problem, though, is not having their own places to breed, so they use the ready-made nests in the rookery. There is rarely conflict between them and the rooks because they choose old nests or wait until the young rooks have fledged before they move in to lay their eggs.

We filmed dwarf irises and dwarf tulips – short-stemmed because of the constant winds – down in the grass along with daisies, green lizards, frogs and insects. Shelduck were inspecting old marmot burrows to see if they would make good underground nest sites, calandra larks and skylarks sang spiritedly from the heavens while their mates sat tight on their eggs. We were even taken out to see one of the remaining stone figures that were sculpted 1,000 years ago and erected as idols by the Scythians, a tribe that once dominated the ancient steppes. There was a time, we were told, when a great many of these two-metre tall figures marched out across flat, open grasslands, the only guiding features for traders negotiating this desolate first stage of their journey from the Black Sea to Scandinavia.

Among its menagerie of animals, Askania-Nova had a herd of Przewalski's horses. There was only one way to approach these original grassland grazers in their large enclosure and that was by pony and trap. This was the only vehicle that the naturally shy horses would allow anywhere near them, associating it with a delivery of food. We took two traps, intending to park one in front of the other so that James could dismount and put his tripod on the ground between them. With a scattering of food, we would then tempt the 30-strong herd to come closer. Even though our backs suffered from two consecutive days in the traps, we were able to tempt the horses to within 20 metres of us.

We also tried to film saiga antelope in another enclosure but they were far too nervous and would not let us even remotely close, not even in the pony and traps that they and the horses had seen so many times before. The few distant shots that James did manage were ruined by heat haze, so this was an important steppe mammal we would have to do without. Our disappointment and aching backs met with little sympathy from our Ukrainian colleagues. They asked us to appreciate that we had ridden the ancient Russian steppes in far greater comfort than the pioneers of the Falz-Fein dynasty 250 years previously. And, they stressed, without the fear of being swooped upon by marauding Cossack bandits.

We returned to the hotel one evening to find that Igor, our snake expert, was in hospital suffering from a bite. He had made a small enclosure for us but the first snake he caught had got the better of him and dipped its angry fangs into his flesh. At midnight Anatoly came to my room to ask, optimistically, if we carried any serum to counteract the snake's poison because there was nothing available in Askania-Nova. He said he had been sent by Igor's wife who was desperately worried about how she and her children would survive if her husband died. Fortunately, Orsini's viper is about the least poisonous of European vipers and Igor was out of hospital within two days. We assured him that we had enough snake footage from that first morning and it would not be necessary to catch another. To celebrate his survival, we were taken out to a club where we found a welcome supply of wine. All of us, including Yuri, Alexander and Anatoly, because they came from Kiev, were stared at and gossiped about by local teenagers. The following day the same teenagers were hanging around the hotel in a confident and mildly threatening way, much to the disapproval of the older hotel staff who 'tut-tutted' and blamed their idleness on the collapse of the Soviet Union. We sensed that they were no more than curious to see visitors in their isolated village.

The day to settle our bill arrived. The weather was amazing, accurately forecast by Anatoly after three days of heavy cloud and wind. Now that it was calm and sunny we had a number of last-minute shots and sounds to get before driving back to Kiev the following morning.

With James and Nigel preoccupied with this last-minute activity, Yuri and I prepared for a financial showdown with Mr Havrylenko. At the Reserve headquarters, with Yuri doing his best to keep the translation flowing, we agreed that I had sent a fax of requirements from England and that the Reserve had carried these out for an agreed sum of money. There was now the balance to pay.

So far so good but, just to make the point again, I added that our first payment had also included a sum to pay for our food on top of the agreed rate of US$50 per person for the hotel. Mr Havrylenko smiled and said that this was perfectly true. Yuri was getting nervous. He wanted to move on from any discussion about how much the hotel was being paid. I didn't pursue the matter. Mr Havrylenko produced a contract which he and I both signed. This was followed by a statement of 'execution of duty' by the Reserve that we also both signed.

Once we were agreed that everybody had done wonderfully well, we were driven to the bank. I was introduced to the manager and he introduced me to his chief cashier who would carry out the transaction. Before that, though, we sat round a large table and went through everything again, examining the contract and the statement in fine detail before the bank manager would add his own signature to ours.

I was escorted by armed guard about five metres to the chief cashier's counter where I signed my travellers' cheques and received US$1200 in exchange. The AK47 then followed me back to the table. I sat down with my money stacked in front of me. After a slightly awkward silence, the bank manager said something to Yuri. Yuri, in turn, asked me to produce the letter from Green Umbrella authorising me to pass the money over to the bank manager. Already humbled by their meticulous approach to bureaucratic formalities, I admitted that, while I did not have such a letter, I was also certain that no such letter existed. Yuri, trying to move things along, suggested that, as even he did not know of such protocol, perhaps the transaction could be completed without it now that the documents had been formally signed by all parties?

This would not do and the bank manager was politely adamant that we could not proceed without the necessary authorisation from Green Umbrella. In its absence, the atmosphere was starting to get a little tense. I made a suggestion: if they could not accept the money from me, then I would just have to keep it.

Their frowned reaction made it quite clear that I was not going anywhere until the money was theirs. Finally, the bank manager came up with the solution: I was to telephone Green Umbrella asking them to fax back authorisation with a signed guarantee that the original was in the post. It was all over in minutes, even though the return fax didn't actually materialise. But they had heard the conversation and that was enough. In return for a receipt, my modest pile of notes was pushed across the table and hurried away to the cashier.

After handshakes all round, Yuri and I were driven back to the hotel where James and Alexander were preparing to go out again, having just come back to look for us. James raised an eyebrow and tapped his watch with an implied 'And where have you guys been, then?' 'It's a long story,' was the only answer he got, at least while we still had work to complete on our last afternoon. Three hours later, as the setting sun took the light from our filming for the last time, James's parting shot of Askania-Nova was of a long-eared owl peering down from the edge of a rook's nest.

On the three-hour flight from Kiev to London, James and I talked about the film. How, he asked, were we going to show the great lowland deciduous forest clearance that took place after the last ice age? Could there be a more interesting way than a standard map with changing colours to highlight modern wheatfields where the great forests once stood? I knew we would have to use graphics to give a European overview of this clearance, but to illustrate it in detail we could hardly go out and randomly cut down a large deciduous tree.

Back in Bristol, assistant Gillian Burke contacted the Forestry Commission and was informed that they did occasionally clear out old trees to make space for younger ones. Better than this, two old oaks – the perfect tree for our film – were due to be removed from Savernake Forest in Wiltshire, just 50 km from Bristol, in a few weeks' time. We would be welcome to go along and film the whole operation. I rang James excitedly.

We arrived in perfect weather, met the manager and his circular-saw crew and set off into the forest. Nigel was with us to record the sounds of the ancient trees crashing down.

The first oak, the smaller of the two, was easy. James filmed the details of the saw biting deep into the base of the trunk and then zoomed out for the final cut that sent the 60-year-old tree tumbling away from him into an open glade, precisely as planned. After a quick tea break under the canopy of the midsummer forest, we moved off in search of tree number two.

This second oak was enormous, over 30 metres tall with a gnarled canopy meshed with the branches of its equally impressive neighbours. Gillian and I walked all round it and decided that James should film it a little more ambitiously than the first. I asked the manager to confer with the cutter and pace out the point from where James could safely film the tree falling towards him. Their combined years of experience gave us the exact mark and James spread his tripod legs wide so that the camera with its zoom lens was just above the ground.

'Can I use the remote control?' he asked optimistically.

'No, sorry, I want you to tilt down with the tree as it comes towards you, gently zooming in at the same time to heighten the feeling of it falling straight at the viewer.'

'Fine,' came the reply', 'On your head be it.'

James, in fact, was quite happy to stay with his camera. He just hadn't been able to resist the throwaway quip. At the last minute, with James in front of us and Nigel stowed in the bushes to our right, having beseeched us not to chatter while he was recording, we were reminded to put those bright yellow plastic hats back on our heads.

James crouched low over his camera, once again a world away through his eyepiece. We stood five metres behind him and waited as the rhythmic whining of the circular saw made the precise cuts that would send the veteran of the forest toppling our way. It was several minutes before the noise stopped and the harsh sound of splitting wood and the distant cry of 'Timber!' finally silenced the birds.

The canopy above us began to move, as though, for a moment, our tree was standing still and the rest of the world was falling gently away. And then their relative movements slowed and changed. Now, the forest giant, groaning and creaking, came crashing towards us through the branches of the surrounding trees, tearing and splintering. The prediction of where it would land was millimetre-perfect but, as it descended, one of its outstretched limbs caught a tall neighbouring birch tree about one third of the way up, snapping it off and hurling it forward like a huge javelin. James was concentrating on the shot through the eyepiece of his camera, and was aware of nothing but the falling oak.

What we saw, in that instant, rooted us to the spot. The birch trunk was flying through the air straight towards James. A moment before the oak tree landed with an explosion of branches and a numbing shock wave, the birch javelin landed and bit deep into the ground. The legs of the tripod kicked violently, the film magazine flew out sideways and James didn't move a muscle. He remained hunched with the birch sticking out of the forest floor right beside his head.

Before we could move, James stirred, stood, and turned to face us. Covered in shredded branches, dust and leaves, his shirt torn and blood dripping from his hand, he looked straight at me and said, 'Well, the camera's f*****, but I think I got the shot.'

The zoom lens, just centimetres from his head as he filmed, was smashed into two pieces. But it had been even closer than that. His yellow helmet, no longer just lip service to a legal requirement, had a scorch mark across its top.

The wooden missile, five metres long and 20 cm wide, had actually gone down in that small angle between his helmet and the lens, making contact with both as it drove into the soft earth below.

We sat for an hour in shock, our relief alternating with the dreadful knowledge of what might have been. No one was to blame; it had been one of those unforeseeable flukes. Later, though, I asked James if he remembered saying,

'On your head be it.'

He didn't, but he did say that, given the choice, he would rather lie down in a field of horny vipers any day. But there was no choice. He had now done both.

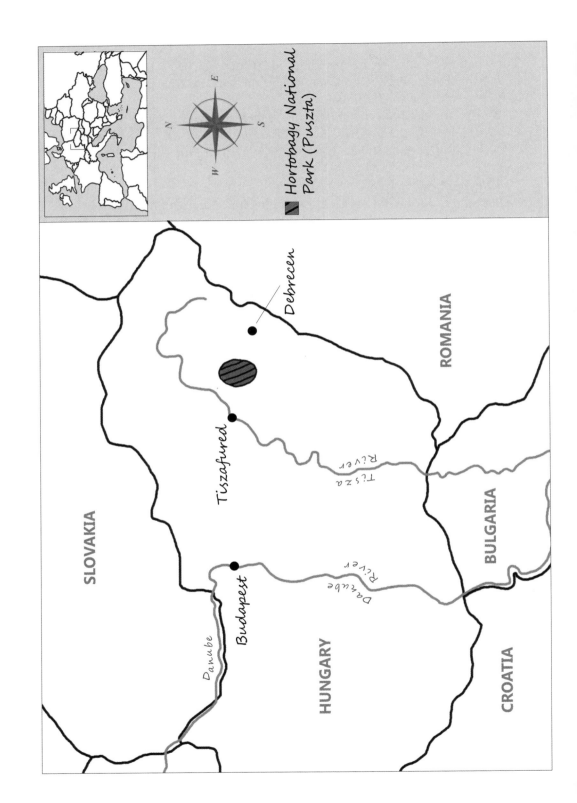

SLOVAKIA

HUNGARY

ROMANIA

BULGARIA

CROATIA

Danube

Budapest

Tiszafüred

Debrecen

Tisza River

Danube River

N
E
S
W

Hortobagy National
Park (Puszta)

13
TO AND FROM THE PUSZTA

Not long after dawn, during late February and early March, thousands of common cranes rise from their dormitories on the grasslands of Africa, Europe and Asia. Thermalling high into the morning sky, they set their internal compasses and head north for an awakening ice-free spring. Their two-or three-week journey, refined over thousands of years, is virtually troublefree.

Once assembled on their breeding grounds, in a swathe reaching east from Scandinavia to the Pacific coast of Russia, the birds begin their ballet-like courtships. Life partnerships are reinforced and bulky grass nests are built on waterlogged ground. From each of these isolated homes two chicks will hatch, and ten weeks later they, too, will be strong enough to fly. After a further two months of bodybuilding, family parties set out on the autumn journey to the warmer south, thus completing their yearly cycle. As on the outward flight, the cranes make stopovers to rest and refuel. One of these is in Hungary's Hortobagy National Park, where we were working in 1996. If we timed things right, we might end our filming with a spectacular sunset sequence of cranes coming down to roost on the famous Great Hungarian Plain, the Puszta.

It took Rachel Pinnock no more than a phone call to establish the difficulty of hiring a four-wheel drive vehicle in Hungary. But, in fact, driving the 2,000 km from Bristol wasn't such a bad idea. If we bought a second-hand Land Rover out of the programme budget, its cost would be partially offset by savings on airfares, excess baggage and at least one hire car on location. Fuel, travel days and overnight stops would need to be taken into account, but at the end of the filming the sale of the vehicle would recoup much of this outlay. And because this was a TV series on the people and wildlife of Europe, we could share the vehicle and its cost between more than one programme. Best of all, we would have our own transport on location and would be able to operate independently of anyone else.

We planned three trips to guarantee 20 minutes of edited footage about life on the grassland of the Puszta. As we planned our schedule early in 1996, the cranes were already on their way to their northern breeding grounds. Not for them the bureaucratic rigmarole of passports, customs and border controls, nor language and currency problems, nor driving on the left or the right. Just the natural hazards of energy and the weather. Some, unlike us, would have made the journey many times before.

The advantage of driving the Land Rover to Hungary was not obvious on the cold April day when Tony Allen and I left England. Arriving at the Channel Tunnel after a 5 a.m. start from Oxford, we asked for the office where we could get Tony's customs carnet stamped. 'You passed it in Ashford,' came the uninterested reply, requiring us to backtrack 25 km to the depot. We queued for two hours before the officer on duty processed our vital document, leaving us with an array of coloured pages to be filled in, stamped, torn out or retained at customs points along the way – out of England and in and out of France, Belgium, Germany, Austria, Hungary – and back again.

Carnets are said to be wonderful things, protecting expensive equipment from import duty as it is transported from one country to the next. The problem is that officials vary in their ability to deal with them. When you are hot and tired, sitting in a shabby little shed at a remote border post and the person in front of you doesn't know what they are doing, the temptation to accept a few random stamps and lose a few coloured pages is quite strong. Or, because they are embarrassed, they play for time, asking you to open two or three of the neatly-stowed boxes so they can check serial numbers against the contents list. The best that can be said for carnets is that they are a necessary evil. It's as well for the bearer to know exactly how they should be treated by Customs. When you finally make it back home, the carnet must be returned to the Chamber of Commerce. If there is anything wrong with the way it has been completed, the door to hefty financial penalties opens wide.

Back at the Channel Tunnel by midday and already behind schedule, we were flagged down, ushered into a little parking bay and told to switch off the engine and wait. Two poker-faced officers came over and, after the usual British pleasantries about the weather, one of them asked if we knew the registration number of the Land Rover. We didn't.

And then, in answer to the obvious question that followed, 'Is it yours?' I had to tell them that it wasn't ours but that it belonged to a film production company in Bristol called Green Umbrella.

One of them ducked into a little cubicle, tapped away on a keyboard and came back with a triumphant, 'Oh no it doesn't. Would you mind showing us your passports and', with a quick glance at the pile of baggage behind the front seats, 'telling us what you are carrying?' The mood had suddenly changed.

We showed them our passports and told them we had thousands of pounds' worth of filming gear, were on our way to film wildlife in Hungary and the possible reason for the vehicle not showing up as belonging to Green Umbrella was because they had only just bought it and the transfer from old owner to new hadn't quite gone through the system. I took a deep breath.

The senior officer looked at us, deadpan. 'Do you have a carnet for your equipment?'

Tony produced the freshly-stamped document with a flourish. It was the turning point. The authority of the much-maligned carnet was already working its magic, though not on its own. The senior officer told us that it had helped, of course, but it was coming up with the ownership transfer idea that had convinced him we could not be telling anything but the truth. He then asked – this was April 1996 – if we had heard about the recent road rage incident on the M25. We had. Apparently, a prime suspect was thought to be considering a sharp exit from England by Land Rover and any of these vehicles turning up at departure points around the coast were being thoroughly checked out. Satisfied that we were not who they were looking for, they allowed us to proceed to a waiting train and the short underground journey to France.

Driving on the Continent is a pleasure. Once you get through Northern France, Belgium and the industrial valley of the Rhine, you enter the highways of Germany. Here, the roads are wide, empty and inviting, used mainly by long-distance European drivers. People travel at high speed in the correct lane, service stations are frequent and the food and accommodation are excellent. Apart from a couple of minor carnet wobbles, the journey was so easy that we didn't envy assistant producer Rachel who would fly to Budapest in three hours and hire her own car.

Our companion landmark from southern Germany to Austria and Hungary was the Danube, the majestic river that defines Europe in the same way that the Amazon defines South America and the Nile defines Africa. We first encountered the Danube shortly after Nuremberg and it remained with us, though not always visibly, for the following 500 km to Budapest. Here we parted company, the river heading south to Bulgaria while we continued east to the Hortobagy National Park and the German-owned equestrian hotel, appropriately named the Epona after the Celtic goddess of horses. The Epona was like a miniature ghost town, with endless rooms and chalets, a huge heated pool, shops, saunas, gym, restaurant and no one to use them but us. It was, they said, too early in the season for serious business. This would pick up in the summer when German schoolchildren came for their big holidays, bringing their own horses to be stabled at the hotel for an endless round of gymkhanas and gallops on the Puszta.

Hungary is famous for this great floodplain, surrounded by the Carpathian Mountains to the north and east, the Alps to the west and the Balkans to the south. The Danube wriggles down between these outcrops to the Black Sea where it ends its 2,500 km journey across Europe. The important river of the Puszta, though, is not the Danube but the Tisza, which flows down through central Hungary to meet the Danube further south at Belgrade.

The Puszta, by nature, is more wet than dry, an area once subjected to such seasonal flooding that it was impossible for people to live there with any confidence. By raising the banks of the Tisza to hold back any sudden or excessive rise in water level and by installing a huge dam near Tiszafured, the flow of water was controlled and the Puszta began to dry out. People moved in to farm its rich alluvial soil.

This ambitious project was undertaken late in the 19th century. When Tony, Rachel and I turned up a hundred years later, it was not only to film the key species of the rich grassland but also the ways in which the flow – more accurately, perhaps, the overflow – of the Tisza was being controlled. Filming the built-up riverbank and farm life was straightforward. The fun began when we turned up at the dam waving our permit to film wildlife, hoping that its official 'National Park' heading would be enough to get us through. Setting out on foot from the car park, we were intercepted by an armed guard who informed us, both politely and persuasively, that the dam was high security and that we should not consider taking another step towards it.

Driving away feeling a little thwarted, we looked at the map and realised that a kilometre or so downriver the road crossed a bridge. It looked as though it might offer a view of the dam. Which it did, but with one problem. It was a single-lane road bridge with a railway track running alongside. Cars queued at a barrier and were let through in convoy, first one way and then the other. There would be no chance of stopping for half an hour to film the clearly-visible dam. We crossed once, returned and then crossed and returned again hoping that the lone sentry on the bridge would not become suspicious of our very foreign Land Rover driving to and fro. The point from which Tony would need to film was just visible from the sentry box but it had the protection of a vertical iron strut that he could partially hide behind. We planned to stop just beyond this strut and let him out with his camera, which he would be able to rest on the flat surface of the side rail. All we needed was to be the only vehicle crossing or the last in the convoy. After pulling out of the queue a couple of times because a car came up behind us, we finally made it over the bridge on our own, offloading Tony as planned. We gave him ten minutes and then returned at the back of another small convoy of cars. He wasn't quite ready to leave, shouting out as we crept past that he had got some safety shots but now needed to get the dam in better light. We drove on to the sentry, who somehow failed to register that this was the sixth time we had passed him in 20 minutes. After a good burst of sunshine, we were the last of three when the green light invited us to proceed. Tony was ready and waiting and we stopped for no more than five seconds to pick him up.

Eliminating the seasonal flooding of the Tisza had a dramatic effect. While it created a place where people could live, it also lowered the water table, killing a great many trees and reducing surface water for cattle and crops. Irrigation and deep wells have gone a long way towards solving this problem but other precautions have had to be taken for wildlife. One of the Puszta's special birds is the red-footed falcon, a small bird of prey that feeds on insects, birds, mice and frogs. Like the lesser kestrels and long-eared owls that also find plenty of food on the Puszta, red-footed falcons depend on the empty nests of rooks and, increasingly, of magpies, to raise their young. But lowering the Puszta's water table killed the trees that the rooks used for their rookeries, forcing them to move away. The birds of prey went with them and the park authorities were faced with the challenge of bringing them back.

The answer was to plant stands of *Robinia* trees, false acacias whose long tap roots would reach the lowered water table during the dry summer months. It took a few years for the trees to become established but once they had reached a certain height and canopy-depth, the rooks returned, and the falcons came with them. It was thrilling to stand at Szalkahalem, just along the road from Hortobagy village, with the first falcons of the year arriving from Africa, dive-bombing the busy rookery, swooping high and low with their penetrating 'keek-keek-keek-keek' calls as they paired off and perched close to the nest of their choice. The adult rooks were largely tolerant of their presence, though the odd skirmish indicated that the falcons were sometimes getting a bit too close. We found a baby rook lying beneath one of these earmarked nests. There was no way the falcons could have been responsible for its death because they wait until the young rooks have flown the nest before they lay their own eggs during May. Perhaps a parent rook left its nest a bit too quickly and inadvertently dragged one of its own feathered but flightless babies out.

Having filmed the arrival of the red-footed falcons, we next needed to find a good rookery where we would be able to film the young falcons being fed by their parents in July. We would try and do this in the same rookery but wanted to look at a few others just in case they offered more accessible nests. Our own estimate was that the Szalkahalem rookery of 200 nests contained ten pairs of falcons. We would have to wait until the adults were feeding their young before we could be certain of knowing which nests were occupied, and even then we had no guarantee of being able to get a scaffolding tower in the right position for filming. We spent the next two days looking at other rookeries with Sylvia Gori, our English-speaking biologist/interpreter from the national park headquarters in Debrecen.

Trudging along the edge of a small stream bordered by willows, having just inspected a rookery whose lowest nests were an impossible 12 metres above ground, Tony and I noticed a dead sheepdog wedged in the branches of a tree, two metres from the ground. It was a macabre sight but we didn't give it much thought. However, the following day while out visiting an even less suitable rookery, we came across another hanging dog and asked Sylvia if this was anything other than pure coincidence. She spoke to the farmer and came back with the incredible story that any dog that wavers in its sheep-tending duties is immediately killed and strung up as a warning to other dogs on the farm. It was evidently the farmer's own brutal form of selection. The dogs were killed to prevent them breeding and passing the laziness gene – if such a thing exists – to any of their puppies.

Back at the Epona one evening, and waiting for the warden Zsolt to bring us news about great bustard displays, Rachel noticed a poster in the entrance hall advertising hot-air balloon flights over the Puszta. Sandor Vegh, the owner of the balloon, spoke good English and said he would be delighted to help us while he waited for the summer invasion of tourists. He had taken film crews before and was much cheaper than a helicopter or a light aircraft. This was encouraging. Our budget was limited and yet the flatness of the Puszta meant the film was in desperate need of aerial views and low-level moving shots over the grass. From high up we would also be able to show how the areas of grassland were hemmed in by agriculture.

The main challenge for a balloon pilot is to follow a chosen route. The wind might be blowing the right way at ground level but this can change significantly the higher you rise. Sandor was experienced enough to know how to play this occupational hazard, changing height to stay in the appropriate layer by burning or releasing gas. Our first flight was exhilarating. The sudden floating lift-off, the silence and the freedom to walk round the basket exposed to the elements. We tracked low over the grass, between trees and then, with a rush of gas-burning, up over the surrounding farmland in the south of the park. Our landing just outside the village of Kunhegyes attracted a horde of children who scampered alongside as we drifted down and bumped to a basket-rocking standstill in a field of emerging wheat.

Sandor, tipped off by Rachel who, unlike us, had flown in a balloon before, produced champagne and matches from his bag and scooped up earth and grass from the field. To celebrate our own first flight, Tony and I were to be adopted as companions of the hallowed group of balloonists. Surrounded by fascinated children, we knelt beside the basket. With fire, Sandor celebrated the burning of the gas that lifted us into the air, with soil he welcomed us safely down to the ground and with champagne, some of which was used to douse the smouldering grassy, muddy, charcoaly mess on our heads, we celebrated the drink of balloonists throughout the world. It was serious stuff and Sandor completed the little ceremony by handing us certificates that confirmed our new status as Nobles of Kunhegyes.

Our second flight was scheduled to take us north towards a remaining mosaic of woodland and marsh at Ohat. But this time, even Sandor struggled to master the layered wind and we were carried further west than intended. I was looking out from one side of the basket when Tony took me by the shoulders and steered me across to the other. There, below us, was the Tisza and straddled across it was the out-of-bounds dam we had managed to film from the bridge. So now, having been denied access to the dam for security reasons, we had both ground and aerial footage to show its significance in the formation of the Puszta grassland. Sandor assured us that it could not really be such a security problem or we would surely have been arrested on the railway bridge and investigated by the Hungarian Air Force as soon as we flew overhead.

We settled on our Szalkahalem rookery as the best place to film the young falcons in July. Zsolt offered to find out the cost of hiring a 7-metre tower and came back with the depressing news that there was just one company able to provide such a structure, with a fixed fee of almost £1,000 a week. As we would need the tower to be in place for several weeks, the cost was prohibitive. We left for England two weeks later with Zsolt and Sylvia promising to do what they could during our absence.

While Tony and I negotiated the carnet-led drive across five borders between Hungary and England, and then back again a month later, the cost of the scaffolding came tumbling down. The exorbitant initial quote was because we were visitors to Hungary, not residents. The park hired the structure for us at a much-reduced rate and lowered this even further because they would continue to use it after our departure. Zsolt had pinpointed several falcon nests in the rookery and had even decided to erect the scaffolding over one of them before we returned. His understanding of our needs was perfect: a nest with two fluffy white falcon chicks, partially shaded but open to good light and even catching the sun in the late afternoon. All Tony had to do was climb up the tower after breakfast and sit in the hide all day while the adults delivered grasshoppers to their insatiable young.

Sylvia, Rachel and I left Tony in his tower hide one morning and set off for Nagyivan in the southwest of the park. We wanted to check on the white stork

141

nests that were perched on nearly every telegraph pole along the road running through the village. We had seen these nests in May when the barely-visible adults were incubating their eggs, and had thought then that we might be able to film the more conspicuous young from the ground on our second trip. One of these bulky nests was opposite the church, which had a wooden door on its bell tower a good 20 metres above the nest and looking straight down into it. That was the angle we wanted because it would also give a lovely panorama of the Puszta beyond the village. Sylvia talked to the church warden but he was very sorry, the door had not been opened for as long as he could remember and it would not be possible to open it now. Besides, it was dangerous: the wooden floor was rotten in too many places.

Now, two months after our May visit, we were driving slowly through Nagyivan with each of the stork nests above us bursting with new life. Every few minutes an adult would glide down, land among its chicks and regurgitate a meal. We stopped beneath the church nest and listened, disappointed that we would not be able to get a better view of the youngsters feeding above us.

Casting a wishful glance up the bell tower, we could hardly believe our eyes. The wooden door high above us was wide open and a wooden platform was protruding beyond it by at least two metres. We went straight across the road. The warden recognised us and smiled warmly. It was, he explained to Sylvia, an amazing coincidence. Two things had happened since our earlier visit. The first was that the replacement for the cracked 300-year-old bell was about to be dispatched from Budapest, and the second was that the Hungarian Air Force, which just happened to be on training manoeuvres at a nearby airfield, had offered the free use of a helicopter to lift the two bells in and out of position. It had taken several days to unlock and prise open the rusted door and a local carpenter had fitted the platform over and beyond the existing rotten floor. The new bell would not be arriving until after the weekend, so we were welcome to use the platform for filming until then.

We climbed the dark, narrow staircase that wound up the tower and emerged into broad daylight high over the road. The view across the open Puszta beyond the village was spectacular, and we could see directly into the nearest nest with its four large young. As we sat on the platform absorbing our good fortune, one of the parent birds glided down to the nest and regurgitated grasshoppers to the squabbling brood. And then, just for good measure, the second parent arrived and the two grown-ups went into a full head-bowing, bill-clapping display that reverberated the length of the little village.

We drove back to the rookery for Rachel to pick up her hire car and drive Sylvia to the park headquarters. I was about to walk round the rookery to look for susliks – Europe's only ground squirrel – when Tony emerged from the hide. It was a bit cloudy, the young falcons were taking their daily nap and he hoped he might be able to nip back to the hotel for a short break, leaving me to keep an eye on things here. He thought there might be another falcon brood about 40 metres to the left of the one he was filming, so I decided to sit in his platform hide and watch.

After a few minutes I saw something unusual. Some of the falcons were flying away from the rookery and landing just beyond the banked-up main road 150 metres away. Through binoculars I could see that the birds, males and females, were in fact going down in front of the road rather than beyond it. They were landing among a stand of reeds growing along a drainage ditch. Keeping an eye on them, I climbed slowly down the tower.

Creeping away until I came to the road 100 metres or so from where the birds were disappearing, I inched towards them under the cover of the bank and its thick vegetation. I got to within 20 metres. Then, through the gently swaying reeds, I saw the most magical thing: in full sunlight, at least 20 red-footed falcons were lined up along the edge of the shallow water, drinking. Birds were coming and going and heads were bobbing, just like budgerigars in the centre of Australia or sand grouse in the heart of Africa, if perhaps not in quite the same numbers.

As Tony swung the Land Rover off the main road, past Zsolt's little house and onto the track to the rookery, the falcons lifted from the mud and wheeled away to the trees. I was so impressed with what I had seen that I suggested we put a hide up over the little drainage pool and film the birds drinking in a few days' time. We positioned a new hide that evening, the final touch being to dress it with reeds, camouflaging it from the birds on one side and from road users on the other.

Over the next two days, we did not see a single falcon visit the little pool. They were all around us as usual, ferrying grasshoppers to their young in the rookery, but the local watering hole seemed to have been abandoned. On the third day I went to have a closer look. The little pool had vanished. It had dried up completely, which seemed a good enough reason for the birds to go elsewhere when they felt thirsty.

Two thousand kilometres from the red-footed falcons, about the same distance Tony and I had driven from England, the young cranes were growing fast in their isolated wetland nests. Three months from now, before their feeding grounds froze solid, they would set out on their southerly migration towards the Mediterranean and North Africa. Common cranes may not have nested in Hungary for a hundred years now, but their autumn stopover is still one of Europe's great ornithological sights and sounds.

For two or three weeks in November, as many as 60,000 cranes gather on the Puszta. The stopover is particularly important for young birds making the demanding flight for the first time. Their parents choose the Puszta for good reason: they can roost in their preferred open country and can also get a little help from their friends. In a joint venture, the park authorities and local farmers contrive to make sure that the late summer maize is harvested very inefficiently, leaving 'wasted' grain to be conveniently found by the birds. They now do the same for lesser white-fronted geese, feeding them inside the park to protect them from hunters.

With Tony and Rachel busy elsewhere, I met James Gray and Nigel Tucker in Bristol to give them the details of the ten-day trip. Sylvia had just telephoned to say that, after a slow start, the cranes were beginning to arrive on the Puszta in reasonable numbers. But she also added that it wasn't quite cold enough in the north for the birds to be heading south on the scale she was expecting, so perhaps in a week or so would be the best time.

James and I decided to set out immediately. We would at least be on hand to respond to anything as it happened and we also had another sequence to film, the small matter of Europe's largest spider, the steppe tarantula. We would film this in Kossoth University in Debrecen where some of these spiders were being held captive. James's task was to film the hairy arachnid catching a Puszta grasshopper in a set. He would be able to get on with that whatever the weather, while we waited for more cranes to arrive. Like Rachel before him, Nigel flew to Budapest and hired his own car three days after James and I climbed into the Land Rover and set off with our carnet to the coast. I was back on the cross-Europe road I was getting to know so well.

With the tarantula filmed, we met up with Zsolt and Sylvia in the Epona Hotel early one evening. The only crane roost they knew of was on Halasto fish pond, just off the road to Tiszacsege.

The following evening, James and Nigel took up a position by the roadside to film and record the cranes gliding in from the surrounding fields. I then drove Zsolt, Sylvia and the Halasto warden a further kilometre or so across boggy and heavily-rutted ground to the very edge of the fish pond. We parked next to a raised platform hide that looked ideal for filming.

But it came to nothing. The cranes came flying into the fish pond all right, low over the last 50 metres in their groups of ten or 20 but, by the time they arrived at the roost, it was too dark to make them out with the naked eye, let alone on film. The birds were also lost against the background bushes and trees as they descended on the last stage of their journey. Although we could hear them – and the massed buglings, honkings and scufflings of 10,000 cranes at such close quarters was spectacular – we could barely see a single bird settling down less than 40 metres away. It didn't help, either, that the mosquitoes from the fish pond were now out in full force. As a final test, the warden insisted that I drive back to the main road with just my sidelights on. He obviously had the cranes' best interests at heart, but negotiating that muddy, rutted, twisting track out of the reed bed in near-total darkness was not something I wanted to repeat.

Back on the road, James and Nigel were also having problems. Their initial excitement at hearing and then seeing the cranes coming towards them had soon turned to frustration. People were the problem. They had come out of the surrounding towns and villages to their traditional 3 km post, the one crossed by the Halasto-roosting cranes every year. There were hundreds of them, slamming doors and revving engines. And they also talked. And shouted. And waved. And the birds were wise to it. They came in over the road late and high, descending quickly once clear of the crane-spotters, and then straight into the bushes and trees around the fish pond as the last slivers of daylight faded away.

The warden shrugged. Sylvia and Zsolt were a bit embarrassed. They explained that, apart from this known location, it was very difficult to predict where the other 50,000 cranes might be roosting. They moved frequently according to disturbances and weather conditions. As for feeding during the day, they could be on farmland anywhere in small family groups. If we did find them, the birds would not let us creep closer than 200 metres before casually lifting off to find somewhere more private. We raised our glasses in the Epona that night and drank to 'Well, don't say we haven't been *here* before!'

We had now used five of our ten filming days, the last two of which were earmarked for the drive back to England. Zsolt promised to scour the Puszta late each afternoon to see where birds might be congregating, hoping, if he found them, that they would return to the same place the following night. Leaving Nigel to his sound recording, James and I set out to try and film the cranes during the day.

We tried early morning, midday and late afternoon, humping maize and heavy gear across ditches, ploughed fields and wet, waist-high grass. Abandoning James to his hide-bound vigils, I drove round and round, sometimes with Zsolt and sometimes with Sylvia, searching for areas the cranes might be favouring. Whenever we came close to a field where James sat patiently in his little shelter, we searched his line of vision with our binoculars and felt the frustration that accompanied him out there in the damp and the cold, and not a bird in sight.

As our time on the Puszta was nearing a disappointing end, Zsolt came into the Epona after dark to say that a large crane roost had been found just off the road to Nadudvar. There were, he had been told by another warden, thousands of birds coming down onto exposed, open grassland, forming the largest known roost of the year. They were visible on the ground and we were still two days away from the full moon. It sounded ideal. But would the birds return to the same place the next night?

It was a private road that took us onto the Puszta at Angyalhaza where, much to our relief, there was not a member of the public in sight. Standing out in the open, the reason for the sudden appearance of all these birds was clear. It was perishingly cold and a bitter chill swept through us all. It was, as Sylvia explained, the sign they had been waiting for. A cold front had pushed many more cranes to Hungary from the north and the current estimate was that there were at least 40,000 of them spread out around the Puszta. They had roosted here, at Angyalhaza, for the first time the previous night. With luck, they would return.

Once again, we heard them before we saw them. But they were distant and the flocks were disappointingly small. True, James could film them all the way down to the ground, which was a vast improvement on Halasto, but they still seemed to be arriving low down just above distant trees and buildings, and the small groups would not look very impressive on film. The setting sun was obscured by low cloud and the opportunity to film the birds flying down in front of it was lost. We went back to the hotel with our fingers tightly crossed for the following night, the night of the full moon and its promise of silhouetted cranes floating down across its radiant face...

We were back in position north of Nadudvar by three o'clock the following afternoon. It had been a clear, cloudless day and we were in for the coldest night of the week. But this was exactly what we wanted. The colder and clearer the better. Tonight we came armed with a flask of brandy to warm us against the elements. We watched the sun sink to the western skyline, getting bigger and redder as its glare faded. Suddenly Zsolt called out. The cranes were coming. The first batch of 50 birds came in beneath the sun, clearly visible as they drifted down and settled on the grass. James picked up all the flight and landing shots he could, the material we hoped would precede our sunset spectacular. We swigged brandy from the flask, jigged from one foot to the other to keep warm, and still the cranes kept coming. Hundreds and hundreds of them, drifting in from the left, above, below and across the face of the sinking sun. Even Nigel, despite the rumbles of a few late afternoon tractors, was happy with his recordings of distant, powerful calls.

We were smiling now. Not because the sequence was complete or wonderful but because it was already a million times better than we had thought possible a day or two earlier.

For an hour we all faced west, spellbound by the mass of birds coming down to the roost. But, as the sun dipped below the horizon and James loaded his fastest film stock to get what detail he could from the darkening shapes on the ground, I turned away to stretch my legs. What I saw took my breath away.

The largest, bloodiest harvest moon imaginable was hauling itself into the night sky. Flocks of cranes were heading straight towards it, apparently on their way to a different roost. I shouted to James and he spun his camera round and set up for the shot that would complete our sequence. The first flight of cranes went above the moon. We would have to wait. Another quick swig of brandy. The image in our minds was the full moon with the cranes, even just a few of them, floating down across its sunlit craters, oceans and seas.

And then it happened. Clear of the ground and still glowing its autumnal red, the harvest moon played host to what may have been the evening's very last flight. The cranes were high in the sky when we heard them over to our right but then, as though on command, they dropped and held a course that took them straight across the face of the distant moon.

All that remained was for us and our carnet to set out on the long road home, while the cranes picked up their seasonal flight path from the Puszta and headed back to where they, too, had begun a momentous journey much earlier in the year.

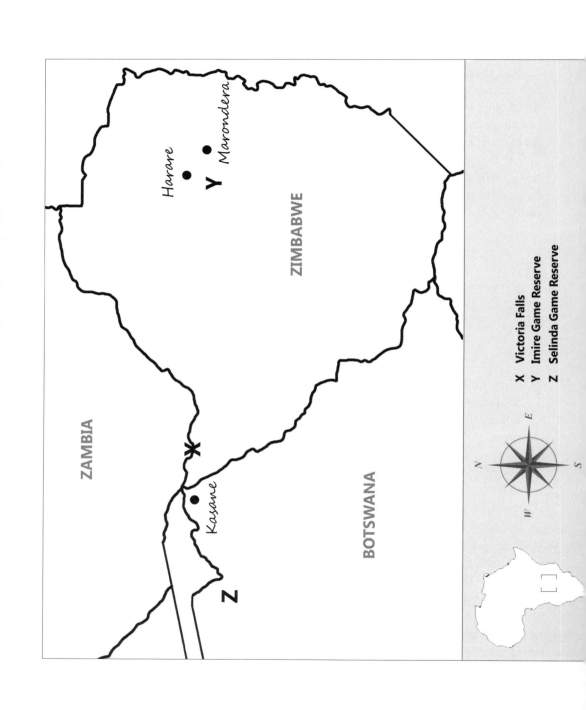

ZAMBIA

ZIMBABWE

BOTSWANA

Harare

Marondera

Y

Kasane

Z

X

X   Victoria Falls
Y   Imire Game Reserve
Z   Selinda Game Reserve

N
E
S
W

14

NZOU OF THE BUFFALOES

Nigel was adamant. There was no way he would touch his meat until it was burnt to a crisp.

'What a waste!' we muttered as we munched through our own barbecued offerings, charred on the outside, perhaps, but soft, warm and pink in the centre. We moved on to fruit, cheese and coffee while our stubborn sound recordist refused all pleas from Ian and the chef to eat his fillet steak before it became a black, carcinogenic lump.

Nigel Tucker's refusal to eat his meat until it had been thoroughly destroyed by fire became a standing joke. He took the laughs with exemplary good humour, a vital quality on any filming trip in the bush. After the daytime pressure of getting the footage to build a programme, a good-natured release of tension in the evening works wonders. It relaxes people and prepares them for a fresh start the following day. We needed this distraction here. A film that had seemed so straightforward in the beginning was quickly turning into a film where cameraman Tony Allen, Nigel and I were having to work very hard to get the commissioned story. It wasn't proving easy because our subjects were very large and very dangerous.

It was the quirkiness of the story that had made it so compelling. I was working in Bristol with Green Umbrella when we had a call from Juliette Mills, hoping to interest us in an idea for a film about an elephant that ran a herd of buffaloes in Zimbabwe. We were intrigued and approached Carlton in London and National Geographic in Washington. They, too, were intrigued though they were not prepared to finance more than 35 filming days to make a 25 – 30-minute programme. That would allow for just two trips, each of two or three weeks.

The first stage was for me to make a quick three-day visit to Zimbabwe to meet the owners of Imire game ranch, get their approval and work out what might be possible round a storyline that, despite the obvious attraction of the star players, still needed the content and shape that would hold for that crucial half hour.

The Travers family consisted of three generations: the old couple Gill and Norman, who blazed the first bush trail from Harare in the 1950s, their children John and Barbara, who managed the tobacco growing and the tourist operation at Sable Lodge, and their grandchildren, who were already looking around the world for a more secure future than was on offer in Zimbabwe. I met them all and they were friendly and helpful, agreeing that the film could be made but that it would be difficult in places.

I had John's blessing to develop a treatment with ranch manager Ian du Preez and, after a few hours driving, observing and talking, it felt as though we had enough to go on. Of particular interest were the strangely predictable mood swings of the would-be star of the film, an elephant called Nzou.

The local people are Shona and *nzou* is their word for elephant. And here was Nzou, a comparatively young 23-year-old living on a game ranch whose owners were devoted to her. Nzou's only problem, if that is the right word, was that she had not seen another elephant since she was traumatically orphaned as a very young calf. The wild herd on which she had relied for everything was destroyed by rapid fire from a helicopter in a culling operation. The little calf was the only survivor, and she was taken in by Norman Travers who was converting part of his tobacco farm into a ranch for wild animals. A little orphaned elephant would fit perfectly with his plans.

The last thing a baby elephant needs is to be deprived of company. So Nzou became the daily responsibility of Murambiwa Gwenzi, the old boy who spent his time walking the ranch with yet another of Norman's schemes. This was a herd of foot-and-mouth-free buffaloes whose male offspring could be sold on to begin similarly 'clean' herds elsewhere in Zimbabwe. It was a new and profitable business and Murambiwa's role was to follow the herd to protect it from poachers, to keep it away from cattle that might be contaminated and to lock it up in a secure corral at night. Murambiwa, always armed with a big stick, kept his distance. The buffaloes were unpredictable and aggressive, especially the bulls.

Nzou walked with Murambiwa, developing a special relationship with him. The buffaloes provided her with a necessary sense of being in, or at least near, a social group. But buffaloes are not like elephants. They are, it seems, expressionless creatures without any emotional glue beyond occasional body contact. Elephants, on the other hand, are among the most touchy-feely of all animals, with mutual dependencies lasting throughout their lives. This is especially noticeable among young male and female elephants that grow up under the leadership of a matriarch, the experienced dominant female who protects and leads through the seasonal whims of the bush. And a key difference between elephants and buffaloes, one that was vital to the relationship between Nzou and her buffaloes, is their different life expectancies. An elephant that should live until it is 60 or 70 has barely reached sexual maturity when a buffalo is already approaching the end of its own 15-year life.

Each morning, these three different mammals – human, elephant and buffalo – set out from the corral on a well-trodden path round the ranch. Murambiwa kept a safe distance from the buffaloes and little Nzou tagged along by his side, occasionally approaching the buffaloes but scampering back when they snorted and mock-charged. She must have felt safer with Murambiwa. Perhaps instinct told her that this was how it would have been with her mother. But the buffaloes still had that 'herd' thing about them. Nzou wanted the best of both worlds, but Murambiwa would not go too close. He was not keen on cuddling half-wild buffaloes.

By the time Murambiwa turned 60 in the late 1970s, Nzou was a sturdy five-year-old with three times the strength of an adult man. She did not abuse this power with her keeper, but she was able to use it to her advantage with the buffaloes, as though her growing physical advantage allowed her to overcome any lurking fear of their aggression towards her. Perhaps she was still driven by the need for close contact in a group and this was still the best on offer. She attached herself most noticeably to older females and even showed something approaching jealousy towards their very young calves. The much shorter life expectancy of the buffaloes meant that successive calves grew up to take her presence among them for granted. Within 15 years, there wasn't a buffalo who had known life without her.

But there was still a major problem, and it had everything to do with the way Nzou would have behaved had she become a dominant female in the wild. In her natural elephant world, young males are raised by their mothers and other closely-related females until they become sexually-active teenagers. The moment one of them begins to show an incestuous interest in a sister or cousin, the matriarch steps in, pushing it away, forcing it to the edge of the herd and denying its genuine need to return to the fold. Avoiding contact and lagging further behind for fear of another beating, the young male eventually sets off on its own until it joins a bachelor herd of similar outcasts. Once established in a male-only society, it will have to wait a further 20 years before it is close to the top of the mating order. And then, lured on by irresistible pheromones, it will periodically return to female herds in search of those who are sexually receptive, though perhaps watching the matriarch with a caution that betrays the scars of adolescent rejection.

Life for a growing buffalo is very different. Herds may number hundreds and there is no obvious emotional struggle going on between the sexes. A three-year-old bull will cut its ties with its mother, who will by now have another calf. The young bull will remain in and around the herd with other males of a similar age. They quickly become independent and within a year or two are living slightly apart from the large herds, which are composed mostly of females and their young. As sexually-mature six- to eight-year-olds, they spar among themselves for dominance before moving among the females to mate with those that are receptive.

When Nzou was strong enough to impose herself on the buffaloes as their leader, she let it be known to sexually-mature bulls that she would not tolerate any mating within the herd. When they mounted, she became agitated, wanting to disrupt what was quite normal buffalo behaviour. Instinct was telling her that what they were doing was wrong. Adolescent male elephants are not allowed to do this sort of thing. She responded to one particular bull's amorous advances by rushing in and physically pushing him off the cow. And when he bounced back, she whacked him all the more. She was the matriarch, and she was already twice his size. Undeterred, the male tried again. She attacked with even more vigour and he, poor thing, simply got up and tried again. Female buffaloes are receptive for just two or three days at a time. He must have been desperate. Nzou, her patience tested to the limit, knocked him down and trampled him underfoot, pressing down on his broken body with her trunk. Within minutes he was dead.

It happened again when the next bull began paying close attention to the receptive cows. And again, and again, until 15 bulls had been summarily exe-cuted. By now it was clear that Norman Travers had a serious problem on his hands. His family loved Nzou but her behaviour was threatening the future of their foot-and-mouth-free buffaloes. With the commercial value of a sexu-ally-mature bull buffalo running into thousands of American dollars and the future of the whole Imire herd at stake if its reproductive effort ground to a halt, what should be done with Nzou? A return to the wild was no longer an option and she was too moody and unpredictable to be used elsewhere on the ranch. Nzou and Murambiwa were inseparable but his job was to stay with the precious buffaloes. It occurred to Norman that Nzou might have to be destroyed.

Fate intervened. One day, while out walking the woodlands of Imire, Murambiwa was attacked by a particularly aggressive male buffalo in the herd.

It knocked him to the ground and was all set to run him through when Nzou charged to the rescue. Barging into the buffalo, she swept it to one side like a limp rag and stood defiantly over Murambiwa's prone body, daring the angry bull to return. It snorted and stood its ground. But Nzou would not leave her friend. It was only when the herd failed to return to their corral at dusk that the search party went out.

The acts of devotion and defiance shown by Nzou, and the certain death that would have been Murambiwa's had she not acted so quickly, set the Travers family and Ian thinking in a different way. Nzou had now demonstrated her value to the ranch, but how were they going to solve her problem with the mating bulls? And how could they be certain that Murambiwa would not be attacked again, with more serious consequences?

Someone had a brainwave. They would build a side pen onto the corral and any sexually-mature bull would spend his days in there. When the herd returned for the night, Nzou would be chained up outside the enclosure and the bull would be released inside to mate with the females. He would then be returned to his pen in the morning before the herd went out to feed.

The only concern was that an incensed Nzou might snap her chain at night and break into the corral with unpleasant consequences for the rampant bull. But it didn't happen. She would have known exactly what was going on not 10 metres from her primed senses but she seemed to accept the compromise as best for everybody. Ironically, the only time she *did* slip her chain and break into the corral, it was to take her precious herd, including the bull, on a midnight feast in the ranch's vegetable garden.

When we turned up to begin filming in September 1998, no bulls had been killed for several years and Murambiwa, now 80, had an assistant named Paul. They could push Nzou around at will but both still kept a healthy distance from the buffaloes. It had already occurred to us that not having the sexually-mature bull with the herd could be a major problem for our storyline. Ian du Preez was adamant that we could not let him out to see what happened. It was too dangerous, for us as well as for the old bull and, anyway, they could not afford the risk of him being trampled like his predecessors: apart from the obvious loss, his life was no longer insured outside the corral.

While there were a number of things we could film on this first three-week trip, like the parts Norman, Murambiwa and Ian played in the story, we had to work very hard to get Nzou and the buffaloes to perform in ways that would provide visual entertainment. A strange herd wandering through the woodland of central Zimbabwe, however eye-catching, would not work for very long. The most important reason for wanting a confrontation between Nzou and the bull was not just to liven up the film. It was to show another remarkable aspect of her personality.

On our first morning at the ranch Norman took us out to intercept Nzou and the buffaloes at a point where tourists are not allowed. He stopped his vehicle within 100 metres of the herd, got out, waved and called. Immediately, Nzou turned to face us, spread her ears wide and let out a loud trumpeting. Norman waved again and hooted his car horn, at which Nzou broke rank and came running towards us, hotly pursued by Murambiwa and Paul, both calling to her excitedly. She stopped ten metres in front of Norman, recognised him and softened immediately, approaching close enough for him to pat her trunk with soothing words of friendship. It was powerful stuff, but it became even more intriguing when, later in the day, Norman led us to the rocky outcrop where tourists were taken every day to have their lunch. We were eating our salad under the pleasant dry-season sun, overlooking the whole expanse of Imire, when word came that they were 'on their way'.

Nzou appeared over a nearby ridge, followed by Paul, the buffaloes and the stick-wielding Murambiwa. Nzou stopped under their favourite acacia 'shade' tree and the buffaloes sat down to chew on their own regurgitated grasses. Murambiwa walked the 50 metres towards the tourists, obediently followed by Nzou. The last 15 metres she covered alone, walking right in among the eager throng who hand-fed her bread and vegetables as though she were some circus-trained zombie without a spark of life. After 10 minutes, she turned and went back to join the buffaloes. Norman turned towards me and winked.

'If any of those silly buggers went over to feed her now, she would kill them. Don't ask me why, that's the way she is.'

It would be easy to film Nzou being fed by strangers at the picnic spot. We could not, however, film her killing one of them 50 metres away! To illustrate her aggression, we needed her to confront the sexually-mature bull in its pen.

John Travers understood that we were prepared to push things only as far as he would allow, and no further. That got him on our side. With Ian's help, he came up with a rough and ready plan. We had already seen how Nzou could be provoked into mock-charging and this gave us a bit of control over her. Ian also knew that Nzou would occasionally break away from the herd and return to the corral to let the bull know who the real boss of the buffaloes was. Perhaps there was some way we could use this to our advantage.

The following morning, Murambiwa and Paul set off with the herd at first light and we slipped in behind them to take our positions at the corral. Tony's tripod was mounted on the open back of a lorry, with both the bull's pen and Nzou's assumed approach route covered by his viewfinder. A zoom lens would give him tight and wide shots of the action without any time-wasting lens changing. Nigel and his sound-recording gear were perched on the back of another vehicle with me at the wheel. Ian stood next to his car with a loaded rifle, just in case. I was never quite sure who it was for.

At a given signal, we began making as much noise as we could. Two helpers were banging jerry cans, vehicle horns were blaring and everybody was shouting and screaming at the tops of their voices. Tony looked over to me and shook his head as if to say, 'This is a funny way to film wildlife', but nothing happened. We tried again, but there was still no response from Nzou. She was, we learnt later, too busy feeding to take any notice, though Murambiwa and Paul could clearly hear the racket we were making.

We took up our battle stations again the following morning, and this time waited until Paul came back to tell us they had finished feeding and were ready to move on. This might be a better time to try. So off we went once more, shattering the peace of an African morning with electronic hootings, metal clangings and human shouts and cries. Suddenly there was a massive trumpeting from within the trees straight ahead, and out she came, Nzou in full flight. On and on, trumpeting her anger again and again, until she reached the far side of the corral, wheeled to her right and confronted the old bull inside, pacing to and fro, leaning into the flimsy wooden barrier but stopping short of knocking it down. The bull buffalo did not move until Nzou had finished her remonstrations and had backed off to leave him in peace.

Tony had got enough action to set the scene for the sequence and we built the rest of the shots around that – close-ups of the bull 'reacting' to her approach, and then, a day or two later, shots of Nzou from the bull's perspective inside the pen. It wasn't dramatic footage in the wild-elephant-charging sense, but we had enough to edit a sequence that became much stronger in the finished film when followed by Nzou being hand-fed by tourists at lunchtime.

It was fascinating to trek around the ranch every day watching for detailed bits of behaviour between Nzou and the buffaloes. Before long, we became aware of many little things that a casual observer might miss, particularly how the physical advances of Nzou were not reciprocated. Her trunk would snake out, feeling, touching, searching and sniffing in the elephant way of greeting and reassurance. The buffaloes took virtually no notice, gave nothing back and moved away when she began to make a nuisance of herself. Was she frustrated by this lack of response? Who knows? She had not seen another elephant for a quarter of a century, so she might not have known that these unresponsive lumps around her were not the same as her.

At many of these dusty gatherings where Nzou's demands for affection went unrequited, she ended up with one old female buffalo standing between her front legs, their bodies actually touching. A cynical view was that it was the shade that pulled the old cow in so close and that Nzou was only too willing to be exploited in return for a bit of physical contact. But if you watched quietly, it was clear that there was some kind of bond between them, difficult to define but definitely there. They would stand like this for perhaps an hour, the

155

old buffalo chewing the cud, Nzou perfectly still apart from her ears and tail which flapped and twitched with the heat and the flies. And then the two would separate, moving with the herd. It may be that in the wild, ageing female buffaloes are increasingly marginalised until they finally die, and that is what was happening here to the old cow. If so, both she and Nzou were content to have each other's company.

After the nocturnal matings between the bull buffalo and several of the cows during the dry season of 1998, Ian was sure there would a birth or two at Imire the following March. Barbara would fax regular reports to me in England. With only two weeks at our disposal to complete the film, our timing had to be perfect. But, once again, it wasn't going to be straightforward. Nzou had developed a problem with newborn calves and was likely to kill them. This made life difficult for the female buffaloes. In the wild, they would retreat to the safety of dense cover, give birth and then, a day or two later, take their calves into the safety of the herd as the best protection against lions and hyenas. But, ironically, the centre of the Imire herd, which was never threatened by lions or hyenas, was actually the most dangerous place for a female to take her calf. Four had already been squashed under Nzou's powerful feet and there were no guarantees that it would not happen again. This time the ranch was powerless to intervene, unlike with the mature bulls.

We wanted to film this first contact between Nzou and any new calves, but it wouldn't be easy. These female buffaloes had adapted their behaviour to minimise the threat. A few days before giving birth, the mothers-to-be would leave the herd and hide in the lush reedbeds along the little stream that wound through Imire. But instead of rejoining the herd quickly, as wild buffaloes would have done, they stayed hidden until their new calves were stronger and able to keep out of Nzou's way. It could be anything up to a fortnight before they went back to the herd.

In April 1999, two pregnant females retired to the reedbeds. One of them had lost a calf to Nzou before and Ian felt she would be very wary this time. She might, he faxed, stay in the reeds for much longer than usual. We faxed back, and we phoned, and all we got was 'They haven't come out yet' and 'No-one has seen a calf yet so we can't actually be certain that one has even been born.'

After two anxious weeks a fax did arrive, to say that two calves had just been seen in the reedbed and that their mothers were getting restless. We should come immediately because they would probably emerge within the next few days. It was a scramble but Tony, Nigel and I were off to London's Gatwick airport within 48 hours.

Barbara met us at Harare airport and we were soon back among the people we had got to know at Sable Lodge. The transformation in the countryside was amazing. Since our last visit in the yellow, dusty, dry season there had been the usual wet-season downpours. Everywhere was bright green and thick with growth. There were colourful flowers and the lodge gardens were alive with male weavers carrying strips of grass to fashion their beautiful nests in the little tree that stood at the centre of an ornamental pond.

We dumped our bags and went straight to the little stream. Only a few months previously it had been no more than a depression with a trickle and the occasional open patch for elephant and buffalo mud-bathing. Now it was flowing boldly through Imire, half-hidden in places by the lush growth of reeds and waterside vegetation. Ian showed us where the female buffaloes were still hiding and Tony spent the next two days filming whatever glimpses of them and their calves he could manage.

We then went downstream to a deeper lagoon where Nzou took the buffaloes each day to swim and bathe. We looked on aghast when something spooked the wading buffaloes, who charged out of the water straight at Murambiwa. Used to such inconveniences, the old boy stood his ground, shouting and waving his big stick from side to side. Like liquid, the mass of horns and hides separated, flowed round him, merged behind his back and carried on up the hill, hotly followed by Nzou.

Each morning we went down to the reedbed at first light. And each morning they were still there. Until, that is, our fifth day, when the calves were nearly three weeks old. The two females had finally emerged and they now stood on the threshold of finding out whether their precious young would live or die. We, with perfect timing, still had nine days to find out.

For the first two days they kept their distance. Nzou knew they were there. You could tell. Her trunk snaked out towards them, testing the air. Occasionally she would advance inquisitively, but no more. The following day we found the 15-strong herd by the river, the two females and calves within 20 metres of it. Ian watched carefully. Murambiwa stood close to Nzou with his stick. If she attacked one of the calves he would be the only person who might safely intervene, though even he was unlikely to risk his own life for that of a baby buffalo. Paul was not yet ready to handle Nzou in such circumstances.

The need to be back in the herd overtook the two mothers. They greeted the other females, sniffing and licking faces with a new intimacy. Nzou stood and watched. Perhaps she was plotting a dastardly act of murder. Or perhaps all she really wanted was to be licked as well. Or perhaps, deeper still, the sight of those defenceless little creatures stirred longings inside her for a youngster of her own. She was already ten years past the age of being able to have her first baby. Did she roll her feet on those little buffalo bodies, knock them around a bit with her trunk, as no more than gestures of frustration or playful friendship?

The two female buffaloes were themselves no more than half-committed to staying in the conflict zone. They were back in the herd where they wanted to be but the danger to their calves was acute. They were nervous, and whenever Nzou made a move towards them they hurried away, their calves clamped to their sides. They compromised by remaining close to the thick riverside vegetation, refusing to accompany the herd when it moved too far away. Ian thought we should be patient, that within a day or two the female buffaloes would take their calves into the herd for good. He thought that the acacia 'shade' tree was where any drama was likely to happen. It would be when the herd was resting and Nzou could take advantage of any heat-induced lethargy that had set in.

Sure enough, the following day mothers and calves went into the herd and stayed put. The river refuge was abandoned and they now followed Nzou on her daily round. Murambiwa knew that if anything was going to happen, it would be soon. The calves would need to be accepted by Nzou and this meant that their mothers had to offer them up for inspection, whatever the risk.

Two lunchtimes later, they arrived at the 'shade' tree as usual. Murambiwa took Nzou off to indulge the tourists while the buffaloes settled down to chew their cud. When she returned to them, Nzou took up a position directly beneath the tree, this time closer to the calves than ever before. They sat obediently on the ground while their mothers stood patiently next to them, as though resigned to their fate. Nzou shuffled closer, her trunk reaching out towards one of the three-week old calves. The tip of her trunk, sniffed, smelt, felt, prodded and poked all round the young head, face and haunches, which remained obediently still. But when the trunk tip came to the young mouth and tried to force itself inside in a typically elephant way, the calf finally, and appropriately for a buffalo, objected. It got up and moved to one side. Nzou did not react. She appeared to have accepted the calf, at least for the time being. She ignored the other one completely. Its turn would come, although Ian was sure that the first hurdle of that initial contact had been cleared for both of them. Neither was safe just yet, but their chances of living in her strange company were considerably increased. And we had all the sequences for our film.

Seven months later, in November, while I was working on an elephant series in Botswana with Wayne and Venessa Hinde, I travelled via Zimbabwe rather than South Africa to deliver copies of the completed Nzou programme to Imire. Barbara met me at Harare airport and the two-hour drive to the ranch was a reminder of pleasant memories. I spent the day with tourists driving around the familiar tracks and had the usual picnic lunch overlooking the wide expanse of the ranch. Nzou showed me no special favours, so perhaps she didn't remember me. Murambiwa and Paul certainly did and it was a delight to be with them again. Could Murambiwa really be so old, in a country with a male life expectancy of around 40, and still walking 90 km to and from home every few months?

I watched again as Murambiwa led the buffaloes back to the corral in the late afternoon, driving them into their pen, chaining Nzou and releasing the mature bull to have his nocturnal way with any receptive females.

As shadows lengthened, I bid Murambiwa and Paul farewell and made my way back to Sable Lodge. I still hadn't been to say hello to Norman and Gill in the main farmhouse. Barbara was waiting for me in her office. She had to go out, she said, but her mum and dad would like me to have supper with them tonight. I could take one of her cars. I had a shower and set off for the main road, drove a few hundred metres to a little shop and turned into the main farm where we had assembled on my first visit the year before. I spent a lovely evening with the ageing couple who were now resigned to an unsettled future in the country they loved. Dark times were approaching for Zimbabwe. Robert Mugabe's popularity was dwindling and he was about to tighten his grip on power by playing the card he had kept up his sleeve for more than 20 years: the designation of the farms, on which the country's prosperity had largely been built, from white ownership to black.

Driving the short distance back to Sable Lodge a bit worse for wear – Norman and I had drunk a fair amount of the duty-free malt whisky I had brought with me – was quite strange. Here I was, all alone, in the middle of the night, in the middle of Africa. I was in someone else's car, just driving along without any idea of what life was really like for millions of people locked into a country that was soon to be plunged into outrageous economic and human suffering. Tomorrow morning, I would be driven at first light to Harare, from where a plane would fly me to Victoria Falls. I would then be driven across the border to Kasane and Wayne would fly me in his little Cessna back to Selinda for two weeks' filming. I would see elephant herds with their matriarchs, mothers, sons and daughters. The life that Nzou never had.

I slept well and was up at 5 a.m. Packed and enjoying bacon, eggs and sausages on the lawn when Ian appeared from the lodge to share a final cup of coffee.

My bags stowed in the minibus that would drive me back to Harare just 24 hours after my arrival, Ian's parting shot had me smiling all the way to the airport. 'Would I *please* tell Nigel that his steak was ready!'

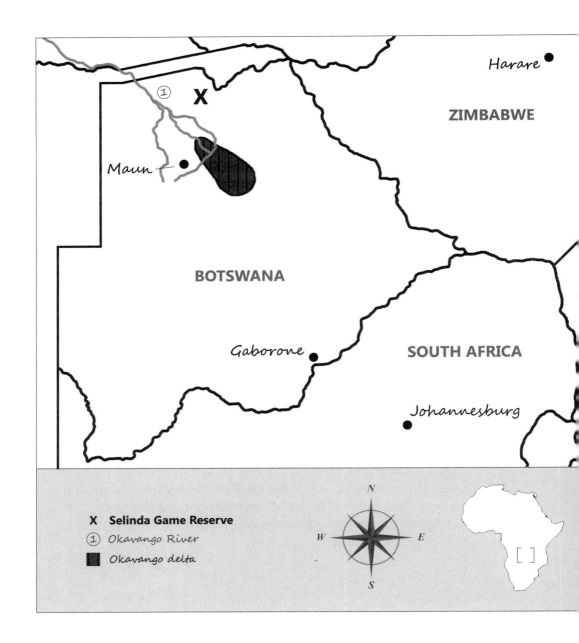

X Selinda Game Reserve
① Okavango River
◼ Okavango delta

N
W   E
S

Harare ●

ZIMBABWE

①
X

Maun ●

BOTSWANA

Gaborone ●

SOUTH AFRICA

Johannesburg ●

15

THE LIONS OF SELINDA

'They won't attack tonight,' someone whispered. 'They'll wait until morning, when they can see exactly what they're doing.' But even now they were less than 500 metres away, somewhere between the forest and the lagoon, their powerful challenge penetrating the trees, our camp and the open plain beyond. And then silence. Were they on the move? Were they coming even closer? Beyond the flickering light of our fire, the dead of night posed questions we could not answer.

How quickly things change. One hour earlier we had finished our evening meal under the stars, the usual flame-scorched fillet steak, salad, bread and beer; unwinding in the warm after-glow of another satisfying day among the elephants. Nearby, hippos grunted, hyenas whooped and unidentified bodies blundered in the dark. I felt exposed and vulnerable, yet safe among people who lived for the bush and knew its ways. And then the roaring started and the mood changed. I could feel tension building in the night.

With just their wits and a steep learning curve to guide them, our African ancestors came to mind. Would they, in this same situation, have been terrified or simply frightened? Would they have had fire? Would they have been able to talk to each other? Would they have even been here in the first place, or would they have been tucked up in some deep hillside cave, waiting for the more comforting light of day? We will never know the details of how these first people lived, except that, whatever it was they did to survive alongside the stomachs of Africa's great carnivores, it obviously worked.

And then it came again, that stark reminder of just one of the problems our ancestors had to deal with in their mosaic land of lakes, woodland and grass. It was 10 o'clock when the roaring began, and it continued deep into the night. Around our fire, the consensus was that large males from a neighbouring territory had come to do battle with the resident males. If this was so, then all ten people I was with were fearful of the outcome. They knew the history.

We were in the Selinda Game Reserve in northwest Botswana. Selinda, until 1995, had been a hunting area and any large male lions that escaped the bullet had moved away, beyond the range of the guns. When hunting was finally banned, three young males, known affectionately to the new tourist camp staff as Tog, Rufus and Blondie, grew up to find themselves prematurely in charge of the territory. They mated with resident females and had three cubs, which wasn't bad for an early start in the mating game. Among lions, though, the mating game can be dangerous. Everyone knew that older, larger males might one day come to claim the territory and the lionesses within it. As we sat under that open, timeless sky sharing the same perils as our ape-like ancestors of two million years ago, the camp staff knew that the moment they feared had arrived. Five large males had been seen north of the Reserve just a few days ago. They had now come even closer and their nocturnal challenge from the edge of the forest could have been to no-one but Tog, Rufus and Blondie.

This spine-chilling event coincided with my first ever visit to Botswana. Wayne and Venessa Hinde were filming *Elephant Empire*, a TV trilogy devised by Adrian Caddy for the Discovery channel in Washington. I had been contracted to produce the programmes.

My first task was to look at 75 already-shot rolls of film with editor Ramon Burrows to decide what would be needed to complete the project with the time and money available. It took just a few days to realise that we would need to tweak Adrian's approach significantly if we were to stand any chance of delivering three different films on elephants. It was this need that saw me board, appropriately enough, a jumbo jet to talk things through with Wayne and Venessa in the bush. After an overnight in Johannesburg I flew north to Maun, from where Wayne and Venessa flew me to Selinda in their little Cessna. We had a really good week getting to know each other, talking over the whole project and following the elephants through the bush, even, once or twice, on foot. And then my last day was upon us. It was 21 May 1999, a day we would all remember.

We left camp at first light in three vehicles, our dipped headlights picking out the track between the forest and the lagoon. A fine mist of rain reduced visibility to just a few metres. We drove slowly, the ghostly, dawn-wrapped trees of ancient Africa slowly taking shape. And then, through the half-light and the drizzle, there they were: five spectacular black-maned lions, probably brothers and half-brothers, together as a team. At six or seven years of age, in their prime and ready to stake a claim, they were standing quietly under the tree where they had spent the night. Two of them stretched and yawned, exposing pink-tongued mouths and yellowing canine teeth. A third stepped forward and uttered the first battle cry of the day, a succession of roars that rose from the depths of his lungs to be expelled from his throat by short, rapid thrusts of his rib cage. He kept going for at least two minutes, pausing only for sharp intakes of breath, until his partners were ready to lend their own voices to his penetrating challenge. There was absolutely no doubting their intentions. As they roared their defiance, they stared straight ahead, out across the withering dry-season lagoon with its cracked, curling mud and reeds that rustled in the wind. By 7 o'clock it had stopped raining, the mist had lifted and visibility was good. The heat of the day was building.

Someone implored Tog, Rufus and Blondie to stay away, to move out, relinquish their tenuous hold on Selinda before it was too late. But lions are not like that. The inexperienced four year olds had heard the calls and were already on their way. Through binoculars, we could see them on the far side of a reedbed 200 metres in front of us. Their approach took them out into an open patch beyond the edge of the reeds and they were seen immediately by the Big Five. With their target visible they bristled, straining their bodies, testing the air, pacing to and fro. Suddenly they launched their attack, sweeping down off the bank and settling into a gentle, easy-paced canter whose economy of effort suggested they were ready for a long, hard day.

It seemed, as we turned and drove after them, that they were travelling in some kind of formation – a central leader slightly ahead of two outriders with two tucked in behind. The roaring continued, those yellow eyes fixed dead ahead, and not one of them taking the slightest notice of the vehicles that bumped and rattled alongside.

The three young lions, which had now come round the reedbed and were in full view, had still not realised what was happening. Perhaps the roaring was designed to awaken their curiosity rather than their fear, drawing them on until they were in sight and vulnerable. The Big Five quickened their pace, leaving us to pick our way through the potholed mud still hard under the sprinkling of early morning rain. And then Tog, Rufus and Blondie saw what was coming and immediately understood its intent. They turned and headed straight for the tall reeds, the only place offering the slim possibility of escape. The response from the advancing males was to fan out and charge headlong after them on an even broader front. Approaching to within 15 metres of the thick wall of vegetation, we cut our engines and listened. The noise was dreadful. Muted roars, yelps and crashings told us that a ferocious battle was taking place. And then, silence.

Five minutes passed before the first lion emerged from the reedbed. He was limping badly and we were relieved to see he was not one of ours. Then another, blood-smeared but still walking. Then two more, and finally the fifth. All had survived the encounter, though two were bleeding heavily and would need time to recover. They walked straight past us, almost through us, as though we did not exist, as though their inevitable sense of purpose gave them no interest, no fear of anything in the world. They oozed such majestic power that it was impossible not to feel a sneaking admiration for them as they made their way back to the forest, fading slowly into the gloom of its densely-packed trees.

Was this phase one accomplished? Had they just retreated to lick their wounds before returning to complete the takeover? It's what lions do. They drive out the resident males preferring, where possible, to avoid physical contact, and then they look for the females, invariably running down and killing their cubs. It makes evolutionary sense for them to do this. Why waste time and effort playing father to other males' offspring, especially if the lionesses will soon be willing to mate again and make a new set of cubs that you can really call your own?

Our distracted gaze after the Big Five was interrupted by the sudden thought of what might have happened to Tog, Rufus and Blondie. The reeds immediately in front of us were impenetrable so we split up, one vehicle staying put, the other two setting off to the left and the right. On the far side we found Rufus and Blondie, walking quietly along the edge of the reeds about 200 metres from where the fight had taken place. Not bleeding, not limping, just walking slowly and purposefully away. But where was Tog? Why was Tog not with them?

We waited five minutes and then drove slowly into the reeds. We used the parked Toyota, which we could just see by standing up in the lurching back of our own, as a bearing. Thirty metres in, we entered a clearing and stopped. It was the battlefield. Three-metre tall reeds had been smashed and flattened over an area 15 metres wide. Blood glistened on fractured stems, fur floated in the air, catching, fluttering, feather-like. A vulture was already there.

We heard Tog before we saw him. A desperate wheezing and panting that we tracked to a far corner where he lay half-hidden on a bed of crumpled reeds. And then he was struggling to stand, but falling, his proud head lolling to one side. It was clear that he had sustained serious injuries in the fight. Lions have a habit of biting at each other's lower backs, aiming to paralyse, to incapacitate while minimising the risk of being bitten themselves. When the decisive blows have been dealt, they pull back, waiting for their next move.

Tog's hindquarters were a mess of deep puncture wounds, his spinal cord obviously touched. But the stabbing 8 cm canines had also got him higher up, penetrating his lungs. Bright red froth bubbled from holes in his chest as he breathed hard and fast to compensate for the loss of oxygen. He must have fought like mad. We will never know the extent to which his siblings joined the fray, though Tog is unlikely to have sacrificed himself to help their escape. More likely, he had been cornered first and the intruders had set about eliminating him from the contest, to make their task with the other two that much easier. If they saw this, Rufus and Blondie might have sensed that it was all over, that even with the battle wounds inflicted on the enemy by Tog on his own, this was a war they could never win. Perhaps they made some sign of surrender or perhaps they just decided to leave while they could, abandoning their brother to his fate. Even if it took the Big Five a week to recover, they would be back. Their mission was unfinished and irreversible, and five against two would be better odds than five against three.

We sat over Tog for the next two hours, bound by the rules of Selinda neither to help him nor put him out of his misery. The camp staff who had known him as a cub were inconsolable. We filmed him dying, getting progressively weaker, struggling to stand less often, breathing less heavily, the vulture still waiting patiently. When it was time for me to go, Wayne left Venessa to film Tog's final breath, reversed his Toyota out of the reeds and we set off for the runway where he kept his light aircraft.

Within 20 minutes we were airborne, sweeping past thermalling vultures who must have known, uncannily, what was happening in the reeds below. Our compass bearing was south towards Maun whose international airport would connect me to Johannesburg and London. We flew in virtual silence for half an hour, consumed by the enormity of the morning's experience. Words, inevitably to do with lions, came slowly. We both had things to say, stories to tell, about one of the world's most impressive and powerful animals.

For a start, I volunteered, how is it possible to sit in an open-sided vehicle just a few metres from a large male lion – or even a smaller female come to that – without him or her appearing to be the slightest bit bothered? Why don't they attack or retreat, rather than just sit there with those superior yellow eyes giving nothing away? What would happen if I stepped out of the vehicle, away from the smell of its fuel, revealing my full human shape? Would I be dead in a flash or would the lion suddenly wake up to what I was and make itself scarce? A bit of both, Wayne suggested, depending on circumstances. Hunger might make a difference. The promise of a much-needed meal often overrides a big cat's natural fear of people.

Wayne told me about a young man in Zimbabwe, a trainee chef at a safari camp, who retired to bed after a long evening in the kitchen, first hanging his smell-drenched coat in a nearby tree. He woke in the middle of the night to the terrifying sound of a lion ripping into his tent, following the promise of food beyond the coat. Alerted, and with enviable calm, the young chef had unzipped the rear of his tent and slipped out. But not far. Other lions were waiting for him at the back, or so it seemed, and he never stood a chance.

Africa's most celebrated people-killing lions were the man-eaters of Tsavo, when two marauding males held up progress on the Nairobi-to-Mombasa railway in Kenya in 1898. For several weeks, these two fearlessly cunning lions terrorised the workforce at the bridge being built over the Tsavo River, killing and eating 28 of them as well as an undisclosed number of locals whose deaths were not officially recorded. It was only after a concerted and abnormally brave effort by Lt.Col. J. H. Patterson, a famous hunter of the day, that these brutish lions were finally laid to rest.

It takes a certain kind of person to track down and kill a large feline on foot, even with the advantage of guns and a deep understanding of the bush. Patterson was one of those people. Jim Corbett was another, though his speciality was man-eating tigers in India before he moved to East Africa in 1947 and was made an honorary game warden. These people, and others like them, knew every precaution a person should take when following the most cunning and enterprising of all animals. Their expertise, combined with an uncanny sixth sense, was the reason they lived long enough to write their memoirs. I was raised on their stories, read to me at first by my father and then on my own until I knew them off by heart. And yet, when I ventured into lion territory on foot for the first time, something was sadly lacking.

It was early in 1972, when I was working on East African birds in Kenya's National Museums in Nairobi, that I was invited to spend a few nights with the expert rhino trapper John Seago. John was already concerned about the future of Africa's black rhinoceros and had set himself the task of relocating any rhinos that were causing problems for villagers outside national parks or reserves. On a trip north to look for the world-famous bull elephant Ahmed on Mount Marsabit, I was able to stop over at one of John's camps at Isiolo, close to Samburu National Park. It was here that I had the encounter with lions that could so easily have been my last.

There were two captive rhinos inside the *boma*, a circular ring of thorny branches erected to protect the camp and its occupants from wild animals, and we spent my first evening talking round a large, warming fire. I turned in at about midnight and slept like a log, which was a great pity. In the morning, the camp was still humming with excitement and several people were repairing a section of the *boma*'s protective thorns. I had missed the night's drama. Apparently, at around three in the morning, everyone except me had been woken by shouts from the guards perched in the trees over the rhino pens. From their vantage point, they could see a zebra being chased round the outside of the *boma* by several lionesses. In a moment of desperation, the zebra had thrown itself

166

against the thorny barrier, which had collapsed under the impact, sending the unfortunate animal tumbling into the middle of the camp. Undeterred by the sound and sight of people and a still-blazing fire, the lions, their minds set on just one thing, came crashing through the barrier, killed the zebra and dragged it back out into the bush. Thirty metres from the broken *boma* they settled down to enjoy their evening meal.

Wayne, perhaps the most bush-sensitive person I have ever met, was amazed that I could have slept through the whole commotion, but he could hardly believe what I did next. Seeing vultures in the trees outside the camp, I grabbed my camera and went out through the still unrepaired section of thorns to try and photograph them. I was 20 metres along a narrow path through the head-high bushes when the stupidity of what I had done dawned on me. If the lions had eaten well, and by all accounts they had, they would be sleeping all around me. I turned to retrace my steps and there, less than ten metres away and neatly covering my own footprints, were the fresh tracks of a lion. Disturbed from its rest, it had turned off the path after following me for half a dozen paces. Overcoming the very real urge to panic, I sneaked back into camp and never said a word.

Wayne's Cessna buzzed along Botswana's airways. We passed low over the famous Okavango Delta and gazed down on hippos, buffaloes and elephants wading through the shallows. Flocks of white egrets. Africa at its very best. With the death of Tog, we knew we had witnessed a power struggle as played out by lions over millions of years, and that nothing will change them or the way they need to behave. It was still upsetting, even for me, who had not known of Tog's existence until a week ago. Perhaps it was also sad because the King of Beasts always kills and is never killed. We feel little pity, though, for the countless animals lions dispatch and eat, often under our very noses. We tell people how 'exciting' it was to have seen a top carnivore chase down and sink its teeth into its prey, yet there was nothing exciting about the lion kill we had witnessed that day.

As we approached Maun from the air, the ancient bush world gave way to suburban Africa. Tin roofs, tarmacked roads, rubbish, football pitches, cars, buses and bicycles. We slid to a halt on the smooth runway and I jumped out with my bag, allowing Wayne to turn and taxi for an almost immediate take-off. He needed to get back to Selinda. The Big Five would return, perhaps even today, to complete their grisly task.

I waved goodbye as the little plane lifted into the sky, and then walked across the hot tarmac to the main airport building. Bright lights, piped music, people in limbo, taxis, tourists, tradespeople. Money. I was on my way to join them once more, loaded onto some alien conveyor belt that would carry me far away from the natural world, back to the more familiar roar of traffic, supermarkets and TV games.

There was still a little while before my flight to Johannesburg, so I stepped out of the air conditioning and back into the warmth of a dusty African street. Across the road was a craft shop, a bit close to the airport but worth a look, just in case there was something local to take home. There were two other people in the little shop, mostly hidden as they pushed through closely-packed stands of colourful clothes. When I came face to face with one of them, it was Chris Tarrant. I was taken aback. With the ancient and compelling behaviour of lions still swimming through my head, I was not ready to engage with the UK presenter of *Who Wants To Be A Millionaire?*

But this was the other side of Chris Tarrant, the side that had to get away, to connect, with his wife, to our ancestral African homeland. And having done just that, they, like me, were on their way back to England, their lives, like mine, enriched by the experience.

POSTSCRIPT

Rufus and Blondie kept walking their slow departure from everything they had ever known. Two days and 12 km later they stopped in lion-free land northwest of Selinda. Without females to attract the attention of other males, they would simply have to sit out a couple of years before they were ready to orchestrate their own challenge on another pride, despite their reduced chance of success now that there were just the two of them. For the time being, a decent meal would be their most important consideration.

The Big Five returned from the forest within a few days, their superficial wounds largely healed, and wasted no time in tracking down the resident females and their young. They quickly killed the two smallest cubs, but also, somewhat counterproductively, one of the adult lionesses.

Another, older cub, named Ziq by the camp staff, escaped death but was still not accepted by the intruding males. They attacked him viciously when he tried to feed too close to them and on several occasions his life was saved only by the bold intervention of his mother. She had not deferred to the new males quite as readily as those females whose cubs had been killed by them. But even so, Ziq's days were numbered. If he was not killed, he would be driven out. A comforting idea was that he should trek north to join Rufus and Blondie, one of whom may even have been his father.

Tog, of course, is dead, his body recycled by the wild inhabitants of Selinda, his flesh part vulture, part grass, part fly. His bones, teeth and claws will take longer to find their way back into the animals and plants within and around the lagoon, but it will happen. His death, in this sense, will not have been a waste.

Back in England, at the end of the filming and with Ramon Burrows breaking the elephant footage out into its three programmes, I prepared a film treatment on the relationship between the Selinda people and the lions that had had such an impact on their lives. Wayne and Venessa, who were central to the programme, had already shot enough footage of the unfolding drama to make the film a safe bet for a TV commissioning editor. With additional filming, a unique story of the human response to a dramatic natural event would emerge, giving us a powerful lion film with a difference. The important thing was to return to Botswana quickly enough to pick up the threads of a fascinating story.

But getting the film commissioned in the post-millennium years that saw so many changes to the wildlife film-making industry proved impossible. After 18 frustrating months, we gave up. And because of the impact of time on people and animals, all of whom move on with their lives, I cannot, I'm afraid, tell you what happened next to the lions of Selinda.

RUSSIA

MONGOLIA

● Hulun Nor

Urumqui
●

Turfan
Depression

Xining
●

Beijing
●

Xian
●

②

Tibetan
Plateau

CHINA

X

①

Lhasa
●

NEPAL

Eastern
Plains

INDIA

Kunming
●

Xishuangbanna

TAIWAN

MYANMAR
(BURMA)

LAOS

HONG
KONG

① Yangtse River

② Yellow River

◼ Poyang Lake

X Wolong Panda Reserve

○ Xiaomenglun

N

W ✦ E

S

16

IN THE LAND OF THE DRAGON

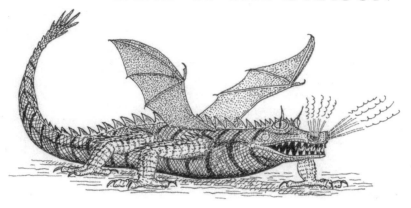

Hong Kong was humid and hot. Under the icy blast of office air conditioning, my first task was to view the existing footage and come up with a storyline to connect eight 50-minute programmes. Then I would have to visit camera crews already working in China, recce new locations and find extra film crews from overseas.

It was no small task and I was glad to have David Reed with me in Hong Kong. I had worked with David before and we were good friends. Now, early in 1987, he had arranged with Patrick Lui, the owner of Sino Films in Hong Kong, for me to join them on a series they were making about China. It would be the first wildlife series on the country and it could not, at that time, have been made by an outside broadcaster. China was still very much a closed country and Patrick had worked long and hard for permission to make his series from Hong Kong. The need for overseas personnel to be involved was accepted. David was the manager and film editor and I was the series producer, replacing a previous incumbent who had left the position after filming had begun.

David and I spent two weeks reviewing footage by day and exploring Hong Kong by night, with a distinct bias towards its food, camera and shirt shops. We finally decided on following meltwater from the high Tibetan plateau down to the sea. Glaciers, snowfields, rivers, lakes, deserts, fertile plains and the coast with its coral reefs and islands would all fit into this series theme. It

171

was such a broad sweep that nothing that had already been filmed would be wasted.

Angie Kwok, a Sino Films researcher/fixer whose father had been a famous Hong Kong jockey, bubbled enthusiastically at the thought of the coastal programme, telling me excitedly of a dramatic wildlife event that would be brilliant for us to film. It was, she said, pointing to the map on the office wall, when millions of fish funnelled down between Taiwan and mainland China in May or June.

Angie then plunged into the most amazing account of how the fishermen put to sea and prepared cauldrons of boiling water which they filled with melons. When the migrating fish were located and the sharks were dashing among them, the melons were scooped from the cauldrons and hurled overboard to be seized by the blindly snapping sharks. As the melons burst, spewing their scalding contents into mouths, throats and stomachs, the sharks writhed in agony, turning belly-up at the surface. The moment the white underside of a distressed shark was seen, a brave and suitably-rewarded deckhand dived in to retrieve it, tying it alongside to be hauled up later. Thousands of sharks were attracted to the frenzy and the fishermen made a fortune selling their bodies, or even just their fins, to hotels and restaurants around the world, particularly in Japan.

It was a gruesome story and yet I knew we should film it if at all possible. The obvious person for the job was Mike de Gruy who was based in Hawaii and currently filming that Pacific island group for the BBC. I sent him a fax to see if he might be able to squeeze our sharks into his schedule in a few months' time.

Mike is an outgoing person, well known today for his skills as a presenter of underwater programmes and raconteur of his own 'and-I've-got-the-scars-to-prove-it' shark encounters.

He cleared things with the BBC and agreed to be on standby, ready to dash to Hong Kong at a moment's notice. It was now up to Angie to find out when the fish were running their narrow channel and to sign up a friendly fishermen whose boat we could use during the hunt. Angie worked away, but her encouraging initial noises were soon being replaced by a succession of reasons why we would *not*, after all, be able to film the sequence. Either the fish were not gathering, or the fishermen were not willing to help, or there were no sharks around just yet, or... or whatever. After a few weeks of zero progress, I had no option but to contact Mike and call the whole thing off. I had learnt an important lesson about Eastern promise of things to come. Angie, whose imagination may have got the better of her, was unable to back down but had neatly prevented any blame from falling on her shoulders. She was, as a result, able to carry on without loss of face, which was just as well, because I needed her. We were about to take off on my first trip to China.

The Beijing Hotel was enormous, on the scale of a grand palace and not far from Tiananmen Square. My window overlooked the wide road on which a lone figure would famously halt a column of tanks two years later. Patrick Lui had lined up a number of things in the country's capital before letting Angie and me loose in its back garden. We spent our first few days visiting more than 20 film production companies – all state-owned and using 35 mm film stock – trawling through their wildlife footage to see what might be useful to include in our series. In with a lot of amazingly detailed material of fairly mundane subjects, there were two things that I felt would be of high viewer interest outside China.

The first of these was a film about a man from Beijing and a removal lorry that had been converted into a mobile aviary. Inside were dozens of beautiful azure-winged magpies. When their owner received the call that someone's pine plantation was being devastated by pine-looper moth caterpillars, off he would go, cheerfully driving into the woods and throwing open the doors of his lorry. The magpies would fly out and spend several hours swallowing every caterpillar they could find, returning obediently when he whistled them home at the end of the day. If he stayed a week, and every bird ate a hundred caterpillars a day, pest populations would be significantly reduced. It was biological control at both its safest and most aesthetically pleasing. And it was delightful because its success relied on the endearing bond between a man and his birds.

The second film, revealing another bond between people and birds, was about 'sky burials' on the Tibetan plateau. Here, in the absence of trees to burn for cremations and with the soil too thin or too frozen for grave digging, scavenging vultures are enlisted to dispose of human corpses. The film I watched was of a young man taking his father's dead body, laid carefully across the back of a donkey, up into the bare mountains. Himalayan griffon vultures were already assembled at the traditional site, which had a large stone plinth at its centre. With their undeveloped nostrils, they could not smell the approaching corpse but would certainly have seen it from a long way off. Gliding down from the cliff tops and the skies, they waited patiently on the ground. When the little party came to a halt, their brown-feathered, bare-necked bodies lumbered towards it.

The young man unstrapped his father's emaciated body, took out a sharp knife and began cutting off strips of dry flesh. These he tossed into the hissing mass of birds. He worked systematically through the corpse, throwing everything into the fray until he was left with just his father's head. Holding it carefully, even cradling it, he approached the central stone, a large hammer grasped in his free hand. The vultures crowded round with a restraint that suggested they knew what was coming next.

They allowed him to place the head on the flat surface of the stone, beginning their frenzied dash only as the hammer swung down through a great arc to shatter the gently-rocking skull. Stepping back, the young man gazed in proud satisfaction as the final, vital part of his father's body, his brain, was consumed by the avian scavengers piled up in front of him.

Both films, I felt, epitomised the China we should try and show the rest of the world, the China whose civilisation pre-dates our own by more than 1,000 years. Its people have established an ancient culture that binds them to plants and animals in ways that we do not fully understand. We do, however, know the difference between right and wrong, and the more integrated China becomes with the rest of the world, the more some of its attitudes and approaches to wildlife – the fur trade and the use of animal body parts in medicine being just two examples – will be challenged. In fact, for the Chinese authorities to address *any* issue, wildlife or human, that might hinder acceptance by the outside world makes, at the very least, sound economic sense. The danger, perhaps, is that China will become such a global force that it will not have to listen to anybody else.

Aware of my interest in the relationship between people and wildlife, one of the Beijing production staff told me about the mulberry tree and how central it was to life on the eastern plains. Typically, a village would have a pond that contained fish. Mulberry trees were grown round the pond and the silkworm caterpillars fed on their leaves, their droppings fuelling a food chain in the water that would feed and fatten the fish. During summer, silk was harvested from the cocoon-spinning caterpillars and the fish would be caught and eaten. In autumn the mulberry trees were pruned, their thinner twigs and leaves combined with grass to make a winter cattle feed, the larger cut and stored as firewood or mashed into pulp to make paper. At the end of each winter, the pond was dredged of its highly-nutritious sludge, which was spread on the fields to encourage the new growth of grass for livestock. And, of course, the pond had ducks and geese whose bodies and eggs provided more food for the village.

When our last day in Beijing arrived, it was reserved for socialising, a vital part of the work process. It meant we were to hold a banquet with some of the important production people who had helped us. While the company and the drink were recognisably good, the identity of a lot of the food dishes left me guessing. Those I didn't trust, I avoided, and I paid scant attention to the fact that Patrick, the host, was tasting and approving every dish ahead of his guests.

The following morning, Patrick returned to Hong Kong and Angie and I, with our official guide and interpreter Zheng Xiao Ting, were left to represent Sino Films on our own. We had a lot of ground to cover, first stop being Xining, 1,500 km southwest of Beijing. Here, I was to oversee the setting up of an enormous enclosure in the nearby mountains so that we could film a snow leopard in 'wild' conditions later in the year. This was a hastily-arranged plan B following the sudden withdrawal of permission to film a genuinely wild female snow leopard and her cub at their den on the Tibetan plateau. We heard later that an underground nuclear test had been carried out close to where we wanted to film, but we never found out what happened to the leopards.

To reach Xining, we had to take a 24-hour train journey into China. Ting and Angie explained the system. First, we needed a ticket that would guarantee us a place on the train, providing minimal space and minimal comfort in 'hard bench' class. Then the two girls would get to work buying our way up to 'hard sleeper' and finally to 'soft sleeper', as long as these luxury carriages were not being used by the military, who tended to block-book them out of Beijing in case any senior staff needed to travel at the last minute.

Our progress through the train's designated areas was memorable. Angie made sure that I was tucked safely into a 'hard bench' corner before she and Ting went in search of the carriage attendant to negotiate our progress to more comfortable seating. It seemed that whole families and their livestock were on the move, bustling and bleating for position while they fed, slept, read and chatted excitedly about the journey ahead. All alone with my inappropriately large suitcase, I was an object of total curiosity. Children crowded round, touching, laughing and scuttling back to their mothers who smiled benevolently as they adjusted to the routine of a cramped compartment. The men kept their distance, watching me silently as they sat smoking, chewing and grinning toothless grins. I felt totally relaxed in their company. When Angie beckoned from the carriage door, I felt almost apologetic to be leaving them, summonsed as I was to the relative comfort of a 'hard sleeper'. Two hours after that, though, and we were sinking into the luxury of a 'soft sleeper' where we drifted away to a deep and welcome sleep.

But not for long. Early in the afternoon, at a fairly anonymous-looking station, we were told to leave the train and wait five hours for an onward connection. And then all that hard work buying us into the 'soft sleeper' would have to be repeated. In a nearby hotel, which I realised had been pre-booked for us, I was advised to get more sleep in readiness for a wakeful night if we failed to get beyond 'hard bench' on the overnight train to Xining.

Not feeling tired, I went for a walk. Grey was the predominant colour, and the rustling of hundreds of bicycles – there were very few vehicles there in those days – brought the centre of a crowded beehive to mind. All those bodies crammed together, busying themselves on their allotted tasks, their eyes fixed on some distant, unknown horizon. After a quick wander round the bus station, I returned to the hotel, thinking that sleep wasn't such a bad idea after all. In the entrance lobby, my eye was caught by a poster of the clay soldiers from Xian, the famous terracotta army excavated in the 1970s and now, in 1987, partially open to the public. The girl on the desk saw me looking at the poster. To my surprise, she spoke English, which was less surprising when she told me we were actually *in* Xian and her job was to look after tourists who wanted to visit the soldiers.

I could hardly believe my ears. Five hours to spare in Xian with nothing better to do than sleep? I rushed upstairs, grabbed my dozing companions and bundled them into the taxi ordered by the girl on the desk. In a little over an hour we were standing outside one of the most spectacular historical finds of the modern world.

We walked round a parapet, looking down on rows of life-sized soldiers modelled in clay. In different pits there were hundreds, perhaps thousands, of them, some even lined up in battle formation with their horses, ready to protect their leader Qin Shi Huang, China's first emperor, whose burial mound was close by.

A thin black line ran round the wall of one of the pits, a metre or two above the soldiers' heads. Here, I learnt, a village had once stood, covering the soldiers, but it had burnt down 500 years ago, leaving a deposit of charcoal which had been slowly buried. Most spectacular of all, though, was when I used my binoculars to inspect the figures in more detail. Life-sized and modelled in clay, they were of different heights, wore precise 'clothing' to denote their ranks and even had their own facial expressions. They were individuals, real people plucked from the masses, their mission in life, and beyond, the safety of their emperor. It would have been no surprise at all if they had yawned, stretched and climbed out of their pit as though no time had passed since their incarceration more than 2,000 years before.

It was time to leave the warriors. Back in Xian, the train was already waiting at the station. In the absence of any military threat to our comfort, our progress towards a 'soft sleeper' was a formality. I slept well that night and woke the following morning to the cold mountains of central China. We had climbed to the very edge of the vast Tibetan plateau. When we stepped from the train and were met by the officials from whom we would need signed filming permits, Angie turned and said, 'Dan, I hope you are hungry, because it seems that, to get these permits, we will spend the rest of the day eating.' As the senior representative of Sino Films, I was to be the host at a banquet.

Inside a small restaurant in the middle of town, a large round table had been prepared for 15 people. We sat down or, rather, I was invited to sit down and everyone else followed. Angie offered packets of 'American' cigarettes from Hong Kong which were gratefully accepted. As everybody puffed away and the drinks began to flow to the call of 'Ganbei!' – the Chinese equivalent of 'Cheers!' but with something of a challenging ring to it – a feeling of warmth and friendship settled over our smoke-filled corner of the restaurant.

And then, out came the food. A large fish dish was placed in the centre of the table. It would be our final course, eaten with rice. Around the fish were smaller dishes, some of whose contents looked distinctly suspect. Something brown was spooned onto my plate, and then onto everyone else's. I waited for someone to begin eating, but all eyes were on me. Ting whispered that I had to taste it first and give my approval before they could begin theirs. Had I paid more attention to Patrick in Beijing, I would have known this was coming. The first mouthful was a bit starchy but was actually quite nice, whatever its identity. I nodded and smiled my approval and the clatter of chopsticks that followed was matched by the clinking of glasses as the clear alcohol – almost neat by the taste of it – flowed. Then the chickens' feet, boiled, soft and yellow. I approved again and more smoke billowed into the air, more glasses clinked merrily and more cries of *'Ganbei!'* penetrated the blue-grey haze that hung over our table like a shroud.

Next came reindeer ligaments, a bit stringy and chewy, but manageable. Then something from a camel, and then something that looked like a cucumber, though that was where its similarity ended. I struggled on. It wasn't the food itself but the strangeness of it. Different cultures, different tastes. And being asked to sample and approve it in front of an audience. What would happen if I turned my nose up at some local delicacy? Despite the relaxed atmosphere, it would, I felt, have been too easy to offend these smiley, chatty, smoky, boozy people. All eyes were on me. Mine were on those unsigned permits.

And then, just before the fish whose flesh would have to be removed without it being turned over, came the almost unbelievable. A large plate of asparagus. Hot, steaming and drenched in butter. My taste buds anticipated the deliciously familiar. As four of the heavenly spears were lifted onto my plate, I gave a hasty nod of approval and popped one into my mouth. A sudden, rude awakening. The 'asparagus' did not melt between my teeth. It was rubbery and I had already passed the point of no return because everyone else was tucking into theirs with glee. I grimaced through the disappointment, chewing mechanically and swallowing with difficulty until my plate was somehow cleared. Our senior guest, smiling, leaned towards Ting who turned to me and said, 'They are glad you specially like giant turtle penis. It is great delicacy here.'

Happily, the banquet passed without further incident and we were soon filing out of the restaurant and up the road to the hotel. In a flashback to the Sundarbans and the unfortunate cow who lost her life to a tiger because of a pair of forgotten binoculars, I realised I had left my own hanging on a peg behind our table. I turned to retrieve them without bothering anyone else. When I entered the restaurant 15 people were sitting at our table, tucking into the remains of the meal exactly as we had left it. Catching up with the group, I mentioned this to Ting. She took my sleeve and led me to the back of the restaurant where there was a queue. She explained that they were the poor people who were allowed to feed on the leftovers when a meal had been finished. And anything left after that, she added, would be taken away and made into animal feed as a legal requirement. Nothing would be wasted.

The filming permits were signed and delivered in the evening, so all we had to do was sort out the snow leopard enclosure. The following morning we went to the zoo to look at the animals themselves. We were astonished to find nine of these beautiful, big-pawed, big-tailed, small-headed leopards, all in wonderful condition and in the proud possession of their doting keeper. We would be pleased to hear the zoo staff had selected a perfect site for the enclosure and that the posts were already in position, just awaiting our approval before the wire netting was applied. All we had to do, it seemed, was look at the map, nod enthusiastically and then go back for another banquet. Well, not quite. We were paying a small fortune for their services and I was determined to see that things were done properly. To their obvious dismay, I asked for an early start in the morning.

We chugged up the hills overlooking Xining in one of the zoo's ageing minibuses, stopping along the way to look at different habitats. In a thickly-vegetated valley I saw my first parrotbills, small finch-like birds with long tails and definite parrot-shaped beaks. Out in the open, there were four or five male cuckoos flying around, calling their familiar 'cuck-oo' notes. We were also shown what our host described as forest, but which was no more than an overgrown gully, a slight embarrassment as it had been listed on our permit as a good place to film. This impoverished patch was a reminder of how densely forested the edge of the Tibetan plateau must have been not so very long ago.

Finally, at around packed-lunch time, we emerged onto rolling grassland above the treeline. The snow leopard site was a bit exposed but would work because it included a steep-sided gully with a natural-looking cave entrance between two large rocks. The only serious alteration I wanted was for the posts behind this gully to be moved so that the fence would not be visible on the skyline in the wider shots. It was just a matter of moving them further back and down so they were out of sight. The fenced-off area was two or three acres in extent, allowing for good following shots and a variety of places for the leopard to explore. There was no chance of filming a natural kill or even of setting anything up that looked remotely like one, so it would have to be portrait shots, wandering around and a bit of feeding on a carcass. The best thing would be if we could film one of the females with a cub, to extend the sequence and give it a bit more appeal. I stressed the importance of the leopards being introduced to the enclosure at least a week before the filming so that they would be comfortable in their new surroundings. These thoughts were received with enthusiasm and our late-afternoon drive back to Xining was accompanied by a feeling of relief.

The bombshell came later in the year when Tony Allen turned up at the agreed time to find that our allocated leopard – one without a cub – was lazing around in her concrete pen in the zoo and had never been taken anywhere near the enclosure. Tony was told to get a good night's sleep after his long journey and tomorrow they would take him and the leopard into the hills for an afternoon's filming! That was all they thought they had to allow and, because of this, it was fixed firmly in their minds that there was absolutely no need for the leopard to stay overnight away from its more familiar surroundings. This was a real struggle for Tony who spent most of his time continuing the negotiations that I thought I had completed several months earlier. After two days of serious discussion, including a banquet and its statutory heavy drinking, he managed to get agreement for the leopard to stay overnight in the hills where he eventually pieced together the minimum sequence for the series. Expensive, not spectacular, but at least included.

Before we left Xining, we were visited by Larry Zetlin, an Australian cameraman who had been driving around the Tibetan plateau in search of yaks and wild asses. The most suspicious of open-country mammals, they have been hunted for centuries and take off when approached to within a kilometre or so, leaving behind them little more than a cloud of dust and another disappointing day. Larry showered and sank down into the luxury of his hotel bed, plugged into a Barbara Dickson tape I had brought with me from Hong Kong. When he joined us on the streets of Xining in search of another good place to eat, his commanding presence – he stood nearly two metres tall, had shoulder-length hair and a long grey beard – gave him instant appeal. The local children followed him down the road like the Pied Piper of Hamelin.

Our next stop was Kunming, the capital of Yunnan, nearly 1,500 km to the south of Xining. This time by air: quick, easy and far less entertaining. Checking into the palatial Green Lake Hotel, we met Ting, who was now looking after Jan Aldenhoven and Glen Carruthers, an Australian couple who had been hired by Sino to make a film on China's southern rainforest. Glum looks all round. They had had a few days, including a quick recce to Xishuangbanna, to get to grips with the problem. And problem it was. The first thing they had seen on a quick walk in the southern rainforest straddling the border with Laos was a fig tree whose fruit lay rotting in great heaps round its base. Jan and Glen were quick to realise that the figs were not being eaten on the tree by birds or on the ground by mammals because there weren't any of these creatures left. Standing by the tree, the forest around them was silent: not a single bird was singing. Despondent, they turned away, but as they were being driven back through Xiaomenglun, they noticed a sign to Xishuangbanna's Tropical Botanical Garden and asked if they might be able to pay a visit. Much to their relief, they found native forest plants visited by nectar-drinking birds and insects, giving them a glimmer of hope for filming at least some of the creatures of the forest. But it wasn't to be. They were politely informed that the Gardens were not listed as a location on Sino's permit and they would not, as a consequence, be allowed to film there.

It doesn't get more depressing than this. Patrick was paying a small fortune to film China's wildlife and here was the country's own bureaucracy doing its level best to stop him. I began to understand the advice – or was it a warning? – that for every 200 km you travel out from Beijing, you go back ten years. A quick glance at the wall map in the Green Lake Hotel showed Beijing to be about 2,500 km away.

Angie suggested we talk to the Professor of Zoology at Kunming University. She arranged a meeting which, at his insistence, was to be at his home in the evening rather than at work during the day. At 7 p.m., we were ushered into a very small flat in a run-down area, where he introduced us to his wife and other relatives. We sat in the living room, made little progress with our discussions on the status of animals and plants in the wild, and soon realised that the whole point of the evening was to show us that he possessed a refrigerator. It sat in the centre of the little sitting room, dwarfing everything else and humming the cooling tune that signalled his status as one of China's important people. The policy of 'a fridge in every home' was still a few years away.

We sat at a table outside the hotel wondering what to do next. It was a pleasant afternoon and, in the absence of traffic noise, we were able to enjoy the novelty of a bird singing in the park opposite. And then it was being answered by another. To take our minds off the frustrations of filming, we decided to look for the birds. We set out across a little stream and into the pleasantly laid out garden whose winding pathway led to the green lake itself. Following the song, we came across an old man sitting on a bench and there, just above his head, was a thrush perched in a cage, singing its beautiful liquid song. Not 20 metres away, another male thrush was singing a reply from its own cage which, like the first, had been hung up in a tree. The singing stopped only as the light faded and the cages were taken down and carried off to their respective homes. The birds were treasured possessions and they obviously brought joy and friendship to a world largely devoid of natural sounds.

While there was something enchanting about the thrushes in the park, they also highlighted our problem in Yunnan: there were very few wild animals. Sino Films was under contract to deliver a rainforest programme which was in danger of not happening, despite the meltwater theme that would allow the inclusion of snow-covered mountains clothed in colourful rhododendrons, azaleas and camellias. We also had some establishing shots of the rainforest and the Mekong River winding through it, taken by another camera crew on an earlier visit, and we had shots of a special forest village dance celebrating the peacock and the endangered South China tiger. But we were still lacking the good selection of rainforest mammals, birds and insects that we so desperately needed. One thing we did know, though, was that Kunming Zoo had a tiger. If Glen could get a big close-up of its face, it would at least give us a living link to the tiger mask in the dance. Angie arranged a meeting with its director.

The next morning we walked to the zoo, taking a shortcut through the back streets. We were touched by the warmth and friendliness of people who might never have seen Europeans before. They were all so busy making and mending, talking and laughing. Birds sang from their cages, frogs squabbled in buckets and snakes were being skinned alive. A small girl from a minority tribe was selling the most beautifully-coloured hand-embroidered bags. I bought six of them and now, more than 20 years later, Becky has just started using her last one.

When we arrived at the zoo, Miss Tshin, its director, assured us that filming the tiger would be easy. It also became clear that she would help in other ways as well. Jan, Glen and I went into a huddle and came up with an idea to put to her. If we could not film animals in the rainforest, we would have to bring the rainforest to the animals. By changing the scenery in a zoo set, and by lighting it for both day and night, we might still be able to save this important film.

Miss Tshin thought it a splendid idea and so, that very afternoon, the tiger was removed to an adjoining cage and we set to work. Glen designed the set – a stream running through the forest with a central pool – and the zoo's builders were soon following his chalk-marks on the floor and cementing the basics into place. When it had set hard, earth was piled in and a lorry sent on its first journey south to collect the leaf litter, tree trunks, ferns, bushes and flowers that would make up the rainforest scene. Three trips later, the set was ready for tree shrews, small cats, monkeys, reptiles, insects and birds to be taken from their respective cages and let loose in what was a perfect replica of their rightful home. Glen and Jan filmed their sequences over two months and came up with enough footage to add to the mountain, river and forest scenes that had been shot earlier.

As Jan and Glen recall:

> 'One of our more intricate sequences was a slow loris, a beautiful, large-eyed nocturnal primate that creeps very slowly, and with great stealth, to nab its insect prey. After filming the loris capturing a stick insect, we hatched a more ambitious plan for it to steal some birds' eggs. We needed to show the loris finding a nest, the startled bird flying off and the loris moving in to claim its meal. So where to get a nest with eggs? We weren't about to sacrifice a real one. No, we'd buy doves' eggs at the market and we could temporarily detain one of the feral pigeons that roosted in the elephant house. With careful silhouette lighting at night it would pass for a rainforest pigeon. But a nest, now that was more difficult. Jan decided to make one, and discovered just how difficult it was. She will be forever in awe of real nest-making! The sticks – and you don't need many to make a dove nest – just wouldn't stay woven together and the end result was a very amateur structure. But, with that shadowy lighting to give just a hint of nest, we'd be OK.'

As always, a 'two-shot' – predator and prey in the same shot – lends authority to a sequence. That was going to be a bit tricky. Glen gently placed the feral pigeon on the nest and held it while Jan put a cardboard box over the top. Daytime birds go quiet when in a dark place. The loris was settled on the branch close to the box. Glen framed the camera on the box and, just as the loris was about to walk into shot, Jan hauled on a string tied to the top of the box. With film rolling, the "rare rainforest pigeon" flew off its nest to give the lurking loris a perfect opportunity. It reached out, grabbed an egg with both hands and stuffed it into its mouth. Now for the close-up shots. No problem – we had put three eggs in the nest and the loris obliged by eating them all, yolk dribbling down its fingers.

'The producer will be thrilled!' we thought, as we shipped the film out to Hong Kong for processing. But how unlucky could we be? Dan knew too much and was quick to put us right. The congratulatory telex we were expecting from him read, instead: 'I can't think of any pigeon or dove that lays more than two eggs.' Our hard-earned, top-drawer sequence was unusable!'

Jan and Glen's most uncooperative animal was the tiger. Offered the chance to return to its former home in the zoo, it had placed one tentative foot on the soft rainforest litter and had refused to take another step out of the next-door cage. It wanted its familiar concrete. What we wanted, though, was a tiger-in-the-forest sequence that would end on a perfect link to the tiger/peacock dance.

Glen decided to clear a path for the tiger through the bushes, trusting it would walk along this until it ran out of concrete. If it worked, it would give him the perfect following shot. The narrow path was cleared and the tiger eventually set off, reluctant at first but then with more confidence. Through the bars of the cage, Glen filmed it creeping through the bushes, with glimpses of face, fur, stripes and feet, until it stopped at the end of its concrete causeway, its whiskered face exactly filling the viewfinder of his camera. The eventual mix from that powerful living image to the tiger mask in the mountain village overlooking the Mekong River was perfect.

I left Beijing the following day and returned to Hong Kong to catch up with other film crews. Jeff Goodman was having an unproductive time along the coastal reefs which had been completely fished out. John Loader was about to disappear into the central mountains to film a short sequence on the strange tarkin antelope. Tony Allen and Chris Catton, joined occasionally by James Gray, were still in Wolong, piecing together a whole programme on giant pandas. Tony had just been attacked by one of the reserve's semi-habituated males that took exception to having another of its bamboo meals interrupted. The viewfinder of a camera gives a distorted picture of the world, reducing the scale and significance of things that are happening. And Tony, delighted to get a bit of action outside the endless munching, kept filming until he realised that the ambling gait towards him was actually directed at him.

Too late to run, his only defence was the sharp end of the tripod legs. An Arriflex camera mounted on its tripod is a heavy piece of equipment but, even though Tony got it up off the ground to protect himself from the angry panda, it was swatted to one side like a fly. Tony lost his footing and fell. Fortunately, the panda was so alarmed by this over-reaction that it veered to one side, doubled back and returned to resume its bamboo meal. Tony was bruised and unhurt apart from a twinge in his back that stayed with him for years. His equipment suffered the most, the critically-important zoom lens taking a sufficient knock for it to have to be sent back to Hong Kong for repair. As soon as the lens was returned, he and Chris set off to film giant gerbils in the below-sea-level deserts of the Turfan Depression, more than 2,000 km away to the northwest. When you get that far from Beijing, around Urumqui, you have travelled a third of the way back to England. China is a big country.

Jim Clare was in the north, filming Siberian tigers in yet another of those enormous and enormously expensive enclosures. This one was knee-deep in snow but he managed to get some wonderfully evocative shots of the largest, most intensely-coloured and hairiest of all tigers slinking around in pursuit of a small deer. Jim also spent time in an isolated lake area where he filmed pied harriers nesting in dense reedbeds, the male a strikingly-patterned pale grey and black. A similar lake, Hulun Nor, lies further west in Heilongjiang Province, close to the Mongolian border, and is almost certainly a nesting site for the rare and endangered Siberian crane. We filmed some of these majestic birds overwintering on Poyang Lake in China's eastern plains after their flight south from their breeding grounds, listed variously as eastern Russia and central and northeastern Siberia. The crane is known to nest in the Hulun Nor region and we heard that it had, in fact, been seen at the lake in the breeding season. The huge expanse of reeds surrounding the lake would easily allow for a few pairs to raise their young in total secrecy.

The footage was slowly coming together. It was with some relief that we were able to meet our first obligation to deliver sufficient material to America for ABC to make up their own presenter-led series of six programmes on Chinese wildlife.

Six programmes would have been ideal for us as well. But no, Patrick's enthusiasm got the better of him and the original eight became ten, based, in part, on the Beijing library footage. But this footage turned out to be technically flawed and the high transfer costs had to be met by Sino Films. By mid 1988, money was running out, Patrick's dream was shattered and contracts could not be renewed. I settled back in England, grateful to have had the opportunity to explore such an amazing country but equally sorry that we were unable to complete the longer series.

It would be a further 17 years before the BBC's Natural History Unit was able to enter into a co-production deal with China Central Television to produce *Wild China*, a long-awaited series on the animals, plants and people of one of the most fascinating countries in the world.

17

Mr Blackbird

He sits, plump and unmistakable, on chimney, aerial and tree. His liquid song, broadcasting the possession of his patch, is among the best in the world. And then he is off, swooping down to the bushes, his chinking alarm call scolding and warning. Danger over, he flips to the ground, pulling worms and flicking leaves. Almost everyone knows the blackbird, one of the commonest and most conspicuous of our garden birds. With five million pairs nesting in the British Isles two or three times a year, hardly a garden escapes their attention.

We met our own 'Mr Blackbird' not long after he emerged from a nest close to our garden in May 2001. Brown-speckled, short-tailed and barely able to fly, he was hiding in a flowerbed and letting out the occasional shrill cry that said 'I'm here and I'm hungry.'

But after one of these calls, he hopped out into the open and began the testing five-metre journey across the short grass to the bushes on the far side of the lawn. The cat – it was the tabby Daisy from next door – was out of her hiding place in a flash, sweeping him up in her jaws and scampering off towards the cherry tree. The little blackbird didn't have time to make a sound. But he was in luck. We had been watching the adult male disappearing into the undergrowth with beakfuls of insects and worms and had heard the youngster's calls for food. When he broke cover, I saw him as quickly as I saw Daisy's mad dash to snatch him from the grass. I was up and after them, shouting and waving, and poor

Daisy, shocked by the onslaught, spat out her trophy and disappeared over the garden fence.

The little blackbird lay where it had been dropped. There was no response from either of its parents. Had they seen or heard any suggestion of trouble, they would have swooped in with their ringing alarm calls. But I was able to pick it up without any protest from either of them. The little bird lay quietly in my hand, its chest heaving up and down. There was no obvious sign of injury, which was a good start, but the possibilities of internal damage or shock were high. My son William carried the little ball of feathers indoors where he and his brothers Michael and Edward made up a bed in a box. After two hours, scrabbling noises from within suggested the fledgling was recovering. We lifted the lid and there he was, bright-eyed and ready to go. The one thing we did notice as we lifted him out was that the first big flight feather on his right wing was bent outwards. A small blood clot at the base of this feather suggested it had been damaged by one of Daisy's sharp teeth or claws, though we had no idea how badly.

Back in the garden, we watched until we knew where one of the other baby blackbirds was hiding, and then released ours nearby. It disappeared under cover, called and, within minutes, was itself being fed. Satisfied that we had helped in at least its short-term survival, we hoped that the baby blackbird with the crooked wing feather would live for very much longer.

By late autumn, surviving young will have dispersed and now their parents retreat from the battlefield of their breeding to lie low, moult their feathers and recharge their batteries. Some will even migrate.

While the blackbirds are away, there are changes in the garden. The delicate winter song of the robin filters out from the yellowing leaves of the cherry tree. Family parties of long-tailed tits clean up persistent greenfly and one or two male blackcaps sing quietly from the shading leaves of the sycamore. TV aerials are perches for collared doves, and woodpigeons sit plumply on chimney pots, keeping an eye on the developing berries of ivy growing on the wall outside our kitchen door.

Around Christmas and New Year, young and old blackbirds return. The battle for breeding rights begins once more. Males chase each other through the bare branches of the cherry tree and along the ridge tiles of nearby roofs. No song, just physical rivalry. Beyond noticing, early in 2002, that they were back again, we paid them little attention. It wasn't until March that we realised that a pair had once again taken up residence and that the female was collecting dead grass for their first nest of the year. She was building deep inside the ivy on the wall, just a few metres from our kitchen door. Her mate, in his first year suit of blackish-brown feathers, dull yellow eye-ring and dull orange-yellow beak, became increasingly prominent. He would perch on surrounding gutters, rooftops, TV aerials, our cherry tree, and accompany his incubating mate with his clear notes of authority. And then he would be feeding on our lawn, pulling out large worms, breaking them into pieces and gulping them down. One day we noticed that the first large primary on his right wing was sticking out at a funny angle. Given that young blackbirds can return to breed close to where they were hatched, we did not question his provenance. The youngster we had rescued from the jaws of a neighbouring cat was back and about to raise his own brood in our garden. For the next three years, he was more than just identifiable, he was almost one of the family.

Our Bristol garden, a typically small city one, is hemmed in by other gardens. All are patrolled by cats, squirrels, rats, foxes, magpies, jays, crows, sparrowhawks and people, any of which might be attracted to a nest site or to a bird preoccupied with singing or feeding. But our pair survived their first season together and, so far as we could tell from the young birds that appeared through the summer, raised three small broods, the first in our ivy-clad wall, the other two further afield. Mr Blackbird's summer was preoccupied with singing and ferrying food to nestlings and fledglings on and around our lawn. We kept cats away as best we could and, as the bond between us and our bird strengthened, there were signs that he was beginning to recognise us as familiar people in the heart of his territory.

Spring 2003 confirmed this. It really did seem that Mr Blackbird, who had now survived his second winter and was in full adult plumage, knew who we were. He would sing from the cherry tree just above our heads as we mowed the grass or put washing on the line. He would come down to a little stone trough which fills with tadpoles every spring, to drink and bathe within two metres of us at a nearby table. He would come out of the trough and make his way under the table, picking crumbs and insects from around our feet, and then he would fly up onto the fence next to us, a stepping stone to our roof where he would sing in the sun. And, wherever he went, that bent wing feather identified him as the bird whose life had become so much part of our own.

The following year, tragedy almost struck. We came down one morning to find black feathers on the lawn, a mix of tail and body feathers beneath the cherry tree. A cat, we assumed, perhaps even Daisy again. It must have taken him by surprise, pinning his tail, which pulled out completely as he struggled to escape. But now he was nowhere to be seen, and we feared the worst. Mrs Blackbird was still sitting on eggs in the ivy and may have been unaware that anything was wrong. We saw her later in the day, feeding on the lawn, as we searched and listened in vain for her mate. We saw her again the following day, when she left her eggs to feed once more. We knew she could manage the eggs on her own, but she would have problems when they hatched and the nestlings needed to be fed and kept warm at the same time. And there was the bigger

Mr Blackbird minus his tail, probably lost to a cat, Bristol, England (Author's collection)

question, too: what would happen to the territory, and therefore to her and her brood, if he was no longer around to sing its defence from on high?

Two days after Mr Blackbird had been attacked, I was in the garden when what looked like a round black ball rolled down the roof opposite and disappeared into its gutter. A small head appeared with a yellow eye-ring and orange-yellow beak. Then the faintest song, perhaps the first notes of reassurance to his sitting mate that he was still alive. He pulled himself up onto the edge of the gutter and sang

188

a bit more. His right wing was clearly visible. It was him. His tail was missing, and with it had gone his balance. He wobbled on the narrow ledge and fell back into the gutter, struggled to his feet and sang a little bit more. It was a valiant effort to reassert his authority. He then flew down to the ground, crash-landed in the soft earth of a flowerbed and immediately took off vertically, landing back on the roof just above the gutter. Mrs Blackbird came out of the ivy and joined him at precisely the moment that another male landed on the roof and launched an attack. Mrs Blackbird intervened, hurling herself at the intruder, chasing him up the roof tiles and back over the road.

Our relief was immeasurable. Our friend was alive and back where he belonged, and his mate, probably a lifelong partner, was looking after him. We could only hope that he would not be ousted by a rival male before he was fully recovered from his ordeal.

For the next few weeks, he was a very sorry sight. His tail grew back slowly, its new feathers emerging in their silvery sheaths like a row of little paint brushes sticking up at right angles to his body. His balance returned only as quickly as his tail grew, so he crash-landed everywhere, unable to steer or brake properly, bouncing on the grass when he dropped out of the bushes to feed. His song, though, was soon as good as it had ever been, enough, no doubt, to reassure his mate and to advise the intruding male to maintain a healthy distance.

Their third year, 2005, was their most prolific. It began like any other, though the first nest was built two gardens away, halfway between us and a pear tree festooned with at least 20 food dispensers that attracted different birds throughout the year. But such a high concentration of finches, tits, sparrows, starlings, doves and pigeons does not go unnoticed by predators.

The early warning came when I looked out of our kitchen window late in April and saw brown feathers on the grass beneath the cherry tree. They were too neatly grouped to have got there by accident, so I went to have a closer look. The feathers were from a house sparrow. It must have been killed and half-plucked on the ground. A few weeks later, there were more feathers in the same place, this time from a collared dove.

At least two babies came out of the nest between us and the pear tree and, while Mr Blackbird fed them in various gardens, Mrs Blackbird was already in-cubating their second clutch in the honeysuckle behind our garden shed. Three young came out of this nest. A third nest was built in early July, about four metres up in a tree alongside our house. This was already late for blackbirds, particularly as it was such a hot, dry summer. In these conditions, the soil is baked hard and worms dig deep. Blackbirds can't get at them so they tend to complete their breeding after two or three clutches, by the end of June or early in July.

Imagine my surprise, then, when, well into August, I found a brand-new nest with four eggs in it. Mr Blackbird was still feeding the three young from the tree nest as well as an older young one from an earlier nest. So now they were looking after the offspring from three different broods at the hottest and driest time of the year – and the youngest of these were still inside their eggs, at least three weeks from being able to fend for themselves.

Beneath our garden wall, close to the kitchen, there was a patch of bare earth where we had grown spinach and potatoes during the year. Although it was very small – measuring just two metres by half a metre – this patch became a lifeline for the blackbirds. We kept it well watered from the dwindling supply in our garden butt and we kept it stocked with worms that we dug up elsewhere. Mr Blackbird was in there most days, pulling out the worms as they hid in their holes close to the soggy surface. The young blackbirds made good progress and we could see them from an upstairs window as they jostled for space in the nest they were so quickly outgrowing. They would be out and about, we guessed, during the first week of September.

Sunday, 28 August 2005 began with the garden bathed in sunshine. I came down at about half-past nine and stood at the kitchen window with a mug of tea, watching the general activity of birds and insects in the garden. Mr Blackbird bounced onto the lawn and began scouring the dry grass for insects and spiders, making his way to the wall which he would then follow down to the worm-stocked patch of earth close to the kitchen window. To reach the wall, he had to pass behind a buddleia bush. He had done this many times before and I knew I would lose sight of him for up to a minute. I turned away to make another cup of tea.

It must have struck the moment I turned, dropping like a stone from the cherry tree where it had been waiting patiently among a thick covering of leaves. I missed it by a millisecond because I had taken barely a single step away from the window when the sound of a blackbird in desperate trouble rang out. It was nothing like the metallic 'chink-chink' that we hear so often, but a desperate, penetrating scream. I was outside in a flash and there, rocked back on its haunches by the buddleia, was an enormous female sparrowhawk. Her wings and tail were spread wide on the grass and, locked tightly in the raking talons of her left foot, was Mr Blackbird. As I approached, he stopped screaming and began struggling. All I remember is his bright yellow eye-ring and gaping orange beak rushing past as the sparrowhawk nonchalantly flipped herself into the air, out through a maze of overhanging branches and across to a nearby roof. She landed on its sloping tiles and, with Mr Blackbird still struggling in her vice-like foot, began ripping the soft feathers from his body – just like the house sparrow and the collared dove she must have ambushed and plucked beneath the tree earlier in the year.

I stood and watched the grim process, shocked and hoping desperately that a third miraculous escape would come his way. But it was not to be. As his resistance weakened and his life drained, the sparrowhawk adjusted her grip on his body and was gone.

Mrs Blackbird, who must have heard the commotion, flew into the cherry tree and sat quietly. Her mate was dead and now she would have to look after four demanding youngsters on her own. She began feeding them immediately but her task was made more difficult the following day when a cat, not Daisy this time, climbed up to the nest and the four well-feathered young exploded out onto the ground and headed for the nearest cover. Over the next few days, Mrs Blackbird kept on finding and feeding them, but they couldn't fly and their new world was full of danger.

Within a week we heard three dramatic outbursts from both young and adult blackbirds that had 'cat attack' written all over them. She was struggling, in September, to finish the task she had been left to complete on her own.

But there was still hope. After a two-day absence, we saw her fly up into the cherry tree with food in her beak. We waited until she had left before going over to peer up through the latticework of branches. There, towards the top and huddled against a main stem, sat a baby blackbird. It had made it this far, could evidently fly and was probably the only survivor, now receiving all the attention and food Mrs Blackbird could muster. It might even survive long enough to face the challenge of a first winter.

Mr Blackbird's death has left many gaps. Mrs Blackbird has no mate, their territory has no defending male and we have lost a special friend. But the blackbird way of life will take care of the first two and we will be delighted if new pairs raise their young in and around our garden in future years.

We may never again know exactly which bird is which, and the relationship we had with the blackbird with the bent wing feather will never be repeated. But perhaps, given the extremes of pain and pleasure we endured during the four emotionally-charged years of his life, that might not be such a bad thing.

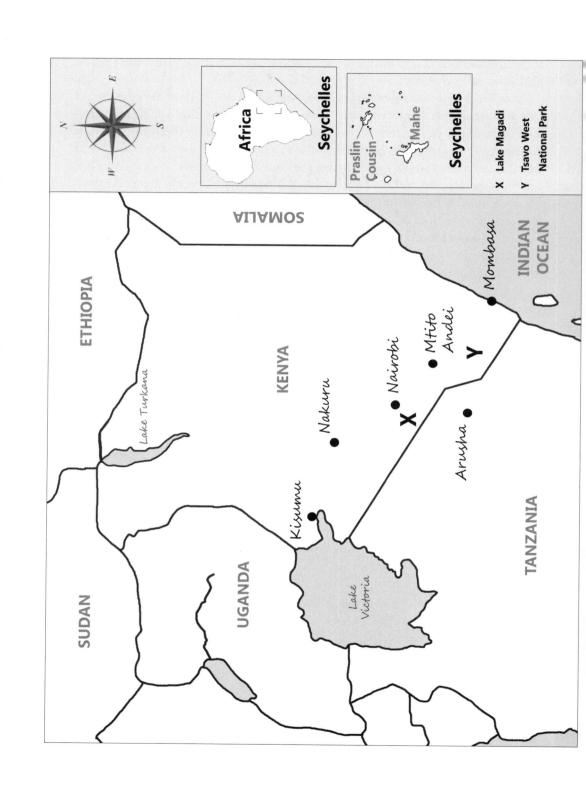

18

Sugar Mike Mike

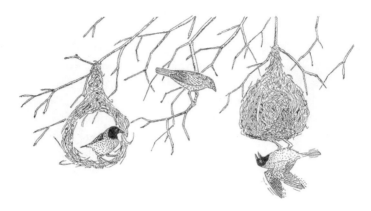

When Cliff and I struggled out of the Point Torment mangroves in June 1968, the pincer of the giant crab firmly attached to his right thumb (see chapter 4), Harry unpacked the radio and sent that 'serious crab bite' message to the Derby hospital. The Kimberley call sign Harry used was 'Sugar Mike Mike' and although that was one of the very few times we used it, the words have stuck in my mind ever since.

Eleven years later, with Cliff now living permanently in Queensland, I joined the Film Unit of the Royal Society for the Protection of Birds (RSPB) in Sandy, Bedfordshire. There I met the two Mikes, whose paired names reminded me of radio days in northern Australia. One, Mike Richards – I will call him Michael to avoid confusion – was a wildlife photographer transferring his skills to the moving image, and the other was Mike Potts, who lived in North Wales and was already a fully-fledged cinematographer.

My first big assignment in the RSPB was to research and write the treatment for a film called *The Masterbuilders*. It was about birds and their nests. I wanted the film to show a progression from virtually no nest on the ground to increasingly-complex nests high up in the trees. To illustrate the point that birds inherited their egg-laying from reptiles, Tony Allen agreed to film baby grass snakes hatching inside a compost heap, and, as a final twist, I thought it would be nice to end the film on a bird that laid a single egg up in the trees without building a nest or using a hole. It would help make the point that our simple-to-complex-with-increasing-height storyline was not exact evolutionary science.

To film the required sequences in England, Austria, Kenya and the Seychelles would mean three things. First, the programme would need to be 50 minutes long, which would be good both for international sales and for the RSPB's commitment to 100 minutes of new film each year. Second, it would mean filming overseas, something the RSPB had not previously done. The third consideration would be cost. It would be quite an expensive film to make but its profile and sales potential would be higher than that of a typical RSPB film about British birds.

I talked all of this through with the head of the Film Unit, John Tomkins. John was ready to fight the film's corner, especially as the title had been his idea. Within a short time it was accepted by Council, the finances were arranged and I was detailed to work with the two Mikes for the coming 18 months. It would take a bit of planning. Both Mikes were committed to other projects. Mike Potts was home-tied in North Wales, filming choughs, but agreed to film the Austrian sequence, while Michael Richards was planning to visit various RSPB reserves to film the secret lives of the birds in our reedbeds. Not the sort of person to flinch from a work overload, Michael committed himself wholeheartedly to both films, even though it would mean, poor chap, a six-week excursion to Africa and the Seychelles towards the end of a very busy year. He would be working flat out from February to August, tearing round the English countryside, putting up hides and watching and waiting while our native birds got to grips with raising their young. It wouldn't be easy. Bitterns do not emerge from the reeds to parade up and down in front of cameras and the birds of field, hedgerow and wood will desert all too easily if disturbed when nest-building early in the season. As the breeding season progresses, their investment in eggs and young increases and they are much more tolerant.

Our commitment, though, was to the vulnerable early stage of nesting. To be successful, the two Mikes would have to approach their subjects with real care. Unless filming in a fixed place like a rookery, which is used by the same birds every year, it is almost impossible to know where a bird will build its nest, even if you do happen to see it sneaking through the brambles with material in its beak. We agreed that we would have to live with this difficulty and just show the nests as early as possible in their construction. It was down to the fieldcraft and patience of both Mikes, particularly Michael, who filmed virtually all the species, that we got as much early-stage footage as we did.

But to imagine Michael sitting quietly in a hide for hours on end, waiting, say, for a lapwing to pick up a piece of grass and pass it along the side of its body to form the crude beginning of a ground nest, is slightly at odds with the Michael we know. His energy is boundless, the sort of person who doesn't do things by halves. His enthusiasm and laughter can be both infectious and infuriating. 'Kestrel overhead!' he would shout while driving through narrow country lanes, twisting his head to watch the hovering falcon in case it made a spectacular dive to the ground. Cars were frequently returned to base with dents and scrapes because he had left the road to get his heavy gear closer to whatever he was filming.

The calmer Mike Potts, with his own special brand of wildlife skills, organised his 1980 chough schedule so that he was free to film in Austria during May. Unfortunately, he would have to drive to Lake Neusiedl on his own, for the simple reason that Becky and I were to be married on 3 May and I would not be free to join him until after our honeymoon. I didn't like the idea of Mike missing our wedding, nor of him having to find and film his bird all on his own. But then we had a brainwave. Becky and I would go to Paris for a few days and then continue our holiday in Austria. We would travel on the Orient Express from France and Mike would meet us in Vienna. I would then be on hand for his final week's filming. It was such a satisfying solution, even though Mike would still miss our wedding, that we completely overlooked the implications of him joining our honeymoon on a mission to film penduline tits!

While Becky and I prepared for our departure to Paris, Michael having taken the photographs at our wedding, Mike was more than 1,200 km away. He had settled into the Hotel Leiner on the northeastern shores of Lake Neusiedl and was able to report the best possible start to his filming. On the morning after his arrival, he had set off to familiarise himself with the reeds and willows surrounding the lake. This, he was assured, was where he would find his bird. Right on cue, a penduline tit flew up into a willow and there, five metres off the ground, was the very first ring of its nest, no more than a day or two old. It was an extraordinary find and, although Mike played it down, I have no doubt that he heard the bird calling its plaintive 'tzee, tzee' and sat patiently until he saw where it was and what it was doing. The luck was not that Mike found the nest but that it was at such an early stage of construction.

Penduline tits are exotic birds, particularly when it comes to nest building. It is the male who creates the basic structure, stripping lengths of stem and binding them like string round the end of a thin dividing branch until that first structural ring has been formed. He then builds out on either side, the rear extending down for 15 – 20 cm to form the egg chamber, and the front narrowing down to form a spouted entrance. It's meticulous work that will take him two or three weeks. Watched quietly by a female – or two, no doubt – so similarly marked that it is often difficult to tell the sexes apart, the day comes when his edifice is ready for inspection.

If her requirements have been met, she reinforces and lines the nest with spider web and soft seed heads collected from nearby plants. A particular favourite is bulrush. When ready for the first of maybe nine eggs, these predator-proof nests are beautifully strong and snug. They have been worn by children further east in Europe as household slippers.

For ten days, Mike tucked himself into the reeds to film the industrious male in the top of the willow. Several times the work of both of them was hampered by strong winds and we had to keep our fingers tightly crossed that the nest would not blow down or be abandoned. On these occasions, when filming was out of the question, we dragged him away for a bit of sightseeing. On the opposite shore of the lake, near the village of Rust, we visited the Haydn Museum and Mike, always happiest when his gear is within arm's reach, was able to film white storks nesting on chimneys in the village. A few days later, when a female penduline tit began her own work on the nest, and the male who built it for her had gone off to build another for a different female, it was time for us to head for home.

Back in England, Michael was also making progress with his own nests. The rooks were done and so, too, were the lapwing, mute swan, bullfinch and wood pigeon. With Michael equally busy on his filming for *The Secret Reeds*, and Mike now back on the choughs in Wales, I settled down to plan the trip to Kenya. I had in mind six weeks from October to December, when there would be enough rain to trigger nest building, but not too much to hamper our work.

Fortunately, I had already spent a year in East Africa, working on birds in the Nairobi Museum and taking people on safari for Don Turner, who ran his own birdwatching company. Don was a good starting point. A further bonus was that Michael's sister-in-law, Helen, was married to Graham Bathe, the warden of Cousin Island in the Seychelles. We could fly there from Kenya to film the fairy tern, the delicate, pure-white seabird that would make the final sequence in our film.

The fairy tern lays and incubates its single egg in a little depression high up on a stump or horizontal branch. Laying just a single egg like this must mean that the chances of it hatching and the chick surviving are good. So why not more than one chick? If predation threats from ground-dwelling skinks and crabs have driven the fairy tern to lay its egg high off the ground, then a single youngster might be all it can manage, particularly if it is not going to build a nest and two eggs would break against each other. Perhaps, though, if the adults have to work hard to catch sufficient fish for just one youngster, then a single egg and no nest, the building of which is energetically expensive, might be the perfect solution.

I remember David Snow, who was head of the Natural History Museum's Bird Room during the last three years I worked there, telling me about the bearded bellbirds he and his wife Barbara had studied in Trinidad. They had wanted to know why the bellbirds raised just one young at a time but, after each

196

visit to the nest by only the female, sat around doing very little before feeding it again. Surely, they thought, two birds could have spent this apparent downtime collecting readily available fruit to feed more than a single youngster? That would at least double their reproductive effort. The reason turned out to be predation. Under intense pressure from monkeys, snakes and other birds, the adult bellbirds could not afford too many visits to the nest in broad daylight. If they went too often, the position of the nest would be revealed and its helpless occupants eaten. A hastily-built nest containing a single offspring fed by just the female was the only way to guarantee the survival of one young at a time. David Snow published some of his and Barbara's observations of South American and Trinidad birds in a wonderful book, *The Web of Adaptation*.

Michael and I flew to Nairobi at the end of October 1980. The overnight flight from London meant that we arrived in time for breakfast, giving us the whole day to confirm the permits that had been provisionally granted in England. Quite remarkably, these formalities were completed by early afternoon, leaving time on our first day to complete our shopping and arrange a hire car. Late in the afternoon, we went to my old haunt, the museum. Mike Clifton was still the resident entomologist and he was delighted to see us. He suggested we went to his club later that evening.

We were not even halfway through our first drink when Mike suddenly apologised for having to leave. But then he was back, bringing with him a mutual friend, Cyril Walker. Cyril was a fossil-hunting palaeontologist and now, with perfect timing, he was passing through Nairobi on his way back to London. Mike, knowing that I would be there on the same day, had kept his secret. The next day was a free day before Cyril's flight home so I suggested, partly to show him a bit more of the countryside and partly to acclimatise Michael, that we drive down to Lake Magadi in the morning.

It is about a 100 km drive southwest of Nairobi to Magadi, the most southerly of Kenya's Rift Valley lakes. Once past Nairobi National Park and the Ngong Hills, we descended into one of the drier and more desolate regions of East Africa. The heat-shimmering lake is fed by hot springs rich in sodium carbonate filtering through volcanic ash. There is no discernible run-off from the lake, so most of its water escapes by evaporation, leaving mineral salts which are exposed during the dry season as a thick white covering, like snow. In some places this Magadi crust is 40 metres deep. Little wonder, then, that there is a flourishing township extracting and exporting soda ash to the outside world. In the wet season, the lake is covered to a depth of about 1 metre in a kind of saline soup, where a single species of cichlid is the only fish able to tolerate the hot alkaline water. Birds visit the lake, and pride of place goes to the lesser flamingos that feed on the algae growing in pink-tinged water. The flamingos do not drink this water – it would kill them if they did. Instead, they suck it into their beaks and pump it out through a thick mesh of hairs which hold back the algae to be swallowed. Some of the alkaline water inevitably passes into the flamingos' guts, so they must make regular journeys to fresh water entry points around the lake for a thorough cleansing, inside and out. A lake full of pink flamingos is a fabulous sight, though their numbers on Kenya's six alkaline Rift Valley lakes can vary from year to year. I don't recall more than a handful of them on Magadi that day.

Michael and I were in Kenya to film black-headed weavers nest-building. They were the pinnacle of artistry in *The Masterbuilders* film, their intricately-woven nests festooning a single tree, perhaps in their hundreds. I particularly remembered a large tree in the compound of Tsavo Inn, about 200 km along the road from Nairobi to Mombasa. Don, who had confirmed in August that the tree was still being used, did some homework while we were visiting Magadi and came up with the frustrating news that the short rains were delayed and there was no weaver activity on my chosen tree. The birds' activity would be triggered by low pressure, the first rainfall and the emergence of green grasses which they could weave into nests. It would be at least a fortnight before we would be able to film at Tsavo Inn.

Time was against us, so rather than wait for rain that might never come, we decided to travel 300 km northwest of Nairobi to the shores of Lake Victoria. At Kisumu, we would be west of the Rift Valley where the likelihood of rain, or at least of persistent wet, was far greater than around the dry, eastern road to Mombasa. This increased the likelihood of our finding weavers building their nests. All we needed, to start off with, was a few males collecting long strips of grass or reed and flying off to their colony. It worked perfectly. I kept in touch with Don by telephone and he finally told me that the weavers at Tsavo Inn were beginning their activity around the tree in the compound. We packed up and headed back to Nairobi.

We climbed the stairs to the flat roof at Tsavo Inn. The first thing we noticed was that the male birds here were different from those in Kisumu. My mistake. Here, east of the Rift Valley, where it was less humid, they were slightly paler and the extent of black and orange around their heads was reduced. In the film we would not be able to cut to one of these males landing on its nest with a strip of grass if the previous shot had been of one of the Kisumu males flying up from the collecting site carrying what, by implication, was the same piece of grass. That, however, was a problem to be shared with editor David Reed at some point in the future. Our immediate concern was to get the finer details of weaving and nest building. In addition to filming the colony from the flat roof with a long lens, Michael also filmed from a hide perched on top of a wooden tower built for us beneath the tree by a local carpenter. With a wide-angle lens, Michael was able to film close to the nests where the industrious males worked and displayed with little concern for his presence.

The approach of a greenish-brown female, however, would send a group of these males into paroxysms of frenzied activity, beating their wings, twisting their bodies from side to side on the underside of their nests and calling for all they were worth. What information could the female, who had yet to choose a particular male to mate with, receive from these antics? The strongest and healthiest males might display more vigorously, but it also seemed that they were declaring their nests to be securely attached, able to withstand the toughest physical onslaught. A nest rejected after inspection by a female was often cut loose by the male, who would then start weaving all over again, paying attention, we hoped, to some detail that had just cost him a vital mating.

If the nests close to the thickest branches were vulnerable to climbing monkeys and snakes, the ones on the thin outer branches would be more open to attack from a bird of prey flying in from the outside. And not just birds of prey. One morning, as I sat watching, a sudden commotion among the weavers suggested that something untoward was happening. Within five minutes, another sudden panic as half-a-dozen bright yellow males careered through the tightly-packed nests, calling excitedly. But what was it? And then a flash of emerald green, followed by another mad dash by the yellow weavers.

This time I saw what the problem was. The flash of green was a didric cuckoo attempting to get its own egg into one of the newly-laid weaver clutches. The nests are tightly-woven, round structures with a short, downward-facing, hooded entrance, making it difficult for snakes to find a way in. The didric cuckoo has to swoop up to its chosen nest from below, lay its egg and get out again under the fiercest opposition from a whole squadron of defending weavers. It may pay the male weavers, who are normally so competitive, to team up like this because none of them knows whose clutch is going to be parasitised and destroyed by the hatchling cuckoo. The adult cuckoos have to be incredibly skilful and quick, launching their assaults while carrying an egg to be laid in a fraction of a second. I couldn't help wondering how many of these eggs ended their lives on the ground below the colony.

A closer look at the tree revealed even more intrigue. Near the top there was a small colony of masked weavers, 12 or 15 nests. Were they cashing in on the abundant black-headed nests, gaining some safety in numbers from predators? Possibly. And then I saw something else. Hanging immediately below the masked weavers was a small group of dark objects that could easily have been overlooked as older, dried-out nests. Through binoculars, though, I could see that these shapes were alive. They were not nests but bats, hanging up after a night's feeding elsewhere. They were quite large, so I assumed they were some sort of fruit bat. I noted their positions and checked them the following morning, hoping they would be in exactly the same places. But they weren't. Perhaps their closeness to the nests they so closely resembled was enough to fool daytime predators, or perhaps they were here for a different reason: above them, in the top of the tree, was a very active bees' nest. The masked weavers and the bats might have been up here because of the added protection of the bees. So now, instead of just one species of bird using the tree, we had three, as well as a mammal and an insect, all of whose lives were linked in some way. No doubt there were others, either hidden from view, currently absent or too small to be noticed.

While Michael completed the filming from his hide, I searched the area within a few miles of Tsavo Inn for different weaver nests. Hanging from a roadside tree was a cluster of red-headed weaver nests with just a single male in attendance; by a little stream, masked weavers had suspended their tight, round nests from the tips of palm fronds overhanging the water; buffalo weavers built ramshackle twig nests in the outstretched limbs of a baobab tree; and the untidy grass nests of white-browed sparrow-weavers were lodged in the branches of small acacias. Some we would film and some we would not.

Our next stop was the Seychelles. We flew southeast from Nairobi out over the Indian Ocean until the isolated group of more than 100 granitic and coral islands appeared below us, like pebbles in a pond. We landed on Mahé. Inside the arrivals building, I handed Michael two letters from my file so he could go to one desk to retrieve his equipment while I went to another to sort out our immigration and suitcases. Five minutes later, we met halfway between the two desks. Michael spoke first.

'They say we need proper import documents for the equipment, not just this letter from the Seychelles High Commission in London saying it will be perfectly all right for us to go filming on Cousin as long as we have written permission from the International Council for Bird Preservation.'

I had been assured by the High Commission that a letter from ICBP, the second letter I had handed Michael, was all we would need. No one had mentioned separate import documents for the gear and, this being my first overseas filming trip, I hadn't thought to pursue the matter. 'And now,' he added a touch accusingly, 'the equipment has been impounded.'

'OK, we can sort that out,' was all I could say as I handed him his stamped passport, 'but there's something else. We haven't got our bags.'

These, I had just been informed, were still in the container that had travelled with us from Nairobi. It was now on its way to Mauritius. Apparently, it was cheaper for East African Airways to unpack the container on its way *back* from 'Dodo Island' in two days' time rather than incur an expensive runway bill unpacking it now. Michael was getting agitated. Penny, his wife, who had been on Cousin Island for a week with their three-month-old son Oliver, had come over to Mahé assuming we would all travel back to Cousin that same afternoon. I suggested that Michael and Penny return on their own and I would follow in two or three days with filming gear and suitcases.

East African Airways, who seemed quite used to the routine, asked me to make a list of the things we would need while we waited for our cases. They took me into the capital, Victoria, and we went shopping. Then I needed somewhere to stay, preferring a family-run bed and breakfast to their expensive hotel. It took me most of the following morning to complete the appropriate forms and secure the release of the equipment. I went back to the East African Airways office to see if they felt a sense of responsibility towards me and my trolleyload of filming gear.

They reacted promptly, flying me to Praslin Island that same afternoon and promising to send our bags on to Cousin as soon as they arrived. By lunchtime the following day, Michael's equipment and I had been ferried across a 2 km stretch of sea to the pure white sands of Cousin, the most beautiful tropical island. Michael was eager to film the fairy terns that hovered over us, chattering in their deep, grunting sort of way. There were also brown noddies, bridled terns and white-tailed tropic birds, all diving and swooping round the island. Inside the mangroves, Seychelles brush warblers flitted through the lower limbs of the trees. Fifty years ago, the small Cousin population was all that was left of this little bird, the main reason for the Royal Society for Nature Conservation purchasing the island. Now, after a successful conservation effort managed by the ICBP, there are between 2,000 and 3,000 of the warblers on at least four of the granitic islands. It's one of the rare success stories.

On my third day on Cousin, I noticed the warden, Graham Bathe, studying the ocean between us and Praslin through his binoculars. He had seen a little boat and couldn't work out why it was heading our way, this not being a scheduled visitor day. When it was close enough, we could see the reason why. Perched in the stern of the boat were two large suitcases, offloaded 30 minutes later with apologies from East African Airways.

Towards the end of my week on Cousin – Michael, Penny and Oliver would be staying on for a bit of a holiday – Penny's sister Helen asked if I would like to go down to the beach with a line and catch our supper. Two hours later I returned with six good-sized fish, but before we could do anything with them we had to comply with a local ritual. The headman of the few people who lived on the island was called in and my catch was laid out before him. He studied the fish carefully, one at a time, then leant forward and pushed the first fish with his index finger. 'This one good.'

The second was also good but the third was definitely bad. And the fifth, but not, to my relief, the fourth and sixth. Helen scooped up the four good fish and led the way to the kitchen. The two bad fish, she informed me, were poisonous and would do us harm if we ate them. The headman would see that they were properly disposed of. She smiled and I understood: there was nothing wrong with those fish – it was a local tax on my catch.

The Masterbuilders, narrated by Jeremy Irons, was an award-winning film. After my two years in the RSPB, I went to the BBC in London to work on Tony Soper's new series *Discovering Birds*.

The two Mikes were also turning their careers towards bigger things. Michael Richards moved Penny and Oliver – Nina and Pollyanne were soon to be added to the family – to Scotland, where he would spend the next two years making a film on the golden eagle for the RSPB. After that he would go freelance and work for the BBC, finally with John Downer for a run of programmes giving unusual viewpoints of animals' lives. The most spectacular of these, for me, was their 2009 series, *Tigers – Spy in the Jungle*, about the survival of tiger cubs in India, some of which was filmed from elephants trained to carry cameras on their tusks.

Mike Potts needed the promise of a lot of freelance work before he would leave the RSPB. I was able to guarantee him a few weeks on *Discovering Birds* early in 1982, and I also introduced him to Partridge Films, who were looking for people to live in Borneo for at least a year to film the ecology and wildlife of mangroves. Mike was one of the camera operators, and the film *Siarau – The Tidal Forest* turned out to be a festival winner.

Working on *Discovering Birds* was similar to working for the RSPB: there was not much money and very little time. In the interests of economy, every day had to be planned to the last detail. For Tony's pieces-to-camera on location, we were allowed just three staff camera crew days per programme and I didn't appreciate until it was too late that, if I wanted the crew to bring any lens on location beyond the standard zoom, I would have to order it in advance and our programme would be charged. But we survived the bed and breakfast rush round England, dodged the weather and had a lot of fun. Tony Soper was a revelation, as nice and professional a person as anyone could ever hope to work with.

We saved money on the series by buying in a quantity of library footage from the RSPB. This gave us the scope to be a little adventurous with our budget, and series producer Ron Bloomfied decided that the final programme could be about a European birdwatching holiday. Thus it was that in the spring of 1982, Ron, production assistant Tina Clarke, Tony, Mike Potts and I drove or flew down through France to the Camargue. We filmed Tony's arrival as a tourist at Montpellier airport and then had him join up with the RSPB's Mike Everett, who was conveniently leading a birdwatching tour while we were there. Bee-eaters nesting in sandbanks and greater flamingos on the salt lakes were the ornithological highlights of our trip.

We were staying in a rather strange hotel in Arles, the Nord Pinus. We came to this hotel by accident because the one opposite, in the Place du Forum, had failed to reserve the rooms we had booked weeks earlier. The two hotels could not have offered more of a contrast. Ours was old, gothic, dark and empty while the Hotel du Forum was modern, bright, noisy and full. Jean-Pierre was our resident manager, a lugubrious, black-suited fellow who wandered around in fear of being summonsed upstairs by the 80-year-old 'Madame' who, we had Jean Pierre's word for it, kept thousands of francs inside her voluminous knickers.

Each evening we assembled in the drawing room of the hotel and Jean-Pierre would solemnly pour us a drink. One evening, the excitement on his face was greater than anything we had seen before.

'We have another guest,' he announced proudly, 'and I have told her she must meet all of you.'

With which he disappeared, to return a few minutes later with a small, darkish-haired woman. She introduced herself as Juliet Stevenson, an actor from London. She was in *A Midsummer Night's Dream* at the Barbican and, during a week's break, had gone to Paris to visit her brother. That not being enough of an escape from city life, she had boarded a train to the south of France, had found her way to Arles and had somehow ended up with us in the Nord Pinus. I don't suppose for one minute that she was pleased to stumble across a BBC film crew but she at least had the consolation of being driven round the Camargue by Mike the following day.

Discovering Birds was a big hit for Tony Soper. Its audiences and ratings were incredibly high. Tony stayed with the BBC in London for a further 18 months to make the follow-up series *Discovering Animals,* while I went to the BBC Natural History Unit in Bristol to begin work on a series about birds of the world. Two years later, Becky and I having by now moved to the West Country, I worked again with Tony on his live *Birdwatch* programmes from the Farne Islands, Bass Rock and Florida.

Mike Potts and his wife Elaine returned from Borneo in 1983. Mike worked for various broadcasters until he finally teamed up with producer Paul Reddish in Bristol. They made a number of films, the highlight with David Attenborough on birds of paradise and bower birds. To film the bower birds, Mike and Elaine spent 10 weeks in Queensland, Australia. The scientific adviser who would help them with the species on Mike's list was none other than Cliff Frith. He and his wife Dawn had spent 20 years becoming world authorities on these two groups of birds, and they were close to publishing definitive works on them.

In June 2009, Cliff emailed me to say that he and Dawn would be coming to England in the summer for six weeks. Wouldn't it be nice if, after far too many years, we could meet up? Nice, yes. Easy, no. Cliff is number six in a ten-child family. When, exactly, did he think we might find time to meet up? The only time, he wrote, would be during their last week, when there might be a spare lunchtime in London. They would be in Cornwall the week before, staying in his nephew's cottage at Wadebridge, but unfortunately they would have friends staying with them: wildlife film-maker Mike Potts and his wife Elaine from North Wales.

I smiled when I read that. Mike and Elaine staying with Cliff and Dawn. Just down the road from Bristol and without the likelihood of a brother, sister, nephew or niece laying last-minute claim to their company. I emailed Cliff to say that I didn't think Mike and Elaine would mind if Becky and I came down while they were staying with him because we knew them very well ourselves. In fact, just to prove the point, I let him into the well-known secret that Mike, already married to Elaine at the time, had joined us on our honeymoon in 1980 and we were all still talking to each other.

And so, early in September, Becky and I set out to drive the 270 km from Bristol to Cornwall. Once there, we spent a lovely day reminiscing and lamenting the passing of both time and friends. We sat at the lunch table for five hours, while outside the sun shone brilliantly. It might as well have been snowing.

Michael Richards is now in his late fifties, while Mike Potts and I have already passed sixty, that final rung on the ladder of youth. Aches and pains are unwelcome visitors. Years of late nights, early mornings and humping heavy gear through forest, swamp and desert, even high up mountains, are beginning to take their toll. Mike Potts has added stills photography to his repertoire and holds exhibitions of his work around the country. I seem to have taken up writing.

Michael Richards pioneers new filming techniques. He continues to work with John Downer on award-winning films that bring animals' lives to our screens from extraordinary viewpoints. His enthusiasm abounds. He is, in his own words, 'still at the top of my tree and finding the digital age...exciting and revealing...There is so much more we need to do now.'

In October 2009, I had an email from Michael just after he had landed in Delhi for a six-week trip into the Himalayas to film some unusual hunting behaviour by golden eagles.

> 'Going to altitude over the next few days so no more emails for now. I am a lit-
> tle nervous but my sherpa has just returned from the summit of Everest, for the
> sixth time! So it should be OK...'

And a month later, from Mike Potts, fresh from photographing blue ducks in New Zealand and about to take to the coast road in Western Australia,

> '[Elaine and I] off up north tomorrow so will be retracing some of your and
> Cliff's footsteps.'

Mike and Elaine were on their way to Exmouth, an invigorating 1200 km drive in its own right, though still no more than halfway from Perth to Point Torment, where Cliff met his giant crab and Harry sent our radio call sign out over the airwaves:

The two Mikes, with Sugar for sweet memories.

Life with Birds
Malcolm Smith

- A compilation of amazing international stories ranging from blood sports to weird bird-eating habits; from religious exploitation to chilling superstition

- Intriguing stories, some bizarre, some told for the first time, about the huge range of connections between people and birds

Life with Birds uncovers the fascinating story of our interdependence with birds. The author weaves an amazing web of inter-relationships, from the Parsi funeral in Mumbai where birds of prey eat the dead; to collecting eider down from nests in Iceland and standing on the once body-strewn battlefield of Agincourt where birds won the day for the English army.

From the earliest days of human existence we have exploited birds; for food, for their feathers, to satisfy our blood lust, to entertain us with their beauty, to inspire our art, our advertising, classical music, popular songs and much more. Cage birds are kept for their beauty and song but this book also investigates the repugnant illegal rare bird trade, and the organised crime it has spawned involving around 1.5 million birds a year. Criminals will go to unbelievable lengths when smuggling rare birds for sale and the 1.5m birds in this annual trade threaten the survival of several exotic species.

Since time immemorial birds have exploited us too. Birds can use our homes to make theirs and can ruin farmers' crops in minutes. Some of the most impressive birds have set up home on high-rise buildings, exploiting the city slicker pigeons that live off discarded fast food and much else.

Life with Birds contains intriguing examples of the huge range of interactions between birds and people. How undercover law enforcement in the US is tackling a cruel and bloodthirsty 'sport'; how birds are being used to smuggle drugs into a prison and across borders; controversial practices such as bird sacrifice in religious ceremonies; and how some Kenyan tribesmen are guided by a bird to find a food they both value. Many myths, magic and religious practices involving birds are exposed such as whether they can predict deadly mining disasters; whether they have killed anyone; and whether the eerie night-time calls in the precipitous mountains of Madeira are the souls of shepherds who have fallen to their deaths.

If you have ever wondered what a nest made solely of bird saliva, considered a delicacy in some countries, tastes like or whether you knew Chairman Mao's 'kill a sparrow' campaign in the 1950s resulted in many millions of Chinese dying of starvation, *Life with Birds* will provide enlightenment as well as a hugely enjoyable read.

ISBN: 978–184995–028–2   240 × 170mm   c.192pp plus 16pp colour section   softback £18.99   February, 2011

from **Whittles Publishing**, Dunbeath, Caithness, Scotland. KW6 6EG, UK
Tel: +44(0)1593-731 333; Fax: +44(0)1593-731 400;
e-mail: info@whittlespublishing.com • www.whittlespublishing.com

The Hen Harrier

Don Scott

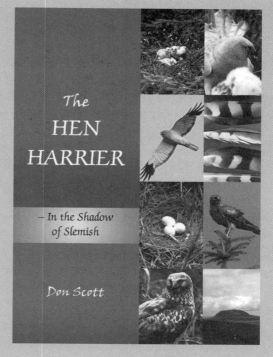

- 'This book is an essential read for all those interested in the hen harrier and the conservation of birds of prey in general, and comes highly recommended'. Professor John Edwards, *Peregrine*, the Hawk and Owl Trust magazine.

- The untold story of the hen harriers' struggle for survival in their former stronghold, the Glens of Antrim, N. Ireland

The hen harrier is one of the iconic species of the bird world and its history is a mix of controversy, persecution, and recent patchy recovery. This dedicated study of the bird in N. Ireland for over two decades reveals previously unrecorded facets of the birds' lifestyle, and provides a detailed account of their life, habits and future prospects. The author presents much new information about the harrier in its continuing struggle to re-establish its hold despite high levels of persecution from man or predation by other species.

Having spent thousands of hours over many years studying these birds, he was rewarded by the discovery that this ground-nesting species was nesting in tall conifers in the forests of County Antrim – the only county throughout their vast European range where this occurs annually.

The author's passion for the bird is obvious as he shares moments of excitement and sadness, and he speaks frankly about the maltreatment and mismanagement of this elegant raptor over the years. This is an unmatched account not to be missed.

A fascinating and detailed study of the Hen Harrier

ISBN 978-1904445-93-7   240 × 170mm   c.192pp + 16pp colour section   softback   November, 2010   £18.99

from **Whittles Publishing**, Dunbeath, Caithness, Scotland. KW6 6EG, UK
Tel: +44(0)1593-731 333; Fax: +44(0)1593-731 400;
e-mail: info@whittlespublishing.com • www.whittlespublishing.com

Kestrels for Company

Gordon Riddle

- A captivating insight into the world of the kestrel through the eyes of an enthusiastic raptor fieldworker
- A comprehensive picture of this delightful falcon is portrayed, based upon almost 40 years' observation in Britain
- Easy-to-read style combining a mixture of facts with entertaining anecdotes and experiences – complemented with a wide range of colour illustrations

Kestrels for Company

GORDON RIDDLE

An appealing book that rightfully raises the profile of the kestrel. It provides an extensive picture of this delightful falcon, including its lifestyle and the factors that affect its breeding success and survival. This is based upon almost 40 years' monitoring of the kestrel in south-west Scotland and further afield by the author and colleagues, giving a flavour of the integrated approach to monitoring and conservation.

As well as the wealth of factual data, there are entertaining anecdotes and stories both from the author's experiences and from the wider media coverage of this raptor over the years. The reader is taken to exotic locations such as the Seychelles, Mauritius and the Cape Verde Islands to see the endemic island kestrels which have always held a great fascination for the author.

Latest figures show an alarming decline of 36% in the kestrel population in the UK, with even more dramatic falls such as 64% in Scotland. The fieldwork techniques which play such an important role are detailed in a composite breeding season. The kestrel is not portrayed in isolation and the bird's current circumstance is tied into the bigger picture of raptor conservation and the struggle against sustained persecution.

The author reflects upon the political, economic and conservation issues that have dominated this field in the past few decades and through this personal and well-informed account the reader gains access to the world of the kestrel.

ISBN 978-184995-029-9   240 × 170mm   softback   c.192pp   full colour throughout with over 150 illustrations   £18.99   February, 2011

from **Whittles Publishing**, Dunbeath, Caithness, Scotland. KW6 6EG, UK
Tel: +44(0)1593-731 333; Fax: +44(0)1593-731 400;
e-mail: info@whittlespublishing.com • www.whittlespublishing.com

Wildcat Haven

Mike Tomkies

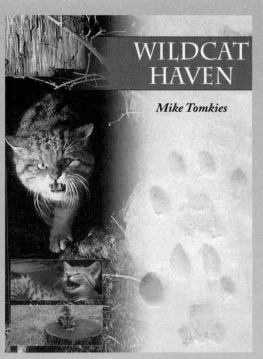

- '...a timely reminder of the need to conserve our only native feline...' **Scottish Field**

- '...His story is illuminating... ...will be deeply appreciated by anyone with an interest in our native wildlife and conservation'. **Habitat**

After abandoning the life of an international journalist for a life in the wilds, Mike Tomkies began a remarkable experiment, rearing the most ferocious animal to roam wild in Britain – the Scottish wildcat.

Now an endangered species, the true wildcat is justly noted for being untameable and is a formidable and fearless opponent of mankind. Mike became the custodian of two spitfire kittens that he named Cleo and Patra. When they were seven months old a spitting and snarling ten-year-old tomcat arrived from the London Zoo to change all their lives. Mike's extraordinary adventures in raising and releasing no fewer than three litters are full of incident, at times hilarious, and deeply moving.

This captivating story by one of the UK's most celebrated nature writers will not only delight Mike's legions of devoted fans but will also bring his work to the attention of a new generation.

About the author: Mike Tomkies is a well-known naturalist, writer and film-maker and has lived for 35 years in remote and wild places in the Scottish Highlands, Canada and Spain. He is an Honorary Fellow of the Royal Zoological Society of Scotland. His other books include *Moobli, Alone in the Wilderness, Between Earth and Paradise, Out of the Wild, Golden Eagle Years, On Wing and Wild Water, My Wicked First Life* and *Rare, Wild and Free*.

ISBN 978-1904445-75-3   240 x 170mm, softback   liberally illustrated,   192pp + 24pp colour section   £18.99

from **Whittles Publishing**, Dunbeath, Caithness, Scotland. KW6 6EG, UK
Tel: +44(0)1593-731 333; Fax: +44(0)1593-731 400;
e-mail: info@whittlespublishing.com • www.whittlespublishing.com

The Storm Leopard

Martyn Murray

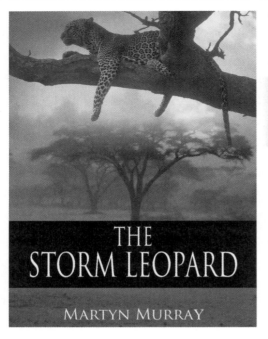

An alchemic blend of travel and nature writing from the Cape of South Africa to the Serengeti Plains

- A journey from the Cape of South Africa to the Serengeti Plains taking the reader through the spectacular, talismanic geography of Africa
- Combines a trek through wild and remote places with a personal investigation into the relationship between people and nature
- A unique book that argues for a new approach to environmental conservation

The Storm Leopard is an alchemic blend of travel and nature writing that explores the primary dilemma of the 21st century the conflict of modern lifestyles with the natural environment.

This is an account of the author's journey from the Cape to the Serengeti Plains and his search for an answer to the Old Timer, a Kenyan who foretold the end of the wild. Martyn decided on one more trip, but this time without an agenda, without a timetable and without preconceptions: with no purpose other than to know, to feel and to understand. The book is filled with insights of African elephants and antelope, and with portraits of a natural world inhabited by Bushmen, game wardens and scientists. Running through it is an outspoken and highly ethical regard for humankind's relationship with nature.

From his first contact with Bushman rock art in the Western Cape, the author is drawn into a spiritual journey as he grapples with the quandary of balancing our lifestyles with protecting the environment. His travelling companion, Stu, a fellow scientist and arch cynic, is nettled by this lack of rationality. Marooned together in their 4×4, the friction, humour and hardship of their journey carry the reader across the continent from one adventure to another, to the final revelation atop an isolated kopje in the heart of the Serengeti Plains.

The Storm Leopard is a unique book that emanates from the author's passionate affair with nature and many years of experience in the field as an ecologist and consultant in conservation – nothing deals with today's environmental issues in the same way.

'Books such as this must be published and reach a wide audience'.
George Schaller, author and world-renowned conservation biologist

'A very-well written meditation on all aspects of travel in Africa'. Peter Matthiessen, American novelist and non-fiction writer, twice winner of the National Book Award

ISBN 978–184995–004–6 240 × 170 mm 272pp, 24 sketches + 16pp colour section softback £19.99

from **Whittles Publishing**, Dunbeath, Caithness, Scotland. KW6 6EG, UK
Tel: +44(0)1593-731 333; Fax: +44(0)1593-731 400;
e-mail: info@whittlespublishing.com • www.whittlespublishing.com

WHAT BECAME OF THE CROW?

THE INSIDE STORY OF THE GREATEST GOLD DISCOVERY IN HISTORY

ROBERT MORIARTY

Also by Robert Moriarty
   Basic Investing in Resource Stocks
   Nobody Knows Anything
   The Art of Peace
   Crap Shoot
   Exposed!
   Entrapped!

First Edition

Editing by Jeremy Irwin, jc9cz@yahoo.com

Library of Congress Cataloging-in-Publication Data has been applied for.

ISBN: 979-8-590-58982-1

DEDICATION

To my beloved wife Barbara, who can never know how much I miss her every day. Life is so boring without her.

A mine is a hole in the ground. The discoverer of it is a natural liar. The hole in the ground and the liar combine and issue shares and trap fools. — The *Detroit Free Press*, 1881

It seems to me that it's up to all of us to try to tell the truth, to say what we know, to say what we don't know, and recognize that we're dealing with people that are perfectly willing to, to lie to the world to attempt to further their case and to the extent people lie of, ultimately they are caught lying and they lose their credibility and one would think it wouldn't take very long for that to happen dealing with people like this. — Donald Rumsfeld, U.S. Secretary of Defense, 2004

The difficult we do immediately; the impossible takes a little longer. — Charles Alexandre de Calonne (1734–1802), Finance Minister to King Louis XVI

As geology is essentially a historical science, the working method of the geologist resembles that of the historian. This makes the personality of the geologist of essential importance in the way he analyzes the past. — Reinout Willem van Bemmelen (1904–83), Dutch geologist

CONTENTS

INTRODUCTION

WITH AN UTTER LACK OF ACCURACY, mining has been described as the art and science of extracting minerals from the ground at a profit. Alas, it's not an art. It's hardly a science. Profit rarely shows its face.

When young students attend school to learn about geology and the mining industry, no one tells them about the idiotic bureaucrats seemingly determined to stop each project at every possible turn. The instructors fail to inform the fledglings about partners who lie, cheat, and steal, or employees who seek at every opportunity to sabotage their company's operations in the hopes of taking it over themselves. Minor issues, such as natives angered at being cheated by every mining venture that ever came along, are not raised. While the locals can't make a company succeed, they can make it fail.

Geology and mining is a minefield where every step you take can blow your leg off. The path forward features booby traps and trip wires, and on each side there are hidden pits filled with sharp punji stakes waiting to pierce the unwary. Competitors will steal your claim markers when they are not stirring up the locals to obstruct everything you do.

Everyone fabricates; everyone deceives and seems dedicated to making sure you fail. Your own technical staff will present their own theories, all of which will disappoint. A geologist without a pet theory that can never succeed is a rare specimen.

When markets go up, everyone wants to throw money at ventures. In an instant those same markets may tumble, and promises of funding turn to dust as demand for the product evaporates.

If geology students had even a lick of common sense they

1

would pursue an honest profession, such as the law or prostitution, where at least someone finds satisfaction.

We are all liars. We tell lies to maintain harmony in our lives. Without the mercy of lies we would be constantly at war with our governments, and even with our significant other. Lying can be a good thing.

Your bride comes home after shopping for a new pair of jeans and asks politely, "Do these pants make my butt look big?"

She's really asking two questions. One has to do with the fit of the jeans; the other is about how much you care for her. There is a correct answer and there is a right answer.

"Darling, the last time I saw an ass that big was on a southbound hippo at the San Diego zoo in 1975." That's an example of a correct answer.

"Darling, the only way your bottom could look better would be if you were wearing nothing at all. Those pants look wonderful on you. Have you lost weight?"

That's the right answer. You may have told a little white lie about the fit of the jeans but you have reassured your wife that you love her.

We want to be lied to. Can you imagine a TV preacher or politician actually telling you the truth or a newsletter writer? They always lie; if they didn't, no one will contribute to their new aircraft fund or vote for them or buy their pulp. We don't want to know the truth. Most newsletter writers specialize in telling people what they want to hear. Their subscribers will never make any money, but the newsletter writers will.

Over the past twenty years I have probably visited about five hundred mining projects. I've seen just about everything. Gold, silver, lead, coal, copper, diamonds, a few oil wells, rare earths, uranium, iron ore. Name a metal or a mineral resource or a country, and I have probably visited a project site and seen it.

Sometimes I've visited the same project twice, with different companies owning it.

I'd guess I get lied to about 75 percent of the time. Many times those briefing me will tell a whopper that even a dunce couldn't miss. The smart ones manage not to mention what you really need to know, if you are to understand whether or not they are sitting on a deal-killer. These are falsehoods by omission. What they fail to mention is what you really need to know.

I like being lied to 75 percent of the time. Twenty-five years ago my wife Barbara and I ran a computer business where we were lied to 100 percent of the time. It's actually quite refreshing for someone to tell me the truth now and again. It's so different.

I read a lot. I'll read anything that comes to hand; I don't much care what the subject is. But naturally, since I am associated with mining, I read as many books as I can about resources, investing, and mining companies. Occasionally I even write a book about those same things.

I've never read a book about mining that told the truth, the whole truth, and nothing but the truth. They all lie, either by omission or commission. I'm old enough now not to really give a shit I can pay my bills, so the purpose of this book is to tell the whole story about mining, leaving nothing out. Well, maybe just a tiny little bit.

This is the mostly true story of the greatest gold discovery in world history. I was a fly on the wall almost from the beginning. The driving force behind it happens to be my best friend.

I'm not sure I'm his best friend, but he's certainly my best friend. I love his wife and his grandkids as well, and doesn't that sort of make us family?

Once upon a time. . .

THE MASSIVE BRE-X GOLD FRAUD [1] peaked in 1997. The price of gold had been fairly steady for years, but gold shares were in the process of completing a long advance. It needed only a pinprick to pop the bubble. Bre-X proved to be more of a whaling harpoon. The fraud ended up affecting the lives of thousands of men and women in the mining industry.

It all began in 1993, when a tiny Canadian company named Bre-X Minerals bought a mining project in Borneo called Busang; it was located on the Busang River. Bre-X did the typical ground survey and basic exploration in advance of drilling, and then started a major drill program in 1995.

At first they claimed to have a gold deposit of two million ounces, which was nice, but not enough to grab the attention of a major gold company. But by the end of 1995 they claimed to have discovered thirty million ounces, making Busang one of the biggest gold projects in history.

They continued to drill, and each hole apparently hit incredible intervals of gold at high grades. By 1996 Bre-X was telling the world that the deposit was really sixty million ounces, and when 1997 rolled around it was up to an incredible seventy million ounces of bonanza grade gold, with every drill hole showing gold.

The sharks moved in for the kill. Placer Dome attempted a takeover but failed. The government of Indonesia announced that the project was far too large for a tiny Canadian junior. Fortunately, a solution was at hand. With the help of President Suharto's daughter, Bre-X could share the project with Barrick Gold. But then it was decided that Freeport-McMoRan Copper &

Gold would run the mine and Bre-X would maintain a 45 percent interest, negotiated by Suharto's son Sigit Hardjojudanto, who would receive a billion-dollar cut for his participation.

It probably should have occurred to someone that in all of recorded mining history, no project had ever gone from two million ounces of gold to seventy million ounces in three years. It also would have been worth having someone look at the gold.

Bre-X's story was that the reported lengths and grades were the result of analyzing diamond drill core extracted from hard rock. But even a cursory examination of the gold particles, with a simple hand lens of the type possessed by every geologist, would have revealed clear signs of it being nothing more than placer gold, mined by local Indonesians and sold to someone at Bre-X.

In short, the company salted the assays. Everyone involved, from the management of Bre-X to Barrick to President Suharto to Placer Dome, forgot to employ their common sense due to being blinded by the idea of instant riches. The reader will find that shortcoming repeated throughout this book; the lure of gold blinds people to reality.

Canadian bureaucrats then came up with an entirely new set of regulations called National Instrument 43-101, designed to set standards on how and when assays and resources would be announced. It was an attempt to put a crimp on the fraud so common in the gold mining business. But wherever there is gold, someone will come up with a new variation on some old tried and proven scam.

Everyone in mining would like to pretend that Bre-X was some sort of outlier that had never happened before and couldn't happen again. But that's hardly true. There are 1,200 to 1,500 junior resource companies in Vancouver and Toronto. In time, ten or twenty of them might come up with a nice economic deposit. The rest are peddling some variety of moose pasture. Management's primary objective most often is to continue to collect paychecks, so they can live the lifestyle they have come to crave.

There are a hundred variations of Bre-X going on at any given time. Bre-X just got carried away, and dragged many others down with them.

Within a year of the fraud being exposed, virtually all gold exploration work worldwide stopped. Dr. Quinton Hennigh, PhD (one of the main characters in this story) was working for Newcrest Mining at the time. The company told its exploration staff that they would be laid off, but they were given severance pay. Quinton left mining in late 1998 and started a new career.

In October of 2001, a young mathematics and science teacher who was new to the job took his sixth-grade class on a school trip, to see an operating gold mine in the foothills of western Colorado. The bus took them to Nederland, where the young sprouts got to visit an authentic gold mine.

It had been agreed that the owner and operator, Tom Hendricks, would give a talk about some subject or other, and then ask the kids questions about what he had just tried to teach them. Those who answered successfully would be given a treat. It might be a sample gold specimen or an old used drill bit or an antique can that had once held blasting caps.

Tom raised a hand and told the kids that whoever could answer his next question would get to light the dynamite. Being twelve years old, the male sixth-graders all thought that would be just wonderful. A few of the young ladies thought it sounded pretty spiffy as well, this being Colorado.

Tom asked the question. A lad named Gavin answered it correctly. So at the end of the tour, Tom said that it was time for

Gavin to light the dynamite.

Tom took an old rubber boot and drew a picture of Osama bin Laden on the side. He went to the explosives locker and picked up a quarter stick of dynamite. Carefully, he crimped an eighteen-inch length of pyrotechnic fuse into a blasting cap. Then he punctured the end of the dynamite stick and inserted the blasting cap. He dropped the now fused and ready explosive into the boot and looked around for his assistant, who was to have the honor of lighting the fuse.

While twelve-year-old boys think lighting dynamite sticks is a great adventure, the first time they try it, they tend to get a bit nervous. Tom reached into his pocket to pull out a Zippo lighter before handing it to Gavin, whose hand was visibly shaking. The young teacher had to steady him so he could start the fuse burning.

Tom held the boot carefully, and as the fuse burned down he continued to lecture the children. It was October, and the weather changed. Snow started to fall. The school was in Longmont, down in the flats east of the Rockies. The kids dressed for fall, not winter. Some of them were now shaking from the cold and wanted nothing more than to get back on the bus and return to school.

Eventually the group dispersed and ran for a warm spot on the bus. When the burning fuse was at the proper length, Hendricks tossed the boot and its explosive over the top of the building behind him.

Twelve-year-old kids have short attention spans. Lighting the fuse was interesting, as was learning how to blow things up. But nothing happened after he threw the boot over the building.

Until it blew up with a massive explosion.

Since 9/11 had been the center of their world for the previous month, the kids were especially sensitive to explosions.

All of them immediately fell to the ground. About half of them proceeded to pee in their pants.

The teacher looked at Hendricks and said, "Tom, those are twelve-year-olds. The dynamite trick might be a little much for them. I think we should wait until I bring out the eighth-grade students. I'm sure they would love the idea of blowing up bin Laden in a boot."

The story of Gavin helping Tom Hendricks to blow up Osama bin Laden in a boot is a tale still often repeated today in Longmont, Colorado, a small town where not much happens. That teacher is still remembered fondly as an inspiration. With the slight exception of the school administration.

Quinton Hennigh was of course that young school teacher. After leaving Newcrest in 1998 he had taken a job at the Twin Peaks Charter School in Longmont. He found himself mostly teaching the children of cattle ranchers. Math and science were hardly a priority, but 4-H [2] was. When Quinton was hired to teach both subjects to grades 6–8, the students were scoring at the 50 percent level in mathematics and a dismal 26 percent in science. Within one year those same students, with their new teacher, rose to 76 percent statewide.

Years later, several of his students went on to attend and graduate from the Colorado School of Mines. One student that Quinton first met and taught when he was in the seventh grade, Jacob Nuechterlein, completed his PhD at age 27. He went on to found one of the world's leading companies designing high-technology materials for 3D printing in the way of advanced metals, composites and ceramics. Quinton and I are both investors in the company.

Quinton likes to tell of the time he blew up the school science lab. As part of his teaching he would get some liquid nitrogen, put a couple of tea spoons of the stuff in a plastic 7-Up bottle and put the cap on loosely. One day he tightened the cap a bit too tightly. Normally the nitrogen turned to gas and made an interesting sight, acting like a gas-powered rocket around the classroom. Alas, on this occasion it blew up, taking part of the classroom roof with it.

I much prefer the story of how he and Tom Hendricks blew up the boot. Anyone can blow up a 7-Up bottle, but it takes a real man to blow up Osama bin Laden. I'm certain the kids who participated still remember that particular adventure.

But while it was interesting for Quinton, his heart was in gold discovery, not the detonation of footwear.

CHAPTER 2
THERE IS NO RUSH LIKE A GOLD RUSH

HOW CAN THERE BE A GIANT GOLD DISCOVERY without an exciting story of just how the gold was found?

What would the California Gold Rush of the nineteenth century be without the exciting tale of James Marshall peering down into the waters of the American River as he built a water-powered sawmill for John Sutter? He spied small flakes of what looked to be gold.

It was gold.

And the rush was on, drawing tens of thousands of would-be prospectors and miners, all believing that the streets of California were surely paved with gold. All you had to do in order to get rich was to pick it up. The California Gold Rush literally settled the west, as fortune hunters from around the world converged on California and the supposed opportunity for rapid wealth.

Bret Harte and Mark Twain left their stories of the California Gold Rush. By reading their tales millions of people still participate in the adventure and heartbreak that accompany every gold rush.

At the dawn of the twentieth century, the world underwent another gold rush as thousands made the difficult and dangerous journey to the wilds of the frozen Yukon in their search for wealth. Many spent as long as eighteen months in their journey. Most, over half, would take one look at Dawson and turn right back around. The trip proved so difficult that the journey itself became the destination.

Jack London wrote a number of exciting books and stories about the people involved, and even their dogs. Robert Service told tales that still entertain today, mixing a few facts, a lot of fiction and a large dose of humor in his descriptions of the Klondike, and what miners went through in their search for gold.

Any form of instant riches will always attract a crowd but gold has a special niche of its own. The first time you see a flake of gold in the bottom of a gold pan you are often hooked for life. And reading about it can be just as mesmerizing.

CHAPTER 3
WHAT BECAME OF THE CROW?

JAMES WITHNELL STARTED THE FIRST Pilbara gold rush by accident in early 1888. All he was trying to do was to protect his lunch from a pesky and persistent crow with a hankering for home cooking.

Mallina Station, on the Sherlock River, has sparse vegetation at the best of times. The boiling summer days of January 1888 had James thinking that the Western Pilbara basin in Western Australia would make a great summer retreat for the Devil, should he ever seek a more temperate clime than that of Hell. The Devil would find the Pilbara in midsummer to be moderate. Mad dogs and Englishmen believe it's hotter than Hell and are not far wrong, with average January temperatures in the Basin as high as 39 degrees Celsius or 102 Fahrenheit.

James wiped sweat off his brow as he and his brother Harding chopped wood for their cooking fire. With one eye he watched a crow hopping about in a dance around his dinner pail; it seemed to be planning to filch his lunch. The crow recognized James and appreciated boiled beef sandwiches.

Raising cattle and sheep in the scrubland of Western Australia, or WA, was proving to be hard work, with little return. It was hot and dusty work. A decent lunch in the field made the task worth doing. But there would be no lunch if James were to turn his back on the crow at the wrong time.

The crow hopped closer to the lunch pail and looked around, as if on a summer vacation on the beach at a crow festival. James saw an iron-stained cobble at his feet and reached down for it, to throw it at the bird. He picked it up, and in an instant forgot all about both the crow and the lunch.

He weighed the rock in his hand; tossed it into the air and caught it. It seemed particularly heavy; far too heavy to be quartz or ironstone. Jimmy fell to his knees and picked up and weighed other rocks, to see if they too were unusually dense. He searched within a ten-foot circle and found half a dozen iron-stained pieces of quartz that were heavier than such rocks usually are.

Could it be gold?

There were rumors of gold having been found in the Pilbara Basin a few years before. And the newspapers were filled with stories of the riches being found in the south of Africa, in the British colony there. If gold nuggets littered the ground about Mallina, he might get rich.

Meanwhile the crow planned on munching Jimmy's lunch as it thought, "Humans are strange creatures, lacking keen appreciation for what is valuable in life. You can find rocks everywhere, but a fine lunch . . . "

James emptied his lunch pail onto the ground and received the crow's gratitude. He piled his newfound stones into the pail, and the two brothers ran home. They quickly hitched their oxen to their wagon and set off for Roebourne,[3] seventy-three miles west on the dusty track, to show their find to their father.

The pair made good time through the heat and were in the tiny town a day later. Roebourne was the nearest town of distinction to the Mallina Station. It possessed scant distinction, but did have the nearest telegraph terminal.

Their parents, John and Emma Withnell, ran a store there and provided cartage services to the other settlers. Their house served as a hub to the two hundred residents of the town, and when necessary provided a venue for church services, until such time as a church could be built.

"Da, Da, I think I've found gold," James exclaimed as he

emptied his pail onto their table.

John Withnell shook his head at the naivety of his son. "Don't be daft," he began. "That's quartz with iron staining." He lifted one of the stones to prove his point. It did seem unusually dense, and there were stringers of something running through the quartz crystal.

"Son, you may just be right. It is substantial. Too stout to be just a simple pebble. We should show this to Lieutenant-Colonel Angelo, and see what he says."

Lt-Col. Edward Angelo [4] had fought with valor in the Crimean War, earning multiple decorations. Later, after retiring from running the Western Defense Force in 1886, he served as the Government Resident at Roebourne. He was said to be a bit stuffy, but would understand what a gold discovery might mean to the Pilbara Basin.

A large gold nugget

The excited trio of Withnell men rushed into his tiny office in the town square. "Colonel Angelo," John began, "I think James and Harding have found some gold. These stones are too heavy to be just quartz. Can you tell us what you think?" James poured his parcel of rocks onto the colonel's otherwise orderly desk.

As the Withnell family had helped settle Roebourne since their arrival in 1863, the colonel always had time for them. The matriarch, Emma Withnell, [5] was the first white female settler in northwest WA. She came from the Hancock family, a later generation of which would start the Pilbara iron mining industry. [6]

Eventually, as a result of the ensuing gold rush, Roebourne would become the largest city between Darwin and Perth, some nine hundred miles to the south, down the coast.

"Tell me the whole story, James. Where and how did you find these rocks? What makes you think they might be gold? They certainly don't look like much," Angelo said as he twirled the mustache that made him resemble a walrus.

"Well, sir, we were cutting wood for the fire about a hundred yards from the cabin. I put my lunch bucket down so I could work. As I chopped, a crow made a move to steal me tucker. I picked up a rock to chuck at it. It seemed very heavy. I looked around and found more rocks that looked about the same. They are all a lot stouter than I'm used to. I thought they might be gold so I came here with them to show Da."

"Hmmm," Colonel Angelo said in the pompous way so common to former military officers, "we must send a telegraph to Governor Broome in Perth and tell them we think you have found gold. O'Malley! Come in here. I want a telegraph sent to Perth at once."

His assistant, O'Malley, came into the office and took a seat, ready to transcribe a telegram to the government seat at Perth.

"Now, O'Malley, it is extremely important to get this off to Perth straight away, so bloody well pay attention," the Resident began. O'Malley had a tendency to wool gather and could be a burden at times.

"Begin. Gold discovered at Mallina. Jim Withnell picked up a stone to throw at a crow." Angelo noticed O'Malley examining his pencil, and stopped speaking.

"Sorry sir, but my pencil needs a topper. Give me a minute." He added, "I have that so far."

He rushed off, believing he had the essence of Colonel Angelo's message. He returned in a few minutes with a sharpened pencil, ready to continue. "Sir, I apologize for the delay, but I have sent that telegraph off to Perth."

Colonel Angelo's face turned crimson as he realized O'Malley believed he had been given the entire message intended for Perth. "You dunderhead," Angelo stammered, "there was more to come." Under his breath he muttered, "Bloody Irish."

At the time the Irish in the colony of Australia were only a few generations old, since arriving on the island as convicts. They disliked the English almost as much as the English despised them. Under his breath O'Malley mumbled, "Bloody pommies." [7]

The Governor of Western Australia, Sir Frederick Broome, in Perth, found himself baffled by the contents of the truncated telegraph.

He quickly responded, "What became of the crow?"

CHAPTER 4
IN THE BEGINNING. . .

WHEN THE GOVERNMENT IN PERTH finally understood that Lt-Col. Angelo was reporting a possible new find of gold, a new Australian gold rush began. Within six weeks, fifty miners had staked claims on the Mallina Station. In some areas the surface portion of the rich quartz vein was twenty feet thick.

Under British law, the mineral rights to a deposit belonged to the Crown. Even if James Withnell had the rights to the station he lived on, he had no more right to any minerals there than anyone else. Within weeks prospectors had chased the gold-rich quartz veins over the entire area, staking claims within the limits rigidly enforced under British law. Every man had an equal opportunity to take a chance on becoming instantly wealthy. Thousands took advantage of the laws governing gold mining.

In 1863, when Emma Withnell and family had settled in Roebourne, the total population of Western Australia was a tiny 12,300. (That's 12,300 people in an area about one-quarter of the size of the U.S.)

In the next twenty-five years, up to the time of the gold discovery in 1888, the population of WA doubled to 25,800. In the next three years it almost doubled again, to 49,790, as men and a few women were drawn to the idea of instant wealth. Naturally, the male/female ratio was skewed. In WA, males outnumbered females by two to one. However, then as now, there were a few female gold diggers around.

More gold rushes in WA followed, at Cue in 1891 and Coolgardie in 1892. The fabulous gold wealth from Kalgoorlie in 1893 drew tens of thousands of adventurers from around the world to Australia, and WA's population had increased to

184,100 by 1901. In just ten years it had almost quadrupled.

The Pilbara discovery was reported in local and national news journals. West Pilbara district drew miners by the thousands, each eager to seek fame and fortune. By August of 1888 the prospectors had searched as far east as the Oakover River, two hundred miles east of Roebourne. They found not only more outcrops of quartz laden with gold, but also numerous gold nuggets in unconsolidated alluvial gravels near water courses.

The gold in the Pilbara region of WA has many similarities to that in the world's largest gold area, the Witwatersrand District of South Africa. You will find hard rock gold in quartz veins such as at Mallina, just as you will find veins of gold in the Wits.

Much of the gold in the Pilbara is located in gravel reefs, some of them consolidated into hard rock, some more loosely packed. Even then, over a hundred and thirty years ago, the miners realized that much of the alluvial gold had its origin in the stacked conglomerate sequences, or reefs, one or two meters thick. In the Wits there are areas with as many as fourteen reefs of varying richness and thickness.

Finally, in both regions there was alluvial gold washed out of the reefs and vein systems and concentrated by both erosion and water.

Clearly, the Pilbara Basin was rich in gold. From the alluvial gravels around Nullagine at the eastern end of the Basin, all the way to the shores of the Indian Ocean, one hundred and forty miles to the northwest, the gravel was rich in coarse nuggets.

But there were problems. Water was always a problem, there being either too little of it or too much. In the monsoon season the entire area flooded. For most of the rest of the year there was a lack of water for processing. With the technology of the day, working alluvial gravels required lots of water to process masses of material. The availability of water proved to be either a feast or a famine.

The technology of 1888 pretty much dictated the type and level of mining. At first, with rich seams of gold at surface and nuggets in every stream and riverbed, a man could become rich within days. There are newspaper reports from December of 1888 about prospectors who found nuggets as large as fifty-eight ounces while using a dry blower to separate the gold from the dross.

The *Western Mail* of March 22, 1890 carried an item about Nullagine: "James Carey has worked out his claim, getting between 300 and 400 ounces of gold. Beaton gold 150 ounces from one load of wash dirt obtained at a depth of 27 feet."

More reports from Nullagine in April of 1890 suggest that several nice patches of gold nuggets had been discovered, with some men taking as much as "20 ounces in a few days, but the majority are only getting a few penny-weights per day." (A pennyweight, or dwt, is about 1.5 grams. There are 20 dwt to a troy ounce.)

In November of 1890 it was reported that at the Turner River, "gold was found at this spot some months ago and it is estimated about three hundred ounces have since been unearthed." Further, "The first few men who arrived hit some very rich patches ... and were getting about 5 ounces per diem each, but that was soon worked out."

While gold rushes are certainly exciting to read about, with tales of instant or near-instant fortunes being made, technology and vision always determine their potential in the end. Northern WA had little in the way of infrastructure and no permanent rivers suitable for navigation. All goods had to be brought in by sailing vessels.

The first Pilbara gold rush made a good story but didn't

produce all that much gold, compared to other areas of the colony of Australia. Official records of Pilbara gold production begin in 1889 and show only 11,170 ounces of gold being produced that year. That increased over the next few years to a maximum of 20,526 ounces in 1899, after which there was a rapid decline.

The primary importance of the Pilbara gold rush was that it acted as a catalyst for prospectors and miners to seek other richer and more amenable areas to mine in the territory. What the tens of thousands of miners wandering the length and breadth of the Pilbara Basin didn't realize was that all the gold there probably had the same origin. It took a PhD candidate from the Colorado School of Mines to realize that the Pilbara Basin almost certainly has a similar endowment of gold to that of the Witwatersrand.

The glory years of the Pilbara and West Pilbara districts were between 1888 and 1900. Production declined continuously after that, although small-scale mining would continue for another hundred and twenty years.

Coins to be used as currency were a big problem in the colony for many years. All of the gold mined in WA would be taken to the offices of the Royal Mint in Perth and exchanged for British sovereigns and half-sovereigns minted in the Sydney and the Melbourne Royal Mints. The Sydney Mint began producing sovereigns in 1855. The Melbourne Mint followed in 1872. It took until 1899 for there to be sufficient gold produced in WA to justify the cost of building a third mint. The Perth Mint opened in June of that year.

The combined production from the Pilbara and West Pilbara districts, from the time of James Withnell's discovery up until 1901, totaled about 190,000 ounces of gold. Compared to the 760,805 ounces from the Murchison district, in the Mid West region of WA, the Pilbara didn't look like much. Mount Magnet, also in the Mid West, yielded 474,100 ounces over the same period. East Coolgardie and the Coolgardie district had produced a whopping 4.5 million ounces of gold by 1901.

Records indicate that only 163 men were working in gold mining in the West Pilbara and Pilbara districts in 1900. In comparison, East Coolgardie counted 5,903 laboring in the gold mines there.

At the time gold was money and provided the liquidity for the economy to function. It's vital to understand just how important gold was to the economy of Australia, and how great the contribution made by Western Australia was, from 1888. The Perth Mint's records show that more than 106 million gold sovereigns were struck between 1899 and 1931, when Britain abandoned the gold standard. That is more than twenty-six million ounces of gold, or nearly $50 billion worth in today's U.S. dollars and at today's gold price. That wealth built the entire infrastructure of the territory and eventually led to the mining of iron becoming a mainstay of the economy of WA.

I'm going to drop in a hint here. The massive iron deposits of WA contributed $4.4 billion in Aussie dollars in royalties to the government in 2018. WA holds an incredible 29 percent of the entire world's iron reserves. It provided 39 percent of the world's iron supply in 2018. The gold came from the same place as the iron, but no one understood that until recently.

The gold was there in the Pilbara, but most of it stayed there for over a century more until a prospector's dream was combined with a geologist's vision and with advances in technology. That formidable combination began to turn a hot and barren landscape covered with spinifex grass and gangly shrubs, with a few brave kangaroos showing themselves at dusk, into — well, let's call it a gold mine.

PRETTY MUCH EVERYTHING IN LIFE, up to and including sex, is easier and more fun with assistance from another person. When you do things by yourself and just for yourself, it's often called masturbation.

The discovery and advancement of the biggest gold project in world history was largely the result of putting two particular people together. Many others took part, including me, but it required the brilliance of (a) the man whom Australians consider to be their greatest prospector, and (b) a somewhat chubby PhD geologist with an interesting theory that he had been gnawing on for years, like a starving dog with a bone.

Seeing them alone, or even together, wouldn't impress anyone. One is tall and skinny, the other not. You could dress both of them for $30 at the nearest Goodwill used clothes shop and make them look sharper and get change back. However, if you talk with either of them for five minutes, about anything, you will soon realize you are in the company of genius. Neither is aloof but neither is keen on fools.

Britain's so-called victory in World War II bankrupted the country. Subsequent mismanagement and poor decision-making by the Labour government in power from July 1945 until its defeat by the Conservatives in 1950 ensured the continuation of the food rationing imposed during the war. Bread rationing began in 1946, after the war had ended! It was 1954 before all rationing ended. Even cheese production took another thirty years to recover to pre-war standards.

The food-rich islands of Australia and New Zealand donated a form of CARE aid packages to Britain in the years after the war. For a young man born in 1944 and growing up in the gray Britain of the 1950s, the Pacific colonies must have looked like a combination of an adventure playground and a food paradise.

Mark Creasy, the young man in question, devoured everything Jack London wrote about the Klondike gold rush. He filled his head with ideas of someday finding a gold mine and becoming instantly wealthy. After graduating in 1964 from the Royal School of Mines, part of Imperial College in London, the twenty-year-old set off for Australia to turn his dream into reality.

At the time, Australia realized it needed immigration, especially of young people with skills and education. The government would pay for their transportation and set them up with housing and a job when necessary.

Mark's first job was in a Queensland coal mine, as a mining engineer. There may be some area of geology and mining more boring than coal, but I have yet to find it.

His interests soon turned to opal mining. For a time he worked at Broken Hill before going into real-life prospecting for opals in New South Wales and Queensland. While opals are exciting for a young mining guy, there are no great fortunes to be made in their discovery. Mark's thoughts kept returning to the tales of Jack London and the riches of the Klondike.

That encouraged him to move in 1972 to Perth, capital of Western Australia, to start a highly successful career as a prospector of gold and then nickel. In an article published by the Australian Prospectors and Miners Hall of Fame, Mark is quoted as saying about Jack London, "It didn't take long to realize that London knew next to nothing about the game."

Mark became one of those overnight successes that are twenty years in the making. "All I know is, I starved for years. Looking for minerals is a fascinating occupation and it gets into you, if you like. It's intellectually stimulating."

His luck began to change in 1976 when, using a metal detector, he found a 46-ounce gold nugget near Mount Magnet. He discovered another monster 86-ounce specimen near Laverton in 1977. The pair of nuggets, along with some of others of smaller size, sold for a $60,000 windfall that would finance his prospecting for the next few years.

Mark believed the Yandal greenstone belt would be prospective for gold. In 1978 he began true prospecting, taking stream samples and doing geochemical surveys to narrow down a potential area. It took many years of hard work and near-starvation wages before he found success in 1991.

Mark tells a tale of how one day he drove his beat-up Land Cruiser from his work in the field into the nearest town for supplies. As he sat in a bar having a sandwich and drinking a beer, he glanced at a newspaper and realized it was his birthday. He had forgotten.

In interviews over the last few years Mark laments how government bureaucracy has changed the nature of mining in WA, and not for the better. When he began fifty years ago, a miner's permit cost fifty cents. When a prospector pegged some potential ground, the exploration rights would be granted in six weeks or less. Those same permits take years to be granted today. That delay tends to kill junior mining companies. Independent prospectors are long gone.

Prospectors prior to the early 1970s largely ignored the Northern Yandal greenstone belt, even though it's but twenty-five miles from the Wiluna gold camp. In 1983 Mark filed claims on what would eventually become the Jundee gold mine. He formed a company with a few others, but they failed to raise the funds necessary to advance the project. They lost the tenements.

Creasy reclaimed the ground in 1985 and went to Chevron

with his idea of forming a joint venture to advance the Jundee–Nimary ground. Chevron took a year to prepare an agreement, and then at the last possible moment decided to pull out of Australia entirely. Mark was left holding the bag, with the required work not completed and a statutory requirement to give up 50 percent of the licensed ground.

Mark had found gold nuggets on the Jundee ground but none on the Nimary property, so he dropped the Nimary portion. The Department of Mines stipulated a ninety-day waiting period before a project could be restaked. A bare week before the ninety days expired, a company called Hunter Resources pegged the Nimary ground from right underneath Mark's feet. Hunter promptly found four gold deposits and built a mine.

During 1989 and 1990 Creasy completed a geophysical survey on the remaining Jundee ground and followed up with eighty-three short rotary air blast (RAB) holes. [8] Only four of them showed a grade higher than one gram of gold per ton of rock. In 1991 he talked Great Central Mines into a joint venture. By 1995 the JV had drilled hundreds of RAB holes and outlined a resource of 1.3 million ounces of gold.

In 1995 Mark Creasy caught the brass ring for the first time. Or should I say the gold ring? Great Central Mines purchased the rest of the Jundee tenements from him, for $117 million. (Mining land properties in Australia are referred to as tenements. It's a block of land, nothing more.)

That payoff rang the bell for everyone from miners to ordinary citizens to the taxman. That most definitely set the fox amongst the chickens. You see, at that time the government of WA understood it needed prospectors to seek and find new deposits. So when a prospector sold a claim, the proceeds were entirely tax-free. Mark collected his check for $117 million and deposited it in his bank account. It was all his to spend.

The Australian public threw a hissy fit. How outrageous that someone might profit by the sweat of his labor and be allowed to keep the money he had earned, instead of handing a big chunk of it to the taxman, so the government bureaucrats could spend it wisely — far more wisely than the person who had earned it. The law was changed, and prospectors could no longer make money without Big Brother taking a giant bite out of it.

The public hated Mark because he had made money and not shared it with them. His fellow prospectors despised him as well, blaming him for causing the change in the law. Sometimes you just can't win. All he had to console himself with was a check with a lot of zeros on it.

Australians will never admit it, but they think Big Brother is just great. They love socialism and hate anyone who makes money through honest work.

There was one little clause in the agreement between Creasy and Great Central that everyone except Mark passed over at the time. It would become important many years later. Mark demanded and got what is called an ROFR: a Right Of First Refusal. If at any time in the future the Jundee mine were to be put up for sale, Mark had the right to match any offer made.

As an asset of Great Central Mines, Jundee later went through a series of owners. But it was the company changing hands, which from a legal point of view isn't the same as the mine changing hands. The ROFR would come into play only if the mine should ever be sold by itself.

Mark now had a lot of money to explore with. I've visited his office in Perth several times. Call it the ultimate man cave. Mark buys all kinds of things related to mining, just because they interest him. He has the greatest collection of gold nuggets in Australia, as well as magnificent specimens of all sorts.

He even has what is the Gutenberg Bible of mining: the copy of *De re metallica* ("On the Nature of Metals"),[9] a sixteenth-century book, that had once belonged to a mining engineer working in the gold mines of Western Australia, named Herbert Hoover.

Hoover and his wife translated that particular book from Latin into English for the first time in 1912 and left many handwritten comments in it. Hoover is remembered in Australia as a mining engineer; in the U.S. he is better known for being its thirty-first president.

Mark's collection of mining memorabilia is easily the biggest and best in Australia, perhaps even in the world. Along with his hundreds of rare and valuable books on mining, someday a museum is going to get a good deal.

In the back of his mind he has always believed that the jewelry box, the big Kahuna of gold as it were, was in the Pilbara Basin. No one had ever made any big discoveries there but Mark believed something giant was hidden there.

Technology and basic metallurgical issues were big problems. Even the first discovery at Mallina Station by James Withnell came loaded with difficulties. A five-stamp mill was brought in to crush the quartz, but the gold was associated with so much antimony that it couldn't be amalgamated with mercury.

Where there was free gold in the form of nuggets — at Egina, Nullagine, and Pilbara Creek — there was either too much water or too little. Mark suspected the Basin was loaded with gold, but it would take someone with a different variation of his vision to figure out how to retrieve it.

CHAPTER 6
IT TAKES TWO TO TANGO

MARK CREASY NEEDED THE RIGHT PARTNER to move the Pilbara forward to where he believed it could go. Da Vinci required someone to block his canvases prior to painting, and to mix paints for him. He might be the master, but even masters need someone to cook their meals. Everything worth doing requires the efforts of multiple people.

Mark continued to prospect, and kept tossing money into prospective juniors. But his crowning glory, of putting together the greatest gold discovery in history, wasn't going to happen until he could enlist the aid of someone with similar vision and intelligence.

That partner wandered into his life in 2005 in the form of a 39-year-old junior geologist, albeit a PhD, working for Newmont Mining. That would be Quinton Hennigh, the other heroic figure in our saga.

When Hollywood gets around to making the movie from this book, the head of casting will encounter a difficulty. The only person in the world even faintly qualified to play the part of Quinton Hennigh, alas, is Quinton Hennigh. Like Mark Creasy, he's unique. Intelligent, visionary, determined, and willing to listen to most, even the idiots. He's a far better man than me. I can't stand idiots and the world is filled with them.

Quinton began his mining career at the age of four, when his grandfather took him into the Colorado foothills, starting west of Denver to search old mining dumps for interesting specimens and maybe even gold. He commuted between his birthplace in Missouri, where his parents lived, and where his father was a university professor, and Longmont, Colorado, where he stayed half the year with his grandparents.

His boyhood was filled with the idea of someday making a giant gold discovery. Maybe even the biggest in world history.

Imagine that.

At the age of eighteen he got his start working on a real gold mine, the Caribou Mine[10] near Nederland, Colorado. Tom Hendricks, the owner of the mine, hired him to scrub the toilets. You might say Quinton got his start in mining at the very bottom. He spent the summers of his high school and college years learning the ropes as he carefully cleaned toilets.

Quinton completed his bachelor's degree, then his master's, and then went to the Colorado School of Mines, the *crème de la crème* of U.S. mining schools, to study for his PhD.

Now, when you undertake a PhD course you should pick an interesting subject for your thesis. Quinton wanted to determine the true origin of the gold in the world's largest gold field.

The largest at that time, that is.

That was the Witwatersrand deposit in South Africa. Goldfields funded his work and he made good progress, but in 1993, due to financial constraints, Goldfields pulled its financing.

One of Hennigh's professors, Richard Hutchinson, found merit in the 27-year-old student and pressured him into working on Hutchinson's theory that black smokers or geothermal vents were the source of gold and pyrite in the sedimentary Wits Basin. Black smokers are underwater hot water pipes spewing out mineral-rich super-hot fluids. The metal-rich fluids resemble smoke coming out of a chimney.

In a moment of serendipity Quinton dissolved samples of the Wits conglomerate in hydrofluoric acid. When he did so, he kept finding particles of carbon. He then came across a paper written in 1978, on the subject of how carbon would trap particles of gold.[11] Not quite. The carbon and carbonaceous material didn't trap the gold particles; it caused the gold to precipitate out of solution. It was Quinton's Eureka moment.

One of the things they teach in geology courses, or which you pick up for yourself on your seventh or eighth site visit, is that gold likes carbon, and will attach itself to carbon under the right conditions. Seawater at the time was reduced, to use a technical term. That means there was no oxygen in it. It was loaded with hydrogen sulfide. Hydrogen sulfide literally eats gold in a reduced environment.

Believing that the gold-bearing solution somehow caused gold to attach itself to carbon, Quinton set out to work backwards. He knew the age of the Wits; it's about 2.6 to 2.9 billion years old. He did some research to determine the chemical composition of water at that time.

Single-cell creatures in the seawater began to produce oxygen and the chemistry and pH of the water changed over time. Since carbon and gold get along so well together, the gold attached itself to the carbon.

I'll backtrack for a moment and give readers the basic chemistry course I was given, I think between site visits 101 and 118.

Hot water deep in the Earth, with various chemical compositions, will dissolve anything. The high-pressure superheated water is always seeking lower pressure, in the same way that winds go from high-pressure systems to low-pressure systems. The fluids generally seek to rise. They find their way to the Earth's surface through fissures in the rock. When the temperature or the chemistry or the pressure of the water changes, minerals will precipitate out.

All minerals are found everywhere. But to have an economic deposit you need a concentration of some mineral.

I mentioned earlier that all geologists have a pet theory. It's like when you were a kid and had a pet hamster. You have your hamster (or your theory) and you come to love it because it's all yours. But you not only love your hamster; you must also pee on everyone else's hamster, or theory. This is for reasons of NIH, or Not Invented Here.

If you came up with a theory, and told everyone about it, and they all loved it, you would inevitably discover that it was a bullshit theory. For a theory to be valid, many people must pee on it.

Quinton found that most people didn't greatly care for his theory, primarily because they hadn't thought of it first. But if you have a good theory, time tends to prove it was right, eventually. You know its time has come when the guys who spent so much time peeing on your hamster now want to steal it in the dead of night, and then claim it was their hamster in the first place.

Those in academia are famous for pinching the ideas they made so much fun of for so long. Naturally there is some of that in our story.

Quinton ended up doing his PhD thesis on ore deposits in Portugal. How interesting. Be still my beating heart.

But he never lost sight of his goal. That was to make the biggest gold discovery in history, and to find it in a basin where the gold had precipitated out of solution. He would also need the right dance partner.

Geologists are a lot like hookers. Eventually they work every corner in town. Quinton learned all that you would ever want to know about scrubbing toilets by working with Tom Hendricks off and on for five years. Then he worked for Homestake for a couple of years, and for the United States Geological Survey (USGS) for just long enough to learn that people take government jobs only because they are too lazy to work and too nervous to steal. He had a short stint with AUR Resources before working for the best of the best, Newcrest, from 1994 to 1998.

The Bre-X scandal in 1997 caught the gold market at a high. As the price tumbled for years, geologists were laid off by the score, Quinton among them. For five years he taught middle school mathematics.

He survived the years spent facing classrooms of teenagers with hormones raging before he signed on at Newmont in 2004. During his period in exile the industry had changed totally. In the 1980s and 1990s the majors did their own exploration work. Gradually, by the early years of this century, the big companies had laid off most of their exploration staff.

Newmont still believed it should be doing basic exploration work. After hearing Quinton's theory about how gold had been precipitated out of salt water 2.8 billion years ago, Newmont hired him. They told him to form a team and go find some paleoplacer deposits. His group spent eighteen months putting together data from all over the planet. They located ninety-two basins of the right age and with indications of gold. The biggest and best was the Pilbara Basin in Western Australia.

It was said of George Washington that he was "First in war, first in peace and first in the hearts of his countrymen." [12] But George married Martha, a widow. Even George Washington wasn't first in everything.

Neither was Quinton with his precipitation theory.

He may have been third or fourth, but writers as early as 1896 were offering the same theory. In a privately published book entitled *Witwatersrand Gold – 100 Years*, printed in a limited edition by the Geological Society of South Africa in 1986, we read this on page 31 (I've put the good bit in bold text).

De Launey in his book 'Les mines d'or du Transvaal' of

1896 also considered a fluviatile origin for the conglomerates. Pretorius (Depositional Environment of Witwatersrand Goldfields) published the following summary of the essential elements of the hypothesis: de Launey wrote that it was possible that the Witwatersrand strata could represent widespread fluviatile sediments deposited on torrential deltas distributed over an alluvial plain such as that which he had observed in Lombardy in Italy however, his preference was for a marine beach origin because of the vast extent of the formations, the frequent occurrence of shingle type pebbles and because the Witwatersrand conglomerates occupied an "intermediate stratigraphic position between two formations known to contain marine fossils — the Bokkeveld beds and the carboniferous limestones.

De Launey agreed with his compatriot Garnier that the gold was secondary and had been precipitated from the ocean waters into the unconsolidated conglomerates.

And on page 32:

Prof August Prister read his paper, 'Notes on the Origin and Formation of the Witwatersrand Auriferous Deposits' to the Geological Society in May 1898. He approved of the generally accepted sedimentary origin of the conglomerates but proposed an entirely new mode of mineralization in these beds. His hypothesis demanded that the Witwatersrand sedimentary epoch be completed in toto, after which the entire supergroup subsided and was flooded by oceanic waters holding gold and pyrite in solution. These waters infiltrated the sediments before final consolidation to precipitate the gold and pyrite in the pore space of the conglomerates. Because the Main Reef group of conglomerates had a well-nigh impervious layer of mud or silt in its footwall, the seepage of this solution ceased at that depth and there precipitated most of its dissolved gold and pyrite. For that reason the Main Reef group of conglomerates were more heavily mineralized with gold and pyrite. The overlying pebble beds received less as the solution was actively passing through on the way down to the Main Reef horizon and the dilutionary effect of rain water reduced precipitation at the shallower depths. Prister stated that the entire process would be repeated as crustal movements brought about renewed uplift and subsidence.

Newmont's exploration manager in Australia, Brian Levet, somehow discovered Quinton at the company office in Denver. They began to talk about the Pilbara and the opportunity there. Brian pointed out that a prospector in Perth held a commanding position in land claims in the Pilbara. His name was Mark Creasy. Quinton suggested to Newmont that he go to Perth and talk to Mark.

There were a few snags.

Actually there were a lot of snags.

By this time Quinton had formalized his theory, passed it around Newmont, and found a lot of support. He just needed the right partner and land package to move forward.

Quinton flew to Perth with Brian. They met with Mark. On behalf of Newmont, Brian signed a confidentiality agreement with Mark that basically forbade Newmont from talking to anyone else about doing deals. It was a dumb move on Brian's part to agree to that because it tied Newmont's hands without getting anything in return. (Quinton knew nothing about the terms of the agreement.)

Mark agreed with Quinton about the potential source of the gold in the Pilbara. He would not only sell all of his Pilbara gold claims to Newmont, but would add a giant land position in South Africa he happened to hold, in another one of Quinton's ninety-two basins of the right age. All he wanted in return was a check for a squillion dollars.

But it would be a cold day in hell before Newmont would write a check for a squillion dollars for kangaroo pasture, no matter where it was located or how well decked-out in gold the kangaroos were.

I mentioned in the beginning of this book that your own people will do dumb things to queer your deal. Brian Levet had signed a confidentiality agreement with Mark, promising that Newmont wouldn't deal with anyone else. Then he went to Wedgetail Exploration (the predecessor of Millennium Minerals, of whom more later) and signed another confidentiality agreement with them in order to discuss doing a deal on the Beatons Creek property.

The Australian mining sector is comprised of a bunch of people who gossip more than a clutch of old ladies with blue hair at a church social. Five minutes after Brian signed the agreement with Wedgetail, Mark knew about it. It annoyed him, to the extent that he was no longer about to do a deal with Newmont.

That first meeting between Mark Creasy and Quinton was in October of 2005, and they got along famously. Quinton traveled up to the Pilbara to take samples, and came back with thirty samples averaging six grams of gold per ton. His fieldwork verified or at least supported his theory. But Mark had first demanded a check that Newmont wasn't about to write, and then didn't want to deal with Newmont at all because in his view, Brian Levet was double-dealing.

Actually, the only dumb thing Brian did was to sign that restrictive agreement with Mark. That was stupid. In mining you have to deal with whoever will do a deal on reasonable terms. Mark was being unreasonable and didn't want Newmont dealing with anyone else.

Meanwhile, Quinton knew he had found the Promised Land. Unfortunately his dance partner wanted nothing to do with his employer. Thereafter, Brian kept fobbing Quinton off with one excuse after another about why the project wasn't moving forward, even though he knew Mark wasn't about to cut an agreement with Newmont. That went on for a couple of years until Quinton accidentally stepped on his dick.

It was the very best thing that he could have done.

Quinton made another trip to WA in April of 2006 to pick up another three hundred samples. This time they averaged four grams of gold per ton. That's economic, and it convinced Quinton of two things. One was that the Pilbara Basin was loaded with gold, and the other was that it was his destiny (that he had been working on since he was four years old) to put it all together.

In May or June of 2006 Quinton happened to run into Pierre Lassonde, the president of Newmont, in the company parking lot in Denver. He gave Pierre the elevator pitch on the project. Pierre asked him who controlled it, and Quinton told him. Pierre is not a big fan of Mark Creasy, and he told Quinton that if he ever ran into difficulty, Pierre would get involved and make the deal happen.

Brian kept stalling Quinton. Eventually Quinton e-mailed

Pierre, asking for his help.

There are two important things to learn here.

One is that people typically will refuse to admit they have screwed up. All business is a series of trial and error. You make a lot of decisions. Some work, some don't. Stick with those that work and chuck the rest. Running a company is nothing more than solving problems one at a time. But solving problems requires first knowing about them, and there's the rub. Because people won't tell you they screwed up. I've run half a dozen projects and said the same thing to every single person who ever worked for me.

I say this to them. "Everyone makes mistakes. It's no big deal. But to solve a problem, I have to know about it. Please, please don't hide shit from me. I hate surprises. If you don't solve a problem promptly it will only get worse. Tell me about every problem you run into and I'll try to fix it."

Of course, everyone I said that to totally ignored me. I could repeat it until the cows come home. It wouldn't make any difference. In the mining business, the people who work for you just love to surprise you with problems they found out about six months ago, and which have grown to giant size in the interim.

Running any business, but I think especially a mining business, means constantly being whacked on the side of the head with a giant problem that was a simple issue months ago when it first appeared, but was covered up and ignored because people just won't admit to fucking something up. They would far rather give you a horrible surprise at the worst possible time.

The second lesson to be learned is that when the big boss gives you his e-mail address with his direct line and tells you to get in touch with him personally if you can't sort something out, he doesn't really mean it. Throw away his business card or you will find yourself in the shit. The boss wants the chain of command firmly adhered to, even if that chain is comprised of blithering idiots who try to hide all their fuck-ups.

Luckily for the discovery of gold in Australia, and for the mining business in general, Quinton needed to prepare to pay for three kids soon to enter college but was dead broke. He had married the lovely Heather in 1989 and they soon started producing infants once they got the hang of it. Quinton had just entered his fifth decade. He had little in the way of savings and frankly wasn't being paid much at Newmont.

Quinton contacting the boss directly annoyed everyone at Newmont. It just wasn't done. Meanwhile, every former taxi driver or drill crew supervisor in Vancouver had now started a junior mining company or two and was raking in the money. You didn't need good sense or any special skill set. You just needed a sufficiently good patter to convince a resource mutual fund to back you and throw some money your way.

I'm not exaggerating when I say that money was being thrown at some of the biggest bullshit mining projects in history. There were areas of Nevada and Newfoundland where there were so many drill holes that it looked like Swiss cheese. You needed to strap a 2 × 4 to your ass in case you tripped and fell in.

Like Daniel before him, Quinton Hennigh saw the writing on the wall and made his exit from Newmont just before the lions got to him. It was a brilliant move on his part to leave in March of 2007. The junior market would surely appreciate his talents far more than the majors did.

But not for a minute did he lose sight of his goal. He knew who he wanted to sign his dance card. Mark Creasy was equally determined to work with someone, but he was free of the financial pressure that Quinton faced. Quinton literally spent the last of his savings to travel to Perth to talk to Mark.

Quinton soon found a job as CEO of a tiny Nevada gold

company named Evolving Gold. He made just enough money to prepare for college expenses for his three children of high school age, and to go pester Mark every six months about doing a deal.

CHAPTER 7
WHERE I FIT IN

I RUN A POPULAR GOLD WEBSITE called 321gold.com. I've done it for almost twenty years now. My beloved wife Barbara was the brain and I was the brawn. Several times a month I would go visit mining projects. She was delighted to have me out of her hair.

Other than having owned and run a couple of small placer projects, I don't have the technical background that many of the other newsletter writers have. That said, I've been to a lot of projects on most of the continents. If you sent a monkey to enough different properties he would develop a pretty good idea of what makes a successful program.

I caught the beginning of the Vietnam War and became a fighter pilot in the Marine Corps at the age of twenty. I spent twenty months in Vietnam, flying over eight hundred missions in fixed wing aircraft and earning a few "I was alive in '65" medals. They made me a captain at twenty-two, and at the time the next youngest captain was twenty-five. I was in the wrong place at the right time. I've written a book about my military experiences with such perspective as I've gained over the last fifty years. You will never find anyone who hates stupid wars more than I do. And they are all stupid.

I did the typical college thing after getting out of the service in 1970 before going to work in computing. Back then computers were bulky, slow, and expensive. My, how things have changed, but the theory is still exactly the same today. By 1974 I had become bored with the pace of writing code, so I found a job delivering small planes all over the world. I made about 240 deliveries, across all of the oceans. I set a dozen international

43

records, won three air races, and even managed to squeeze in a flight under the Eiffel Tower.

One trip in particular is noteworthy. In 1976 I flew a Rockwell 685, previously owned by Wayne Newton, from California to Melbourne. For once I had a passenger: Greg Hayward, the son-in-law of Lang Hancock of iron ore fame. That was a true adventure. We barely made it after a few hair raising emergencies including a engine out landing, a catastrophic oil leak with a full load of fuel on board and ten hours of flying with no battery. Just your average ferry flight.

During the seventy-five hours or so of flight time, as we hopped from one island to another on our way to Australia, I listened to Greg's tales of how rich the Pilbara was in iron, in the form of banded iron formations. Thirty-five years later, that little kernel of information would prove very valuable.

I've seen photographs of Gina Rinehart, the richest person in Australia and the daughter of Lang Hancock, standing next to a painting of her dad. That's my 685 in the background of the painting.

In October of 2008 Barbara got a call asking if I was available to go to Colorado to meet with the president of Evolving Gold, and then to go with him to Wyoming to look at the Rattlesnake alkaline gold project. Since I was just across the border in Nevada, looking at some projects for another company, it was easy to fly to Denver and drive to Longmont and spend the night there.

Barbara's caller from Evolving Gold was Quinton Hennigh. He would pick me up at the motel the next morning and we would be on our way to Wyoming.

He did so. I didn't waste a minute in getting to the point. "I put a bunch of money into Evolving at ninety-five cents, six months ago, on the basis of you guys having the extension to the

Carlin Trend. That's in Nevada. We are now in Colorado, on the way to another project in Wyoming. My shares are now worth fifteen cents. What the fuck happened to the Carlin Trend? I just came from the Carlin Trend and no one wanted to show it to me."

Quinton looked a bit sheepish as he explained. "The financing guys got a little carried away with the Carlin extension story. That's a great place to spend a lot of money and not have much to show for it. We are going to see an alkaline system that I found when I was working at Newmont. We can see a lot more progress there for less money than we could in Nevada."

I was hardly mollified.

The journey from Longmont to Wyoming took about four hours. Over the course of the site visits I have made, I have noticed that on the occasions when I could actually sit and chat with someone for hours, I could learn far more about the project and the character of the company than in any corporate briefing, even those of almost unendurable length. The same was true when I was making a living as a ferry pilot. Most of the time I learned far more in the bar than in a plane.

Mining execs must constantly pitch their company and whatever project is on their agenda. This is because the process of resource exploration consumes money like a school of starving piranha gnawing on a wayward cow.

They might as well record the pitch and play it to you while they go and have a drink. I never learn anything from the programmed pitch because they are telling you want they want you to believe, and are deliberately preventing you from asking questions in order to learn what you really need to know.

Quinton is a little bald and slightly portly; call it pleasantly plump. But then, he lives in Longmont, Colorado. It gets a bit chilly there in wintertime. No doubt wife Heather appreciates a little blubber keeping her warm and snuggly on those bitterly cold nights.

There really isn't all that much you can say about an alkaline system. You have to go see it and do the typical pointing and arm-waving to get any feel for it. So if you happen to be on a four-hour car journey, you have to fill the time with chatter about various other things.

Being in a vehicle alone with Quinton Hennigh for hours was a gift. He is one of those very bright guys who can talk intelligently on a wide variety of subjects. There was his work for Newcrest of Australia, and Newmont in the U.S. After the Bre-X fraud was exposed and the industry shut down for all practical purposes, there was his time as a middle school mathematics teacher. It showed; he was always a great teacher.

We sorted out world peace, and how to fix the financial system. We finally got into a discussion of how the gold got into the Witwatersrand Basin. While working for Newmont in 2005, Quinton wrote a paper about a pet theory of his that he had been working on since his PhD studies.

According to his calculations, 2.8 billion years ago seawater would have had the capacity to absorb gold at a concentration of between four and forty parts per billion (ppb). At a concentration of 10 ppb, one cubic kilometer of seawater could have contained 300,000 ounces of gold. Today, seawater still contains gold, but at a vastly reduced concentration of about 322 ounces per cubic kilometer, or ten parts per trillion.

Where did the rest of the gold go?

He identified a number of basins of similar age around the world. The area with the greatest potential seemed to be the Pilbara Basin, in Australia's WA province.

I was intrigued by the theory. I knew of the two main competing arguments about how the gold got into the Wits, one being that it was some form of modified paleoplacer, and the other that it was hydrothermal in nature. You can take any two geologists in the world, sit a case of beer between them, and they can fight for their preferred candidate until they run out of beer or fall off their stools.

His theory was interesting.

We arrived at Evolving Gold's Rattlesnake deposit. We were just in time to see the drill crew dragging the machine out of the field as snow started to fall. Had we visited one day later we would have been either snowed out or snowed in. The site visit was about as close to useless as any trip I ever made. I saw some dirt and a lot of snow. There was a lot of fog as well. Who knows, there may well have been some gold underneath all that dirt.

Snowfall at the Rattlesnake deposit

So there wasn't much to see. Evolving Gold had made some interesting hits, but as with all big systems, it takes a lot of money to define a large resource.

Quinton's work at Rattlesnake was the success he believed to be possible. The stock went a lot higher as the market recovered from the crash in 2008. But he wanted to follow up on proving the Pilbara gold theory. He just knew it was the next giant gold discovery.

Quinton and I soon became best friends. I had three placer projects, one in British Columbia in Canada and another in Tanzania, and a small operation in Sonora in northern Mexico. Quinton advised me on all of them. In the dozen years since that first meeting we have spent a total of several months together, and have always found something interesting to talk about.

On one occasion, Heather managed to drag him down to Grand Cayman to spend a week with Barbara and me on vacation. I already knew Quinton could find gold, but while swimming thirty meters out from shore he found five one-dollar bills. I can imagine finding one bill, but how on earth do you manage to find five?

He would be even more successful finding gold in the Pilbara if he could only conclude a deal with Mark Creasy. Alas, that's a high hurdle to jump.

CHAPTER 8
PLACER GOLD ON THE SOUTH ISLAND

AFTER THE DUST HAD SETTLED from the first Global Financial Crisis in 2008, I returned to my pattern of making two or three site visits a month. In May of 2009 I got a call from Quinton asking if I could block out three weeks, to make a trip to the South Island of New Zealand for ten days or so, and then to go on to Perth, where he would try once again to make a deal on the Pilbara with Mark Creasy.

It was an unusual trip for me, as until then I had never paid my own way. After all, I was going to write about and bring attention to the companies whose projects I was visiting, so they paid airfares and expenses.

Quinton mentioned that he too would be paying his own airfare, since the purpose of the visits was eventually to do a deal with Mark Creasy. That deal would be with Quinton's own company, not with Evolving Gold. And he needed some help from me in the matter of expenses. I readily agreed to that. Every day spent with Quinton was as good as a semester or two in geology class.

I had delivered a couple of airplanes to New Zealand in the 1970s and 1980s but, other than a short hunting trip to the South Island, I had never visited as a tourist. I suppose that was true of everywhere else, too. While I have traveled the entire world and lived outside the U.S. for years, I never made the time to be a sightseer. It's one of those traps we all fall into. The only time you ever explore where you live is when someone comes to visit and you need to entertain.

New Zealand has always been a favorite of mine. When I first landed there, in the 1970s, it was like visiting a country on a commercial flight and hearing the captain say, "We have just landed in New Zealand. Please set your watches back fifty years." What can be bad about a country with five times more sheep than people?

A few of New Zealand's main inhabitants

Quinton and I flew into Christchurch, on the east coast of the South Island. Someone he knew named John Youngson and his wife or girlfriend, Sue Attwood, picked us up. I was interested in finding a placer project if I could.

I never figured out just what the relationship was between John and Sue, but at the very least they were living together. I liked Sue at once and that never changed. I didn't much care for Youngson and that also never changed. He was a hell of a lot more impressed with himself than I was.

In the southern half of the South Island, a rock type can be found that has tiny quartz veinlets running through it, with particles of gold in the quartz. It is very low grade, but New Zealand is a rain forest with a lot of rain, so the rocks are constantly breaking down and the gold is moving toward every possible body of water. After many travels, visiting different kinds of deposits, I never saw as much gold as I did on the South Island, albeit low grade. But you could go to every river and every stream and pan gold. It was everywhere. It was heaven.

Low-grade quartz stockworks in the rocks

John had a plan for the week. We drove around the entire southern half of the island. He briefed us on all that he knew. Every time we saw water, we stopped and panned. Anyone driving by would have thought we were idiots; grown men on their hands and knees, butts in the air, trying to find a pennyworth of gold. Perhaps we were idiots, but when you catch the gold virus your outlook changes.

For a portion of the trip we were accompanied by Donna Falconer, who had been a master's student financed by Newmont while Quinton was still with the company. She shared his belief that gold precipitated out of salt water in the increasing presence of oxygen. Indeed, many of the attributes of gold precipitation and the bacterial remobilization of gold were (and are) taking place in New Zealand.

Part of the reason for the trip was to prepare me to accept the theory, but I was already halfway aboard. I remembered my hours of discussions about banded iron formations with Greg

Hayward while flying Lang Hancock's 685 to Australia.

John Youngson shoveling under a giant boulder

We stopped at one area that had been a stream until the

water had changed course. There were giant boulders. The water traveled down the river and encountered these big rocks in the way. The water would slow and gold would be trapped under the rock. I took a panful from under it. Sure enough, I found some tiny bits of gold in my pan. But we found the same thing everywhere, including gold trapped in moss.

Bob looking for gold under a boulder

It was easily the most interesting ten days or so I ever spent on a working holiday. Quinton could and did pour information into anyone willing to listen.

With so many sheep in New Zealand, the economy is based around livestock and farming. Gold was discovered in commercial quantities as far back as the 1860s, and attracted tens of thousands of prospective miners. Mining drew people to New Zealand and financed much of its early infrastructure, but it was never as important an industry as it was in Australia, particularly Western Australia.

You can't do much with the land in Australia, while New Zealand has water and a mild climate perfect for sheep rearing. And a long history of great sheep jokes.

Quinton Hennigh in New Zealand

Lord of the Rings was filmed on South Island. As we drove west from Christchurch, we passed through the area where the filming took place. I may be biased, but New Zealand is one of the most beautiful places on earth. And having lots of gold never hurts.

I was well aware of the economic catastrophe on the horizon and wanted my own gold mine. The purpose of our being there was to look for placer projects. I am of the belief that wealth springs from an excess of savings, not from an excess of spending. You cannot spend your way to prosperity. Nor can you borrow your way to success.

Bede Falconer panning for gold on South Island

Our party was comprised Quinton, John Youngson and Sue Attwood, along with Donna Falconer and her son Bede, aged ten. We also had a German stock promoter named Werner Ullman, plus me.

Of all the people on the trip, it was undoubtedly Bede who got the most out of it. What an incredible adventure for a ten-year-old boy. He's probably still telling all his friends about standing in six inches of freezing water in a creek near Queenstown, where he discovered the first tiny flakes of gold in his own gold pan. Every kid should learn how to pan for gold.

Parts of the trip were especially memorable. We stopped at a tiny beach on the west side of the island. The wind was blowing like stink, June being midwinter, and the cold made it through right to the bone.

Under the right conditions of tide and wind, this particular beach would be covered with tiny gold flakes right on the surface. People living in the area would bring snow shovels with a wide blade and scoop up the top layer of sand. They would run that through a sluice box with the right flow of water and collect gold every time it showed up.

I was quite familiar with the concept of gold flowing down rivers and out into the ocean, but less familiar with the concept of the ocean bringing back some of the gold on occasion. That's what also happened on the beach in Nome, Alaska. The gold went out in the river water and came back on the tide. Same thing with this beach in New Zealand.

The few small-scale gold miners in New Zealand are adapted perfectly to the conditions. Their mining equipment is so well matched to the country that even at a time when gold was just under $1,000 U.S. an ounce, they could profitably mine at 0.2 grams of gold per cubic meter. That's right, at $6.50 a cubic meter. (We would need this information later, when trying to determine the profit potential of Egina in Western Australia.)

Tide and winds will cover this beach with gold

The South Island portion of this trip is hardly important to the overall tale, but John Youngson would play a key role in the attempted *coup d'état* nine years later by the Toronto Mafia, when they tried to steamroll Novo. Luckily I had kept my trusty quill sharpened, and intervened at just the right moment.

I did arrange to put in a claim on a project where someone had overspent terribly. They had this giant, floating plant. But they were dumping their tails right into the same pond. Eventually someone realized that all they were doing was processing material they had already processed. As you read on you will find more examples of project planning of similar quality and rigor.

Abandoned gold dredge in New Zealand

There was a wonderful, combined gold mine and tourist attraction for sale. It would have made a decent living for a couple thirty years younger than Barb and me. It needed a bit of investment but could have made an interesting project. The owners could then make money from tourists from both sides of the island, because it was located right in the center of South

Island, and could go pan for gold all day if they wanted a break from working for a living.

A gold tourist attraction

Werner Ullman would end up taking over my mining project a couple of years later and then blowing it sky-high. He never paid what he owed to either Youngson or me.

Such is how the mining industry really works.

CHAPTER 9
SWILLING $1,200-A-BOTTLE WINE

AUSTRALIA IS ONE OF THOSE COUNTRIES where the government never runs out of projects to dump money into. Consequently it is constantly trying to raise more money via taxes, even though the country is one of the most important mining jurisdictions in the world. The government takes a bite out of every ton of iron ore, every gram of gold, and every trainload of coal. And still doesn't have enough.

Australians should be the richest people in the world, even richer than the Swiss, but they are not. It all goes for taxes, including some really stupid taxes.

Like the tax on wine.

In Australia they tax wine every year — the same wine. If the tax on a bottle of red wine is X for one year, and if the vineyard ages it for five years to improve quality, then the tax due is 5X.

So, think of Australian wine as similar to bathtub gin. The winemakers pick the mature grapes, toss them in a tub, stomp on them, pour that into a bottle, slap a label on and sell it immediately. It is loaded with sulfites and every bottle comes with a money-back hangover guarantee. No winemakers age wine in Australia. They make it and flog it to the nearest chump. Because of all of the sulfites, it also makes a great "green" weed killer.

Mark Creasy has a fix for this. The first time I saw him use it, I was convinced he had lost his mind.

Quinton and I finished our New Zealand trip and headed for Perth to meet with Mark. Quinton had arranged for a few people to join us there, mostly financial people. We had Craig Roberts, a mining engineer who worked closely with Cal Evert in

Vancouver, along with David Eaton, chairman of the Baron Group out of Vancouver. If Quinton was to form a company to develop the vast gold resources of the Pilbara, he not only needed to cut a deal with Mark Creasy, he required financing. Quinton had worked with the Baron Group since becoming president of Evolving Gold.

The entire mob went out for dinner with Mark, his chief geologist George Merhi and his consigliere, Steve Lowe, at the Old Brewery on Mounts Bay. Mark brought along a bag containing various bottles of wine.

Mark handed over the bag to the wine steward, who evidently knew him. A few minutes later we all had a fresh glass of red wine in front of us. Mark pulled out a small spray bottle from his pocket and spritzed his wine with a couple of shots. I had never seen anyone do that before.

Most of the wine I ever drank with friends came with a screw-off top and cost $3.99 a gallon. We also believed Ripple to be a deluxe wine, if you bought the right year and were satisfied with Night Train Express for everyday swilling.

I leaned over to Mark so he could hear me speak and asked, "What the fuck was that?"

He looked at me for a long moment, as if I had just fallen off a truck filled with green bananas. Then he said, "I sprayed my wine."

"I know that, I could see you. But what did you spray it with?"

Now he knew I was daft. "Hydrogen peroxide. It offsets the sulfites used as preservative."

I thought about that for a minute. Hydrogen peroxide is water (H_2O) plus an extra oxygen atom per molecule: H_2O_2. He was oxidizing the sulfites in the wine in the same way that light and water oxidize sulfide minerals.

I wasn't real sure just how much sense that made. I figured out the next morning at zero dark early, when I awoke with a splitting headache from the wine. I just had to get a little spray bottle and some hydrogen peroxide.

But we had to be on our way to the Pilbara. Soon we were packed and driving to the airport.

George Merhi worked his magic when he learned that we planned to spend a few days inspecting various sites in the Pilbara. He and his two assistants, Rob and Morgan, made a run to a sporting goods store to buy tents and the supplies we would need. He also raided Mark's wine cellar for a case or two. He sent the lads on their way to Newman to await our arrival.

Main Street, Nullagine

We would spend almost a week out in the field. We flew first to the airport at Newman, picked up our vehicle and drove to our first stop at Nullagine, to look at the Beatons Creek area.

Anyone who doesn't understand the importance of mining to the economy of Western Australia needs only to fly into or out of Perth at any hour of the day. The airport and terminal will be jammed with hundreds of mining people on their way to a two-week work period or returning from a two-week work period.

George Merhi, chief geologist for Mark Creasy

George Merhi knew Mark's projects in the Pilbara better than anyone. He had worked for Mark for years, and did most of the exploration and drilling. Rarely would you find him without a big smile on his face and hair going into all directions. I suspect he may have had a run-in with a light socket, and lost.

Quinton knew George and the Beatons Creek area, having visited a few times. But for Craig Roberts and me it was brand new. Here he is, pointing out where the gold would be found.

Quinton shows us the gold conglomerates

We hadn't been at Beatons Creek very long, looking at all the reefs of conglomerate, when Quinton reached into some old workings from over a hundred years ago and handed me a piece of carbon. At that moment I became a believer in the story. Nothing since then has dashed my belief that Quinton had found exactly what he was looking for.

Bob is handed a piece of carbon

While carbon has an affinity for gold, it's unusual to find plain carbon or graphite with a gold deposit. A key to understanding Quinton's theory of gold precipitation from salt water is the matter of where you find it. In the Witwatersrand, one of the fourteen reefs featured a thin sheet of carbon with tiny particles of gold within. Some geologists believe the carbon seam captured the gold in the same way that moss did on the riverbanks in New Zealand that Bede had panned. But it was the chemical attraction of gold to carbon that put the gold in the carbon leader in South Africa, not friction from the carbon catching the gold. It's a subtle but key difference.

Of course, Craig Roberts and I were like kids in a candy store. There was gold everywhere. You could see where miners a century before had taken hand tools and carved out a tiny cave to dig out a few ounces of gold.

We were in Nullagine in the late days of fall. It was still hot. In summer the area would be hell on earth. A nice cool cave to work in might be just the right thing.

Craig Roberts looking at the conglomerate layer

Everywhere we found a dried-up waterway, we would take a sample and pan it. I won't say it was the richest ground I had ever seen, but you could find gold just about everywhere. That is exactly what Quinton believed would happen when he pitched his theory to Newmont four years before.

Bob preparing to pan scrapings

In hindsight, it's interesting how things worked out and what I saw from the very beginning. We were standing on top of a small hill at Beatons Creek. Quinton was doing the typical geologist's pointing and arm-waving. He pointed at some reefs on the side of a hill perhaps half a mile from us. "Do you see those layers of conglomerate outcropping?" he asked. "Each of them has the potential for gold."

I replied, "That's all very interesting, but a lot of the reefs have eroded away. Where did all that gold go?"

Viewing the layers of conglomerate

Quinton laughed. "Much of it was in the creek beds, and where the streams run during the rainy season. That's where the early miners back in the 1890s made their fortunes. The rest of it all went into the Indian Ocean."

For some strange reason, that comment about the gold all going into the Indian Ocean struck me as important. I wouldn't forget it. Later in our saga, readers will understand why it turned out to be so vital.

Not all of the gold made it all the way to the ocean. Some of it dropped out en route, at Egina and the flats to the west, covered with loose gravel.

We spent about five days in the field. All day we would be out in the field on quads, running around banging on rocks and panning for gold. At night we would return to our campground to cook dinner out of cans. And drink wine at $1,200 a bottle.

Bob goes panning

No kidding.

The first night, George and Quinton discussed the wine they should drink. Should they tipple the $750 wine or the $1,200? I just looked back and forth between them, feeling very stupid.

Clearly it was one of those male inside jokes that you had to be part of the group to understand. They understood it. They were insiders. Craig and I being on the outside, plainly didn't get it. Neither of us wanted to look silly by asking what they were talking about. So we just glanced back and forth between George and Quinton.

That first night they compromised, and we sipped the $900-a-bottle wine. I couldn't tell the difference between that and Ripple. I didn't feel like advertising my ignorance so I pretended I was also an insider and understood it.

We could cover a lot of ground on quads

Next evening my curiosity got the best of me when they were comparing the bouquet of the $1,200 wine to that of the cheap stuff, the $750 wine.

"OK, guys. You have to let me into the club. What the hell are you talking about, twelve-hundred-dollar wine? It couldn't possibly cost that much."

George pointed out the obvious: "That's in Aussie dollars, so it's a lot cheaper than you think. It doesn't cost twelve hundred U.S."

I suppose he thought he was making things easier to understand but it was hardly clear to me.

I responded, "We are in the middle of the Australian outback sleeping in tents in a campground after pounding on rocks all day. What the fuck are we doing, drinking twelve-hundred-dollar wine? Or nine hundred dollars, or seven-fifty, or wine at five bucks a gallon. Have we lost our minds?"

George smiled and said, "Mark gave it to us. Anytime we go out into the field we take a case or two of wine. Some of it is pretty expensive."

That was what you might want to term an understatement.

"Mark seemed *compos mentis* when we were out to dinner. I don't care how rich he is, he won't be rich long, handing out cases of wine at those prices," I said.

George explained, "Mark likes wine a lot. He will come across a vintage he favors, and buy cases of it. Eventually he tires of that stuff so he puts it in his warehouse, and we take it with us on our trips to the field. It's a treat for us."

"I wish someone had said something to me in Perth," I began. "I would have been happy to take the cash instead. I could have handed my camera to you. You could have brought me the pictures back in Perth. I'd love to take the money and run."

Frankly I couldn't taste the difference between the $750 plonk and Night Train Express, except Night Train Express doesn't leave you with the same hangover.

Tent City in the Pilbara

The gold Quinton and George showed us was in a thin layer or reef of conglomerate material. You could look at the surrounding hills and see where the conglomerate reefs came to surface. Quinton was in his element. George had had far more time on the ground but he didn't have the benefit of Quinton's theory.

Even though Quinton had been in the field several times since his first visit in 2005, while with Newmont, he seemed to know the deposit so well that you would have supposed it had taken him years of exploration to get to that point. But he didn't need it. He understood it better than anyone, right back to 1888. But even Quinton would be surprised a few times in the future.

Left to right: Quinton Hennigh, George Merhi, and Bob

The richest gold in conglomerate

75

We saw the project at Beatons Creek, which had oxide gold of relatively coarse size. Quinton wanted that project. Millennium Minerals was in the process of spending $100 million to construct a mill in Nullagine and would soon be in production, but their material was mostly sulfide and across a fault from Beatons Creek. Quinton saw a lot of potential at Beatons which Millennium failed to see. Over the five days we covered a lot of territory, including a visit to the Millennium mill.

The Millennium Minerals mill at Nullagine

On the last leg of our tour we visited Marble Bar, reputed to be the hottest town in Australia. The original settlers back in 1891 believed the rock formation was marble. Actually it was jasper, but beautiful nevertheless. There were several sites around Marble Bar that Mark Creasy had controlled for years.

Quinton wanted them too.

He dressed up in his finest outfit to take us to see the Marble Bar. The primary thing that I got from this visit, and everywhere else we went to, was the enormous scale of the gold area. It could only be explained by something as simple as Quinton's theory of gold deposition, but everyone else had missed it. There was gold everywhere we went. And it was all connected.

Quinton Hennigh at Marble Bar

On our last night in the field we were staying in a campground near Marble Bar, sipping on a nice mild $750 wine, when a bus pulled in. A seeming herd of young kids flowed off the bus. They looked to be thirteen or fourteen years old, right at that dreadful time when their hormones kick in and at the same time they discover their parents are stupid. Several hundred of them came down the stairs, each clutching a cellphone.

We all laughed at the thought of them, determined to spend

as much time as they could chatting on their phones with their BFFs, who just happened to be sitting three rows behind them. Where did they imagine they would recharge their phones?

Even in 2009, kids needed a phone to travel

Clearly the tour director was wise to the ways of teenagers. The kids got out, set up their tents, and by the time they completed their work the matron in charge had set up a table and a connection to a generator on the bus. I guess they had run into this issue before, and maybe most adults were just as dumb as their parents.

We made our early flight back to Perth from Newman and drove into town to sit down with Mark, to chat about our visit to the Pilbara and possible terms for a deal.

Mark Creasy is one of those people hard to define. He would probably be a great poker player as he keeps his cards close to his vest. He's interesting, intelligent, and rarely misses a trick. Quinton, Craig Roberts and I sat down to meet with Mark and his accountant, Steve, in Mark's office.

I will give him full credit for knowing how to build the ultimate man cave. For twenty years I delivered small aircraft all over the world. In the U.S. at that time it was possible to buy a nice, fairly new light plane for about the cost of a new car. In the rest of the world, with their incredible taxes and duties, for all practical purposes all of the people to whom I delivered aircraft were rich.

Mark is rich, but he spends his money on what is probably the world's finest man cave. We live in strange times, where young people no longer know what sex they are. And it seems there are twenty or more types to choose from. I didn't know that. When my kids were born, we just looked down and knew what we had on our hands right away.

I've never heard of a woman cave. Owning and outfitting a cave seems to be a man thing, not a woman thing or a thing pertaining to any of the other variations of sexual orientation. Mark does a man cave right. He had a large round table in the middle of the room for us to sit around, so everyone could speak and see the others. In one corner of the room was a giant stainless steel tank of some sort. The rest of us, all being male, naturally had to know the story of the strange tank.

It seems that in July of 1979 the first U.S. space station, Skylab,[13] was about to tumble to Earth. The boffins who controlled the vehicle attempted a course change that they thought would bring it down in the Indian Ocean. Alas, parts of it landed in Western Australia. Mark determined the approximate path of the ship and drove out to the desolate area where he supposed it would impact Mother Earth. He located the stainless steel tank and a few smaller pieces of Skylab.

I said that he rarely misses a trick. While he was searching the Fraser Range for those pieces of the ship, he also found indications of nickel.

When he returned to Perth with his new and old treasures he immediately got on the phone to NASA, and asked them how much they would pay for a couple of slightly used oxygen tanks, previously owned by a little old lady and only driven on Sunday. NASA laughed and said he could keep them, with their blessing.

I have cast-iron proof that not only is there a God but also that she has a sense of humor, I offer his subsequent research into the nickel in the Fraser Range. He came across documents showing a joint venture between Newmont, WMC and Anglo American from twenty years before. The JV found intercepts of copper and nickel. At the time it wasn't economic because nickel was still in the discount sales bin, but it later increased in price.

Mark staked a large nickel deposit that became the basis for Sirius Resources' Nova-Bollinger mine. He later sold his 30 percent stake in Sirius for hundreds of millions of dollars. As if he really needed it.

That wasn't quite good enough for him, as he was out the

cost of the gas to search for the Skylab bits, so he did a deal with Legend Mining on another nickel prospect in the Fraser Range [14] and kept 28 percent of the shares. That stake is worth $75 million today.

Mark always wants to be sure of the names of the people he is dealing with. He asked Craig Roberts again what his name was so he could firmly implant it in his mind. Craig told him. Mark looked as if he didn't understand and asked once more.

And pondered again after a pause before saying, "Oh, Craaayyyghhh" with a broad Australian pronunciation. Then they both laughed. While Craig had grown up in Australia, he had lived overseas for many years and had literally forgotten how to pronounce his own name in Australian.

The meeting was mostly a rehash of what Quinton knew from his previous trips, plus some thoughts from Mark. In private, Quinton and Mark would discuss terms, and what Quinton wanted in the way of ground. Mark still wanted someone to write a check with his name and a lot of zeros on it, but in June of 2009 his holdings in the Pilbara were best suited for raising kangaroos.

The Pilbara offered sniffs of gold. Mark had drilled a few holes showing gold, but in 2009 it wasn't a deposit with a known or even a predictable resource. No one was about to write a big check without doing millions or tens of millions of dollars' worth of exploration.

At the time, Quinton knew that he needed Mark but Mark hadn't yet realized just how much he needed Quinton. No man is an island. No kangaroo pasture is a mine. Mark had had it way too easy in the past. But he was under no financial pressure, so Quinton was going to have to drag him kicking and screaming into a deal.

As we concluded the meeting, someone — probably Steve

Lowe, the number cruncher — asked if anyone had any other questions or comments before we concluded. I hadn't contributed much to the discussion prior to this, so I spoke up.

"Well, you have most of what you need to make a giant gold discovery," I began. "You are missing only one key element."

I nodded to Mark. "He's got the ground."

And then towards Quinton. "He's got the interesting theory."

On to Craig: "And Craig will help find the money to move it forward. And you, Steve, are in charge of all the accounting paperwork bullshit."

By now I had their full attention. What on earth was I talking about? What could possibly be missing?

I spoke sagely, as if I knew very well indeed what I was talking about. "You can't have a giant gold discovery without someone to tell the story. If you don't tell the story, there is no story."

Everyone in the room looked at me as if I had lost my mind.

"I'm not doing anything else right now. I'll tell the story, unless we can find someone better." (I didn't mean that last part, of course.)

I got some strange looks. It was something no one else in the room had considered up until that point. They all pondered it and sort of mumbled, "OK."

I still think it needs someone good to tell the tale. How can you have a gold rush without a Mark Twain or Bret Harte or Jack London? I wonder who could write it? We need someone really good.

CHAPTER 10
GALLIARD AND THE EARLY YEARS

QUINTON HENNIGH'S GOAL was to turn the Pilbara into the gold producer that he was convinced it could be. There was this one obstacle. Mark Creasy controlled much of the Basin, and dealing with him was a slow business.

The $117 million Mark collected from Great Central Mines made him rich. When you have that much of it, money tends to lose its ability to motivate. But he never ceased being a prospector at heart. He had, and has, pieces of dozens of Aussie juniors, including several that have multiplied his fortune. He certainly wasn't a one-trick pony. I don't think he has made it to billionaire status yet but he'll get there one day soon.

So Quinton was holding a burning candle, but neither time nor money moved Mark. Quinton presented his case and made an offer in June of 2009. Time passed and the candle became shorter.

Quinton formed Galliard Resources in October of 2009 from a shell controlled by Robert Bick. They raised $400,000 at twenty cents a share to keep the door open and the lights turned on, in anticipation of doing some sort of deal with Creasy. Quinton was eager and ready to follow his dream.

One part of Mark Creasy's deal-making technique is to leave the other party hanging. Quinton had no doubt about the potential of the Pilbara. He had lined up financing, taken a shell company public, and was ready to do the mundane day-to-day exploration necessary to turn scrub ground into a gold mine. Mark wasn't in such a hurry. He knew it cost $1 million a year just to file the financials, keep the stock exchange happy, meet payroll and pay the light bill. But it wasn't costing him $1 million a year so he could take his sweet time in doing a deal, knowing that the other party was under pressure.

Galliard did a second, more serious financing in January of 2011 to raise $1.6 million at twenty-five cents a share. I didn't participate in the first placement but bought as many shares as I could in the second: just under 10 percent of them. The reporting requirements and rules for selling blocks of shares change at that level, so unless you plan on being in the shares for a long time, most investors will chose to stay under 10 percent.

By accident, Quinton then trumped Mark. Quinton had all his ducks lined up in a row before we went to Perth and the Pilbara in June of 2009. He wanted to get moving. His goal wasn't just to make a giant gold discovery; he had higher aims. But first he wanted to find an economic deposit, put it into production, and expand it until he was running the largest and most profitable gold company on earth. But he had to do a deal on that kangaroo pasture before he could do anything else, and Mark proved hard to budge.

So Quinton began talks with Millennium Minerals, which was in the process of building a mill and had dozens of satellite gold deposits near Nullagine. We had been banging rocks at Beatons Creek, which Millennium also controlled. They saw no potential in the project, but Quinton did. He was on the verge of completing a deal with Millennium but also wanted the adjoining ground owned by the Creasy Group. When Mark realized that Quinton was about to do a deal with Millennium on Beatons Creek, he came back to the table and became serious about doing a deal with Galliard.

Galliard and the Creasy Group signed a non-binding memorandum of understanding in February of 2011. It would give Mark Creasy 6.6 million or 43 percent of the shares in Galliard in exchange for 70 percent of his land position at Marble

Bar and Nullagine. It was essentially the deal that Quinton would have signed in June of 2009. Mark had dragged his heels for eighteen months in the hopes of an even better deal for himself.

But Mark wasn't through dragging his heels just yet.

In April of 2011 Galliard completed the deal with Millennium Minerals on Beatons Creek, agreeing to fork over $500,000 worth of shares and to spend $1 million on exploration to earn a 70 percent interest.

All of this time Quinton was the brains and most of the brawn behind Galliard, but he was also still president and CEO of Evolving Gold. From the dismal days of October 2008 at the bottom of the market, when Evolving shares dropped to a low of fifteen cents, Quinton worked his magic at Rattlesnake and increased the value of the shares by 1,100 percent over the next year. He turned Evolving Gold over to William Gee in March of 2011 and left entirely that November.

Quinton is a great consulting geologist but not quite as good at picking his own replacement. Gee took over a good company making solid progress, which had done a joint venture with Agnico Eagle on Rattlesnake in mid-2011. But then Gee got sick, and a fellow named R. Bruce Duncan took over.

Now it may be terribly unfair and indeed unkind of me to point this out, but every person I have ever met, going right back to H. Ross Perot, who used an initial for his first name was a pompous blowhard. Duncan was no different. He took a good company, in bed with one of the top mid-tier gold companies in the world on an excellent project, and he ran it right off a cliff. I didn't have all that much to do with him, but after every conversation with him I would contemplate just how pompous he acted and how dismally the stock performed as a result.

He became CEO in early 2012. The stock price never again

touched the height it had under Quinton, and Evolving Gold finally evolved into bankruptcy.

I have nothing against arrogance, but if you insist on being arrogant you should at least have done something worthwhile in your life. Duncan was a legend only in his own mind. He took a solid company with a good partner and destroyed it.

Quinton took over formally as president, CEO, and a director of Galliard in April of 2011. In June they changed the company name to Novo Resources. I hated the name Galliard; it was like listening to a cat trying to climb a chalkboard. The name just grated.

With the change of name came an increase in the speed of the work being done. Quinton was finally achieving the initial phase of his dream. Creasy remained stubborn, but Novo could forge ahead as a result of doing the deal on Beatons Creek with Millennium. Novo did a private placement in late 2011 and raised over $5.6 million for exploration.

Novo started a 5,000-meter reverse circulation drill program at Beatons Creek in the fall of 2011. By February of 2012 they had completed 80 percent of the program of forty-five vertical holes, ranging from 50 to 250 meters deep.

But Novo began to run into a problem that would later turn critical, and I'm going to illustrate it with a picture.

Half the gold ever mined in all of history has been mined since 1965, when low-grade open pit heap leach deposits as employed in the Carlin Trend in Nevada came into common use. Gold fetched only $35 an ounce then, so the industry needed to find a cheap way of producing it.

Unlike most of the gold taken from the Earth's crust up until that time, the gold in Nevada was microscopic. You can't see anything, even in the richest Carlin-style gold. The rock is butt ugly. However, the tiny size of the gold particles makes it perfect for a heap leach. More mining vocabulary: a heap leach is where the ore is stacked up in big piles and a solution of cyanide is literally dripped through it. That pregnant solution of cyanide loaded with gold is further processed to produce bars of gold (or often with some silver) called doré. Only then does it get sent to a refinery for further processing.

The size of thirty grams of gold

Fine, microscopic gold is desirable for two important reasons. Generally, it tends to be found spread throughout the rock containing it, so drilling the rock gives a consistent indication of grade. Secondly, the smaller the gold particles, the easier it is for the cyanide to dissolve them.

Let's go back to the picture above. I show three ten-gram pure gold cubes, about a quarter of an inch along each side. The British gold sovereign and the U.S. dime are for scale, so readers will understand how tiny those gold cubes are.

Imagine that you possess two cubic meters of rock that you

believe contains gold. If you have a dining room table for six people, you have about two cubic meters of air underneath the table top and between the table legs.

One cubic meter of rock weighs about 2.5 tons, depending on its density.

Now imagine those three gold cubes are located somewhere within your two cubic meters of rock. The grade would be thirty grams of gold divided by five tons of rock, or about six grams per ton. At today's price of gold, a ton of that rock would be worth about $360. So the rock is quite valuable. Any geologist or mining engineer would be happy with six grams of gold from each ton of near-surface rock.

But if you have those three little cubes in a block of two cubic meters, how do you find them? You could drill that rock under your dining table twenty times and not hit even one of the cubes. That's called the nugget effect. It's why geologists always want gold to be microscopic, firstly so they can find it, and secondly so they can process it. You could drop one of those cubes into the cyanide solution used all the time in heap leach deposits, come back six months later, and it still wouldn't be dissolved.

Quinton was aware of the nugget effect. What it meant was that his drill results were never quite as accurate as they would be if the gold was both microscopic and uniformly distributed within the rock being drilled.

The gold at Beatons Creek proved to be coarse from a placer miner's point of view but small enough to give Novo a fairly accurate measure of grade. In what would be one of the very few times I disagreed with Quinton, I took a sample and panned it. I told him that his theoretical grade was off by about 50 percent. He had more gold than his tests indicated. It would take almost two years for me to be proven right.

Quinton has a knack for understanding geology and his

explanations almost always make complex things simple. In this case, his method of measuring gold was inherently inaccurate because of the nugget effect.

Relatively coarse gold from Beatons Creek

Novo and Quinton hit the ground running in the fall of 2011, starting a 5,000-meter drill program at Beatons Creek to earn the agreed 70 percent interest in the property.

Those outside the mining industry probably believe that the hard work in mining is the discovery of the mineral in question. Actually that is just the beginning. Permitting and building a mill might take several years and cost anywhere from $50 million to a billion dollars, or more. And that is after you convince the mines department that you have an economic deposit that is safe to mine. The discovery is the quick and easy part. Millions of dollars will have to be spent on surface exploration, environmental studies, heritage investigations, and lots of drilling.

Actually, Quinton and the Novo crew had it easy. Quinton's theory was that the gold was in a certain sequence of reefs that could be identified by eye. His theory held that there would be gold throughout the reef but not above or below the reef.

When a company drills for a mineral, it is just as important to know where the mineral is not as where it is. Running a ton of rock through a mill costs the same, no matter the economic value of what it contains.

Galliard's (later Novo's) agreement with Millennium required it to hand over $500,000 worth of shares and to spend at least $1 million in Aussie dollars within two years to earn the 70 percent interest. Millennium was busy finishing its $100 million mill and couldn't be bothered with the gold from tiny Beatons Creek. In hindsight, that was a massive error of judgment that would eventually put Millennium into bankruptcy.

The gold from Beatons Creek was particularly suitable for the mill Millennium had designed, while the deposits Millennium planned to mine were low-grade and generally unsuitable for the mill. For all the paperwork and all the testing that government agencies require in Australia, or anywhere else, no one ever insists that a mine plan or mill design must make sense. Quinton and other observers in WA watched in wonder as Millennium thrashed around like a headless chicken for almost ten years, trying to make a silk purse out of a sow's ear before falling over dead.

Junior mining companies rarely go out of business because they have a poor project. What kills more small companies than anything else is making a poor job of managing cash flow. Mining in general, but especially exploration, requires a constant infusion of cash. From the market crash in 2008 until late 2011, the cash spigots opened wide as the price of gold climbed higher and higher, to a new record. Quinton was especially well connected and respected in the tiny mining community. He had little problem raising cash. Others broke their picks and crashed.

The newly-named Novo Resources began releasing assay results in February of 2012. Those numbers certainly justified the deal with Millennium. Almost all of the holes intercepted the reef, with mineralization widths of as little as two meters and as much as eight meters, with an average of about four meters. Four meters is a nice block of material to mine, and it was all connected.

Quinton had drilled within an 800-meter by 800-meter square. If you assume an average of two grams of gold to the ton and an average thickness of four meters, that multiplies to a little over 400,000 ounces of gold. For a 5,000-meter drill program, that is a lot of gold, developed cheaply. For an initial drill program, that would be a home run out of the box.

Quartz veining in drill core

There are two kinds of drills that are most often used in junior resource exploration. When it is necessary to understand the structure of the rocks being drilled, core drill rigs are used. They have a drill head shaped like the end of a straw, lined with industrial diamonds, and will cut through anything. The extracted material maintains its shape and goes up what looks like a tube, all the way to the surface. The drill crew pulls up that piece of drill stem, carefully removes what is called the "core" and places it in boxes identifying the hole it came from, and the depth from which it was lifted. Core drilling provides the geologist with a lot of information, but is expensive.

The other common type of drill rig performs what is called "reverse circulation" or RC drilling. An RC rig grinds up the rock into tiny pieces and blows those pieces back up the drill stem, where they are caught and split into smaller fractions for assay.

Since Novo was drilling conglomerate at Beatons Creek, there was no need to understand the structure — only when they entered and exited the targeted reef. RC drilling doesn't tell the geologist anything about structure, but it's a lot cheaper. When you don't need to know about structure you get a lot more bang for the buck with an RC drill program.

Quinton signed the memorandum of understanding with Mark Creasy in February of 2011. It then took until July of 2012 to drag Mark across the finish line to sign an agreement. He was used to calling all the shots when he did a deal. One of his most effective techniques was simply to delay and delay until the guy on the other side of the table finally gave up and bent over.

By doing the deal with Millennium and delivering an extremely successful drill program right out of the chute, Quinton proved that he was serious, and showed that he was willing to work with somebody else if Mark wouldn't belly up to the bar and sign that piece of paper.

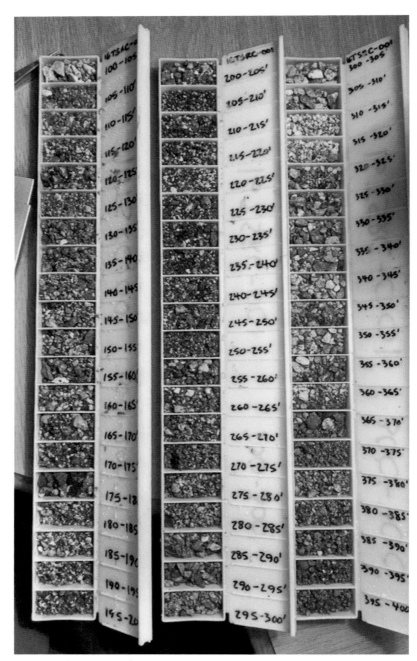

Chip trays from reverse circulation (RC) drilling

Quinton managed to put together an impressive land position in the part of the Pilbara that he thought offered the most potential. On the other hand, Mark Creasy's side of the deal meant that he was now by far the biggest shareholder in Novo. Keep that fact in mind, because two years later it would become a giant issue.

Both were happy for now. Quinton had his teeth solidly into a proven gold property. Mark in effect had Quinton doing all the heavy lifting for him.

Quinton continued work at Beatons Creek into late 2012, with a second drill program of another 107 holes. Results continued to be excellent.

I waited for three years to write my first piece about all this for 321gold.com. It took that long to get Mark to agree on terms reasonable to Quinton. While Novo's deal with Millennium proved to be the catalyst to move Mark off of top dead center, or TDC,[15] it was the ground that the Creasy Group controlled that would turn Novo from just another Canadian gold junior with an interesting project into a gold powerhouse. Quinton wasn't seeking just one medium-sized deposit. His theory was that the Pilbara had the potential for millions or tens of millions of ounces of gold.

My first of dozens of articles on Novo Resources appeared on the website on August 15, 2012. I titled it "698.3 Square Miles of the Wits"[16] since most of my readers are American and hopeless at anything as complicated as the metric scale. I include myself among them.

In the article I explained the basic geology, and Quinton's theory, and the proof we had put together so far. I talked about banded iron formations, since they seemed to me to be such an important clue, although ignored by the geological world.

At the time the company had a $14 million market cap and

an enterprise value of $5 million. I went on to say, "I rarely give a pure buy signal. This is a buy signal, the most important that I have ever made. The stock at $.44 is like stealing." A couple of paragraphs above that I pointed out that investors were getting nearly seven hundred square miles of the Wits. "It is easily a ten-bagger. It could be a 100-bagger. It's going to be big."

So far it has gone up only twenty-fold. But give it time; Quinton has big plans still.

About the same time, in mid-August of 2012, Quinton had Novo put some money into a private placement with Evolving Gold. [17] He knew the potential of its ground and, should it advance the work he had begun, he wanted Novo to have a piece of it. Novo bought two million shares at thirty cents. People in the industry vary in their opinions about just how wise that was.

The other company I visited in Nevada just before flying to Colorado to see Quinton in October of 2008 had about $12 million in cash at the very bottom of the resource mining stock crash. Since it had partners funding all its projects, I suggested putting some of that money into juniors that were being given away. A year later most of those stocks were up a thousand percent or more. The company ignored me and ended up blowing all its cash. That's how the mining business really works.

I happen to agree with Quinton's theory. And over the years he had Novo invest in half a dozen projects and companies. Some turned into giant home runs. Some, like Evolving Gold, turned into a duster.

Novo released the last of the initial 5,000-meter drill program results on August 21, 2012. [18] The company had drilled forty-three holes and found significant gold in forty-two of them. In the two weeks after I wrote my first piece, the share price shot up from forty-five cents to eighty. While more and more juniors pulled in their horns in preparation for a long cold winter as gold prices dropped, Quinton continued to charge ahead.

The holes were interesting; many had a short interval of maybe one or two meters of high-grade gold within a longer, low-grade interval. Quinton always planned for production so he needed to know the distribution as well as the size of the gold.

Again, most junior resource companies don't fail because of their project or projects, but rather from poor management of cash. Drilling eats money. The money is there to make progress with, but investors hate paying big checks to support the lifestyles of the stock promoters and management.

Quinton was one of a handful of geologists (the others include Peter Megaw, David Lowell, and Keith Barron) who everyone in the industry knew and trusted. There's a balance to be found between issuing too many shares, to the point where an investor couldn't make money if the company drilled into Fort Knox, and running out of cash but having the world's tightest share structure.

As long as I have known Quinton, he has been able to tread that fine line between too many shares and too little cash. Investors like results. No, investors *demand* results, and it's like going into a Chinese laundry in San Francisco: "No tickee, no laundry."

Quinton was on a roll. He had the ground to be tested. He had money in the bank. The first drill program was followed up in September of 2012 with a 7,500-meter program. There was another two thousand meters of drilling at Golden Crown Hill, where historic drill tests had found gold-bearing conglomerate beds less than fifty meters from surface. Then at the end of 2012 Novo conducted another two thousand meters of RC drilling on the new Creasy ground south of Grants Hill.

Frankly, most of what goes on in the mining business is important but boring beyond belief. Quinton was drilling conglomerate beds that you could see outcropping on the hillside. Ninety-five percent of the holes reported some gold. At times the results were incredible but, to the average investor, boring. The first red meat as opposed to the far more common tofu came in May of 2013, when Novo announced its initial 43-101 resource.

I'll backtrack here for just a few lines. As a result of the Bre-X fraud, the Canadian government prepared and issued a series of regulations that it calls 43-101. [19] The full name is National Instrument 43-101 Standards of Disclosure for Mineral Projects within Canada. It's a set of rules about what you can and cannot say about a deposit. While the government functionaries love their 43-101, I'm not a big fan. It's more petty worthless rules, often ignored. When someone really does bust the regulations they are given a small and meaningless fine for their misadventure.

Novo's press release talking about the new resource estimate was interesting because it allowed investors some insight into Quinton's thinking.

The initial resource was only 421,000 ounces of gold at a grade of 1.47 grams of gold per tonne. That number was derived from 16,100 meters of reverse circulation drilling. Novo also drilled eight core holes, and could use the core to determine average specific gravity for their 43-101 report.

Quinton also did something interesting that I hadn't seen done before. I found it extremely valuable. He mentioned the cost of the drilling. It had cost about $2.5 million for 16,100 meters. So, according to Quinton's numbers, it had cost about $6 to find one ounce of gold. That's an excellent price of discovery.

A couple of years later, when Novo did a bulk sample, Quinton found that the drill results had downplayed the actual results. He was in fact finding gold for about $4 an ounce, and that is nothing short of brilliant.

Gold and gold shares reached their peak of popularity in 2011 after a dozen consecutive years of higher prices. The shares peaked months before the gold price touched $1,922 in September. For the next four brutal years, investors got beat up if they had not been sharp enough to take some money off the table. Gold and the mining shares hit bottom in January of 2016.

During that dark era for investors, few writers were interested in covering new stories. Jay Taylor of *Turning Hard Times into Good Times* [20] was an exception; he recommended Novo for the first time on August 9, 2013. More about Jay in a moment.

I had visited Novo several times by now and was fully conversant with both the theory and the story. I had been part of the Pilbara story since a year or so before the company was born, so it's natural that I covered it early on. But even with my half-dozen or so pieces there on the website for anyone to look at, no other writers took any notice for a year after I began telling the story.

I had almost five years of thinking about the theory and the ramifications, so that gave me an easy advantage. But how many writers are ever in a position to talk about a Witwatersrand lookalike?

Jay Taylor hadn't visited the project, but in his own way he did a great job of telling the tale. I repeat that Quinton excels in explaining complex matters simply. Jay not only understood the theory; he glommed onto a vital issue.

In his piece he said:

"Without speaking to management about this issue, the fact that management mentioned the fact that Millennium is operating a mine and mill may be of importance in the longer run. Indeed, Hennigh told the *Northern Miner* in October of 2012, "If we can get away from having to build a mill the permitting for a smaller operation is really straightforward and could be very quick.""

Quinton had known about the characteristics of the gold at Beatons Creek ever since his first visit, when he was a Newmont employee. It's coarse and hardly suitable for a heap leach. It takes time for cyanide to dissolve gold, and the bigger the gold particles, the more time it takes.

Jay understood that. His story continued:

"Hennigh may have meant to imply the possibility of an open-pit heap leaching operation and thus no need to build a mill or the possibility of working a milling deal with Millennium Minerals. But for now, that's getting ahead of the story. First we need to know more about the size and grade of the deposit and a host of other issues before we think in those terms. Still, the fact that Millennium's milling operation was mentioned in a press release can't help but get you thinking along those lines."

Jay Taylor nailed it with those comments, even if it had taken several years to arrive at that point.

I've been on dozens of trips to various projects with Quinton. He thinks further ahead than anyone else I know in the mining business. Five minutes after he found the first nugget at Beatons

Creek, he would have been pondering just how to process it cheaply. When he learned that Millennium Minerals was in the process of building a mill, he probably thought he had died and gone to mining heaven.

In September of 2013, Newmont Mining surprised both Novo and the market with its announcement of the purchase from two large shareholders of 17.76 million shares of Novo at a price just above market. [21] That would give Newmont control of 35.7 percent of the shares, if it were to exercise all of its warrants.

Newmont was well aware of the potential of the Pilbara through Quinton's work years before. Some self-dealing on the part of its Perth staff queered the agreement with Mark Creasy and left a bad taste in Creasy's mouth. Newmont figured that if you couldn't get in the front door, perhaps the back door was open.

With the exercise of about eight million warrants by September of 2013, Novo brought in $4.8 million, bringing its cash on hand up to $8.26 million. It had about fifty-eight million shares outstanding and a market capitalization of about $48 million at the time of Newmont's announcement.

Juniors that make deals with majors are often left at their mercy. Making a deal with a big company with lots of expertise and money sounds attractive, but the majors enter into deals for their own benefit, not that of the junior. It's OK to get into bed with the pretty lady but it's also important to not catch the clap.

To give Quinton full credit, for as long as I have known him, he has always tried to structure deals such that no single party could take control. On the Beatons Creek ground, Novo had a 70 percent interest and Millennium Minerals had 30 percent. On the ground around Beatons Creek, controlled by Mark Creasy, Novo had the same 70/30 split. Remember that it was Novo's side deal with Millennium that prompted Mark to conclude his deal with

Novo, after dragging his heels for years.

For Quinton, Newmont would be an especially valuable major to deal with. First of all, he had worked for them. There was mutual respect. Second, majors have large staffs and have all the various technical areas covered. If you know what a major is especially good at, having a big company as a partner can be a marriage made in heaven.

Newmont had a strong group doing bulk leach extractable gold (BLEG) sampling. [22] That was ideal for Novo. Novo had a lot of ground with potential; the Creasy portion alone was about 1,800 square kilometers, or 700 square miles. They needed a cheap, fast way to determine where to concentrate their efforts. BLEG is interesting because it determines only the presence of gold in a location, not its grade.

A BLEG survey team will go out and take fine material samples of decent size from streams, both flowing and dried-up. When you have thousands of samples, you can get an idea of where to look for gold, and the distribution. One big problem for juniors with large land positions is deciding where to expend their energy, and more importantly, where to spend their limited money.

The Newmont–Novo partnership would prove to be valuable to both partners. Newmont's large share position would become a problem for them later, however, when someone made an agreement to sell just before very valuable news was released. But that is a tale for the future.

In late 2013 the Department of Mines and Petroleum of WA notified Novo that it had been approved for two grants, each of $200,000 Australian, for the purpose of drilling deep holes further out in the Nullagine embayment. [23] The plan was for one deep hole at Beatons Creek and another at what Novo called Contact Creek, part of its Marble Bar project.

Gold and silver began a perfectly natural and normal price correction in 2011, such as happens in every market. It took gold bugs by surprise. Quinton knows that no markets ever go straight up. He had always sought to keep a healthy cash position in the bank. Investors tend to want instant gratification and scream for companies to do things faster. But often, slower works better.

In March of 2014 Novo concluded another deal with Mark Creasy on an additional 18,000 square kilometers of new tenements. [24] That's another 7,000 square miles to cover. BLEG sampling would be an important technique to use, to narrow down the search for the mother lode. Novo would be the manager, and this venture had the same 70/30 split that Novo had concluded before. Remember, all of this could have taken place four years earlier, but for Mark being dilatory.

Newmont had completed a baseline BLEG survey over Novo's ground at Beatons Creek as early as November of 2013, shortly after its purchase of Novo shares. In March of 2014, after Novo had added the additional 18,000 square kilometers of ground, Novo and Newmont realized it was time to get serious about carrying out a BLEG survey over the entire new property. [25] That began in March and was scheduled for completion by the end of winter, in July. The BLEG process belongs to Newmont, and its team of technical people is the very best in the industry.

The terms of Novo's deal with Millennium called for the completion of a bankable feasibility study (BFS) by August of 2016, in order for Novo to complete its earn-in of its 70 percent interest in the Beatons Creek project. Quinton had seen enough disasters brought on by missed milestones, and didn't intend to be mousetrapped by the terms of the deal. Under the "shit happens" rule, everything that can go wrong will go wrong.

Novo planned to complete all the requirements of the BFS as early as July of 2014. [26]

In August of 2014 the BLEG work was completed and the results obtained. [27] The results around Beatons Creek and Marble Bar demonstrated that Novo had correctly identified where the gold was coming from. Novo immediately began an aggressive shallow drill program at Beatons Creek to expand the near-surface oxidized gold resource.

Millennium Minerals was now in production but had done a rotten job of designing the mill. There would eventually be a resource of a couple of million ounces of sulfide gold, but the mill was designed for oxide ore. Once Millennium ran out of near-surface oxide material, they were screwed. Their dismal recovery of the sulfide rock meant they were really just running a hiring facility, not a mining company. They produced about a hundred thousand ounces of gold a year but never made money. They existed to provide employment and to repay their bank loans. And to allow management to collect paychecks they otherwise would have had to work to get.

But an escape route was at hand. Millennium's partner in Beatons Creek had five years' worth of ore that would be perfect for their mill. Novo had done all the work of locating and defining the conglomerate gold. All Millennium had to do was to agree some kind of arrangement with Novo. They could buy out Novo's interest. Or they could perform toll milling for Novo, whereby Novo mined the gold and Millennium processed it for a fee. Or they could sell the deposit and the mill to Novo and retain a net smelter royalty.

The only really stupid, head-up-the-butt thing they could possibly do would be to do nothing.

So they did nothing.

It seems clear that Millennium Minerals never saw the value

in Beatons Creek, even after Novo proved a nice five-year resource. Since Millennium saw no value in the project when doing the deal with Novo, they weren't about to rethink their position, no matter what new facts emerged. Millennium was convinced that Novo had bitten off more than it could chew, and would eventually walk away from Beatons Creek.

Quinton Hennigh, on the other hand, could not come to grips with the fact that he couldn't interest Millennium in any sort of deal. He didn't need Beatons Creek; he had nearly twenty thousand square kilometers of other ground where he could go find more gold. He knew it was there. He had perfected a way to locate the best areas and he knew how to mine it. And in the worst case, he could define enough gold to justify building a new mill for Novo's use alone.

What baffled him was a company so stupid that it would risk bankruptcy rather than change a business model that clearly didn't work.

Over the years he came to realize that there were no average mining companies in Australia. There were some of the best-managed and technically most astute mining companies in the world, such as Newcrest. He had worked there and knew they were brilliant. There were also dozens of tiny, poorly-financed juniors run by idiots who would have been on welfare and living under a bridge, if they'd ever had to compete on their own merits. In short, Australia has a few great companies and a number of others run seemingly by dodo birds with nothing better to do.

Millennium was in the dodo-run category. Alas, the Australian dollar kept going down as the price of gold slowly went higher, bailing out the idiots who should have been in bankruptcy court wearing nothing but an old pair of jeans with patches on the knees.

By the end of 2014 Novo had drilled off more of the oxide resource and had completed recovery testing, using both a gravity process and cyanide leaching. Tests showed as much as 91.7 percent recovery via a gravity circuit, with an additional 7.6 percent recovery using cyanide leaching, for a total of 99.3 percent recovery of the gold. [28] That's both brilliant and rare.

CHAPTER 11
2014 BRINGS PRODUCTION OPPORTUNITY

IN LIFE AS WELL AS IN MINING, serendipity often plays an important part. A golden opportunity dropped into Quinton's hands in February of 2014, only to slip away three months later.

As I mentioned, Mark Creasy made his first fortune selling the Jundee gold mine to Great Central Mines in 1995. In May of 2014, Newmont Mining announced its intention to sell Jundee to Northern Star Resources for $82.5 million. [29]

You can think of this as tossing a hungry cat into a pen of chickens. It was an amazing episode, complete with self-dealing, Novo's second-biggest shareholder screwing its second-biggest shareholder (that is not a misprint), and a lot of simple old skullduggery.

I keep saying that Quinton was quite close to Newmont. He had worked there only a few years earlier, and now Newmont was conducting the BLEG survey on Novo's tenements, after having injected a fair bit of cash into Novo.

Newmont (and my readers) may or may not have remembered that the deal between Great Central and Creasy included what is called an ROFR; a right of first refusal. That ROFR would come into play in the event that the mine was put up for sale, as distinct from the company that owned it being sold. (And Jundee had been part of a series of corporate moves since 1995, operating under a number of flags.)

If the Jundee mine itself was put up for sale, Mark Creasy had the right to match any offer made, and his bid would be superior to any other. He hadn't forgotten this, and neither had Quinton, even if Newmont had forgotten or had never realized it.

Mark had given that ROFR to Novo in February of 2014 because he recognized it had value to Novo. No conditions were attached to the transfer; he just handed it over. Novo now had the right to match the Northern Star's offer to Newmont, and Novo's money would be deemed better than Northern Star's.

Jundee began as an open pit mine. Then, when the strip ratio got too high, it was converted into an underground mine. Production varied, but was usually between 250,000 and 400,000 ounces of gold a year.

The deal would catapult Novo into the status of a mid-tier mining company literally overnight. Quinton was standing at the goal line. All he had to do to complete his life's dream was to step across the line.

Any time life hands you everything you wanted and it looks just like a beautiful red rose, just be careful and avoid the pricks because they are always there.

An interesting book came out in January of 1971 called *Games People Play*.[30] The author, Eric Berne, discusses the three main games that people play with each other, often without knowing or understanding the potential dangers.

One very common game is the negative sum game. In this game, one or both of the players want to make sure the other loses. Divorce proceedings in the United States frequently turn into negative sum games, because lawyers know that the best way to maximize billable time is to stir up the pot. If they can enrage both sides to the extent that each starts to focus on destroying the other, the lawyers will make a lot more money.

Wives often enjoy this game, because it is an opportunity to right every real or perceived wrong ever done to them. They are quite prepared to beat the soon-to-be ex-husband to death, even using their children as weapons. The lawyers benefit, the husband and wife both suffer financially, and their children wonder why and how they lost a parent. And two people who once promised before God to honor and obey each other until death did them part now despise each other.

A better game is one in which each party realizes it is in its own best interests not to screw the other party. Divorces don't have to be acrimonious. It's pretty stupid, especially when there are young children. Likewise, if you are planning to enter a business arrangement or a marriage, why would you want to beat up your prospective partner? In a positive sum game, each player makes sure the other side benefits. They realize that working together is far more productive than fighting over pennies.

A perhaps more common game is the zero sum game. Here, one player wants only for the other player to lose. Ross Perot, Steve Jobs, and many businessmen actually believe that making sure your partner loses means you win. In a zero sum game you have one winner and one loser.

Mark Creasy always plays zero sum games. He is totally focused on making sure the other side loses in any business deal. In this deal with Newmont and Jundee, that created a giant problem.

Mark handed the ROFR to Novo when rumors began to fly in Perth about Newmont selling Jundee. Frankly, he wasn't in a position to do anything with the ROFR, while it fit Quinton's plan for the future perfectly.

Quinton was familiar with the geology at Jundee. Instead of the million or so ounces of gold that Newmont believed it had, Quinton believed Jundee had a good five to ten years of production potential. He got on the phone and rapidly amassed commitments for the money. After all, if Northern Star could raise the money, why couldn't Novo?

That's when it began to get interesting. Mark Creasy's accountant, Steve Lowe, was part of Chalice Gold. Chalice had money in the bank and was certainly a candidate to take over the mine. So, all of a sudden, Mark Creasy tossed a monkey wrench into the works. Instead of the easy and smooth road into production for Novo — which would certainly benefit Newmont as Novo's largest shareholder, and Mark Creasy as the second-largest shareholder — the deal was blowing up.

Steve would have known how to push Mark's buttons. I think he did. All he would have had to accomplish was to convince Mark that he would get a better deal with Chalice. And I have a sneaking suspicion that some of the Newmont people in Perth may have been playing both sides against the middle just to stay employeed.

Newmont had a major stake in Novo, so why wouldn't Newmont want Novo to go into production at Jundee? Just because Jundee no longer fit its plans didn't mean Novo wasn't a better home. I find it curious that a number of former Newmont people from the Perth office found a new home at Northern Star.

Creasy demanded Novo return the ROFR to him. He was going to cut Novo out of the picture and do a deal either with Chalice or with Northern Star.

At the time this took place, I was driving an early Ford Mustang from California to Texas. As I sat in a motel room in some podunk town, Quinton filled me in.

"Mark wants the ROFR back. He's gotten Northern Star to agree to hand him ten million dollars in shares of the company."

"Quinton, I love you like a brother, but in this case you need to tell Mark to go fuck himself," I said, and added, "He's a big Novo shareholder. If he screws Novo he is also screwing himself. Why would he screw himself?"

The answer, of course, is exactly the same as why a dog goes

110

into the center of the road to lick his dick. Because he can.

Quinton is flawed. Not in a major way, as is the case with most of us, but nevertheless he is flawed. He is much too nice a person. At times that gets in the way of doing sensible and reasonable business. I don't have that problem.

Quinton gave the ROFR back to Mark. Mark got $10 million in shares of Northern Star. By Mark's logic, he did well. His $10 million in Northern Star is now worth almost $100 million. He didn't make that much, because over the years he sold off his shares. But he did well.

But Novo would have been producing at least 300,000 ounces of gold a year, and today would have only one-quarter of the number of outstanding shares it actually has. The guy who got screwed the most in the deal was Novo's second-biggest shareholder. Mark cost himself $500 million in gains. I think the term for that is "penny wise, pound foolish."

It was a fleeting opportunity to leap forward. There are always lots of people doing everything in their power to queer any proposed mining deal, provided it brings them short-term benefit.

Overall, Novo had done quite well in the thirty months since it went public. The deal with Millennium had worked out well and looked to have potential for a couple of million ounces of gold. Two deals with Mark Creasy had been signed. Newmont had come on board with some money and all its technical expertise. Novo had assembled a great team and a series of projects in the Eastern Pilbara that had good potential. It wasn't going to be some giant area of twenty million ounces, but what Quinton had put together was well worth having.

But there was this one clog in the machinery that hampered Novo for years. Millennium had spent well over $100 million and several years building its mill. And while everyone else in

Australia and the world understood it was the wrong mill for the feed, Millennium's management (to use the term very loosely indeed) sailed blissfully along, ignoring the fact that it had turned the company into a sort of home for unwed mothers.

There wasn't a chance of Millennium ever making money processing the rock they owned in the mill they owned. In a masterpiece of self-delusion, they also convinced themselves that Novo was stupid and they were smart. They refused to even consider the one solution that would turn them into a mining company instead of an employment service.

Quinton Hennigh and Novo were caught in a trap of their own making. Had they made a similar discovery a couple of hundred miles to the west, they could have taken the conventional route of defining a resource, raising money, and building a mill to go into production. Quinton never made any bones about it. He wasn't running a lifestyle company that existed to pay management large salaries for doing little until the day that shareholders woke up and booted them. He wanted production.

One reliable rule of mining management is not to spend a few years and $100 million building a mill right next to another $100 million mill that desperately needs, and is suitable for, your sort of ore. That would be mad. In short, by succeeding, Quinton was failing.

The shareholders didn't mind much. Novo has always been one of those resource companies with great liquidity. For some reason the theory of gold precipitation seemed popular. You could always buy and sell its shares. That's a lot more rare than you would think, with the penny mining stocks. For years, there was a 200 percent or greater change in the value of Novo shares every year. The stock would double or triple, dribble back down to where it started, and then repeat. To a wise investor who recognized the pattern, a 100 percent annual return was possible nearly every year no matter if you were a bull or a bear.

CHAPTER 12
WHEN ACADEMICS LOOT AND PLUNDER

IF BY NOW THE READER IS NOT CONVINCED that mining mostly consists of lying, cheating and stealing on the part of many of the participants, then I as author have failed. However, the chicanery in the mining business does not compare to the sheer ruthlessness that exists in academia. Professors tend to make pirates look like pussycats.

Whenever someone comes up with a new theory, perhaps the best measure of it actually being true or working is when no one, including the "experts," agrees with it. In fact, the more they make fun of the theory, the greater the chance of it being correct.

In 1982 two Australian doctors came up with a theory that a bacterium caused most cases of ulcers and stomach cancer. After years of study and gathering proof, they presented their case to an audience of surgeons in Sydney at a medical seminar. Literally, they were laughed off the stage. You see, the surgeons made money by cutting into people due to the belief that ulcers were some sort of lifestyle disease. If you could cure a case of ulcers with a simple course of antibiotics, they would lose patients. And, more importantly, income.

In 1985 one of the doctors, Barry Marshall, had himself tested to prove he was free of the bacterium before deliberately infecting himself. Soon he tested positive for the Helicobacter pylori bacterium and came down with a mild case of gastritis. A short course of antibiotics caused both the gastritis and the bacteria to disappear.

Doctors Barry Marshall and Robin Warren eventually won the Nobel Prize for Medicine in 2005, twenty-three years after their discovery. Once everyone else in academia realized the theory might in fact be correct, they rushed to jump onto the bandwagon. By 2005, some twenty-five thousand articles had been penned on the subject of bacteria causing ulcers. No doubt many of the authors claimed to be the first to come up with the concept.

Professor Hartwig Frimmel from the University of Würzburg crossed paths with Quinton in May of 2005, at a meeting of the Geological Society of Nevada, in Reno. Quinton had been assigned by Newmont to form a team to find volcanogenic massive sulfide (VMS) deposits. [31] Think of an undersea black smoker; that's a VMS deposit. They can be very rich in valuable minerals.

By this time Quinton had developed his theory on gold precipitating out of solution from seawater. He showed his paper to Frimmel, who found it faintly interesting but said it was basically rubbish.

In May of 2014, Frimmel published a paper [32] setting out his new and brilliant and original theory of the deposition of gold from seawater through precipitation in the presence of carbon, as the oxygen content in water increases through the action of single-cell bacterial colonies, the first life on Earth. His theory was Quinton's theory, from nine years before, that he had mocked.

When Quinton formed Galliard and began moving forward in the Pilbara, I wrote and published on 321gold.com a dozen or so pieces, talking about the theory and the Pilbara. As the market accepted the idea, Frimmel began to pass it off as his work.

When Frimmel's paper was published, Quinton contacted him and mentioned that it was his idea and work, and that Frimmel had mocked it years before but was now passing it off as his own. Frimmel tossed Quinton a bone and agreed to share credit as authors on a paper to be submitted to an academic journal, *Mineralium Deposita.* [33]

The full name for *Mineralium Deposita* is: the International Journal for Geology, Mineralogy and Geochemistry of Mineral Deposits. That sort of rolls off your tongue.

That paper was published in January of 2015, with Frimmel as the primary author and Quinton in attendance. They called it "First whiffs of atmospheric oxygen triggered onset of crustal gold cycle." [34] It won the best paper of 2015 for the organization.

At the time, while Quinton was a bit miffed, I thought it had its amusing side. After all, I was writing more about Quinton's theory than he was. While professional journals are far more inclined to publish papers written by academics talking about interesting theories than by the guys in the field actually doing the work, Quinton had taken the theory and was putting it to work to produce gold.

It was humorous right up until the time that two minor newsletter writers pulled the same trick on me, two years later. Then it wasn't even a bit funny.

CHAPTER 13
COMING UP WITH A NEW WAY TO MINE

QUINTON HENNIGH'S PRIMARY STRENGTH is his uncanny ability to think laterally; to view things from a different perspective.

Most of the mining activity in the Pilbara, all the way back to the nineteenth century, revolved around hard rock, vein-style mining. That continues even today. Few companies seem to care about the conglomerates or to be able to figure out how to mine at a profit.

The essence of where and how the gold was found in the Pilbara has been known for a hundred and fifteen years. Government documents as far back as 1906 make references to gold in various layers of the conglomerate, just like the Witwatersrand. Vein deposits present problems of their own, such as trying to find them, but the conglomerates were everywhere. Yet were ignored by most of the mining companies.

It has climate issues as well as geology: hot, dry, and nasty, with either too little water or far too much; and wildlife, with the top ten deadliest creatures on earth. Australia has snails and spiders that can kill. The constant barrage from flies is enough to make you consider suicide just to rid yourself of them.

And don't even think about summers in the Pilbara, at forty degrees Celsius (or a hundred and four Fahrenheit) in the shade. Not that there is any shade.

The conglomerate gold at Beatons Creek was in nugget form, which made it difficult to measure accurately but also meant it would be easy to process. However, it was located in the matrix, between cobbles and boulders. You needed the matrix for the gold but not the rest.

If Quinton decided to use the typical mining method of drilling and blasting before moving the material to the mill for processing, 90 percent of what he moved would be waste material. Further, in doing so he would shake a lot of the free gold to the bottom, where it couldn't be removed by heavy equipment.

I have worked in two very different fields where the people around me were often very smart. One was computer operations and programming in the early 1970s, and the other, obviously, geology and mining for the last twenty years.

Computer programmers are often very bright. They are also commonly self-taught. I have never taken any programming training but I was a systems programmer. Everything I knew, I learned on the job. Steve Jobs had a smattering of college, as did Bill Gates. I'm not sure even today that college is any real advantage in learning programming. The industry and languages change far too quickly. What is popular today will be old hat six months from now.

Geology and mining, on the other hand, require years of study. Nearly everyone in mining or geology has at least a bachelor's degree. Master's degrees are common and doctorates not uncommon. I have met more overeducated people in mining than in any other occupation I have ever been associated with.

The flaw in this emphasis on education is that people tend to become too specialized. They are very good in some narrow skill but lack general knowledge. In the course of the many trips I have made, I have often noticed (and often commented) that most geologists believe mining to be about spending money rather than making it. They would all do far better if, while earning their doctorates, they spent a semester working in a 7-11, learning how to sell a quart of milk at a profit. Profitability seems to be the last thing that most mining folks consider in a project.

Here's an example. My first visit to an underground mine was in 2002; it was the Ken Snyder gold mine in Nevada.[35] At the time it was one of the highest-grade gold mines in the world. Our escort was the senior geologist on the project. He had been there for several years, as I recall. We went down to the working face and saw just how narrow the vein was. Their working width was two meters, but it was ultra-high grade. In the week before we were there, they had mined 28 ounce gold measured across the mining width, but the vein was less than 0.5 meters.

We were suitably impressed, picked up all the samples we could carry and returned to the surface. As we exited the portal, one of the skip loaders followed us and proceeded to dump his load on a stockpile just outside the portal entrance. I turned to our guide with a couple of questions.

"You find some extraordinary high-grade gold now and again. The average is much lower, though still pretty rich. Do you need to blend the ore? Or how do you handle the high-grade stuff to get the best recovery rate?" I asked.

"I don't know how they work with that." He pointed up the hill to where the mill was situated. "They do all the processing there. I've never been in the mill."

I was nonplussed. How could you work on a mining project for years, in a senior capacity, and not have some idea of how things worked? How could you not want to see how ultra-high grade gold was processed?

That's not Quinton. He's the most inquisitive bugger I have ever met. He wants to know everything. He thinks about everything. He has come up with a practical solution to a problem before anyone else has even figured out there might be a problem.

Mines rarely run out of ore. They just get mined to a point

where, for one reason or another, the company running the project cannot make a profit any more. So they shut it down and walk away. Likewise, it's common for a project to be bought by half a dozen companies, consecutively, before someone works out how to run it at a profit. There's that gruesome word again, that keeps popping up.

Political risk, commodity prices, and labor and energy costs are all factors as well, but plain old human stupidity is more important than any of them. Ten thousand prospectors and miners had walked all over the Pilbara seeking fame and fortune, while it sat at their fingertips and right below their feet. Quinton went into lateral thinking mode, and communicated the results to Novo's shareholders. [36] Here is what he said.

Recent trench sampling revealed a potentially important behavior of oxidized reef material. While collecting samples, the matrix of many reefs was seen to break away from cobbles and boulders as rock was moved during sampling (see Figure 1 below). The Company believes that it may be possible to significantly upgrade material for processing by first removing largely barren cobbles and boulders. If so, this could mean the tonnage of material that will require grinding may be reduced, a potential cost saver. To test this possibility, further test work will include a component of "scrubbing" whereby material will be tumbled to cleanse cobbles and boulders of gold-bearing matrix. A new bulk sample (250 kgs) has been collected from representative reef material on Golden Crown Hill and is being submitted for bench scale scrubbing, gravity and floatation testing. Results from this new test work are expected back in the first quarter of 2015.

(Figure 1: Left, 1m high gold-bearing conglomerate horizon exposed in a trench. Hard siliceous boulders and cobbles occur in a dark brown sandy matrix. Gold occurs in matrix material. Right, loosened conglomerate material collected during trench sampling. Water bottle is approximately 20 cm tall. Note that most of the siliceous boulders are free of matrix material. Screening these out may potentially help upgrade matrix material and help reduce costs of grinding in a mill.)

When faced with the matter of how to mine the conglomerate, Quinton looked beyond conventional drilling and blasting from the start. It would cost too much money to move what was mostly waste rock. Processing it would be inefficient and expensive. He needed a better solution.

He was attacking a problem that had been known about for a hundred and twenty years. Everyone knew there was gold in the conglomerates, and that the conglomerates came to surface in a number of areas. Any prospector could go to a gold-bearing portion of conglomerate and start pulling out gold. In the early days, many did.

But they couldn't extract much gold by hand, and even after

bulldozers and dump trucks arrived after World War II, the gold was too low a grade to be economic. That is, if you did it the same way every other mining engineer or miner would.

So Quinton devised his own solution. It was both simple and elegant, and it worked. He wrote about his theory and how it would work in a press release dated February 9, 2015. [37] He also talked about a deep drill hole. Here are two illustrations and an extract from the text.

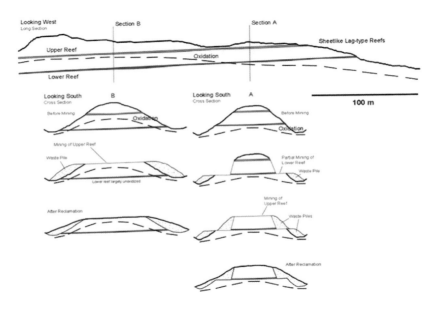

(Figure 1: Conceptual mining plan at Beatons Creek. Overburden above the lower reef, blue, will be pushed aside into adjacent low areas and this reef will be selectively mined. Overburden above the upper reef, red, will then be pushed off to the sides, and this reef will then be selectively mined. This technique reduces the need to haul waste material and could be very cost effective.)

Quinton envisioned stripping off the waste rock and mining only the gold-bearing ore. This was a low-cost mining method, and easy to gain a permit for.

Using simple equipment such as an excavator with a flat

edged bucket, gold bearing conglomerates can be selectively mined thus reducing dilution and, thus, helping maintain higher grades. Costs of this type of mining are anticipated to be low. Figure 2 (below) is a schematic illustration showing this selective mining technique.

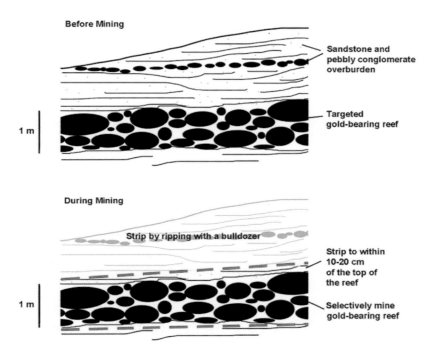

(Figure 2: Schematic illustration showing selective mining concept at Beatons Creek. Using a bulldozer, soft sandstone and conglomerate overburden can be ripped and pushed away exposing the underlying gold-bearing conglomerate horizon. The targeted conglomerate can then be selectively mined utilizing an excavator fitted with a flat edged bucket.

To succeed at Beatons Creek, Quinton needed more land and more gold. After hearing tales of Millennium's desperate financial condition, Quinton went to them with an offer. He would pay them $3.8 million Australian for the remaining 30 percent of the Beatons Creek project. While Millennium owned that 30 percent, Quinton was trapped. If he was too successful developing ounces in the ground, it gave them a life ring. They were too stupid to consider doing any sort of deal with Novo, as they were still convinced Novo would fail. By buying the remaining 30 percent, Novo could take Beatons Creek out of the equation.

Quinton announced Novo's offer late in March, 2015. [38] Millennium jumped on it. They couldn't wait to get their hands on the money. Six days [39] after agreeing to purchase the remaining portion of the property, Novo sat down with the management of Millennium to sign the transfer papers.

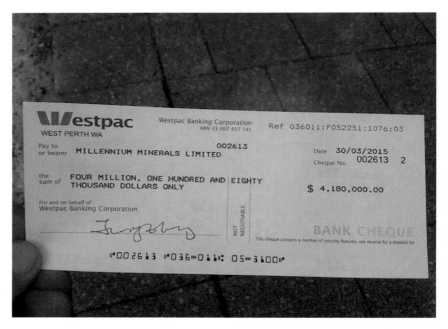

The final payment for Beatons Creek

The meeting concluded with Novo handing over $3.8 million Aussie, plus 10 percent stamp duty. The Millennium lawyer grabbed the check, quickly checked it over and then literally ran out the door to go to the bank. They were in that bad a shape. A few minutes later the bank called to ask if they should honor the check. They weren't used to issuing a corporate check and having it presented in person two hours later. They wanted to make sure the right person had the check.

That was a major coup on Quinton's part. In any mining deal, the health of your partner is important. People get stupid when they have their backs against the wall.

Quinton was still in the position of having enough gold to mine but no mill. Financing and building a mill for the small amount of gold Novo had defined didn't make sense.

But the price of gold kept going down. Eventually Millennium would tip into bankruptcy and Novo could take over its mill.

In May of 2015 Novo moved from the C-Exchange in Canada to the TSX Venture Exchange. [40] That would give the shares more liquidity because in general, small investors are hesitant to trade C-Exchange stocks, believing that TSX Venture shares are safer. I'm not sure how true that is, but the liquidity issue is one that hampers trading in all the companies using the C-Exchange.

About this time, Quinton went to Toronto on business. Customarily he would meet with Eric Sprott, another big Novo shareholder. Quinton wanted to keep him abreast of things.

It was an important meeting, seen in hindsight. It would put over $1.5 billion in Eric Sprott's pocket. It would make a multi-millionaire of Tony Makuch, president and CEO of Kirkland Lake Gold, along with Greg Gibson, Eric Sprott's man Friday. Alas, jealousy would ensure Quinton was stiffed for his contribution.

The highest prices for gold shares were seen in mid-2011, while the top in gold was later, in September. Prices for the metals and resource shares had been tumbling since then, after a twelve-year advance. Many stocks were down by 90 percent and more. The very bottom was at the end of 2015, with the Sprott funds down as much as 93 percent.

A meeting had been held earlier in 2015 to determine just whose head should roll. The verdict was unanimous: Eric Sprott was dumped as the lead portfolio manager on the Sprott equity and hedge funds. This was after being replaced as CEO of Sprott Asset Management in 2014.

While the press releases mumbled about how Eric was in his seventies and wanted to spend more time with his family, and while it soundly praised him for his contribution to mining around the world, the fact remains he was fired. Ultimately, in April of 2017 he was removed as chairman of Sprott Inc. [41]

Eric's old office was on the twenty-eighth floor of the RBC Building in Toronto. Lest he forget his new position, he was demoted two floors to new, smaller quarters on the twenty-sixth floor, also losing much of his view of the city.

When Quinton saw him that day in May of 2015, Eric Sprott was downcast and disheveled, to say the least of it. Normally well dressed and natty, his changed appearance matched his demeanor. He was dejected at being removed but it was impossible to dim his belief in the ultimate fate of gold and gold resources. But he wasn't interested in hearing about Novo or the progress Quinton had made in Western Australia. After all, why should he be? He had little to do with Sprott now.

He hadn't lost his optimism. He still had hundreds of millions of dollars of his own. Eric asked if Quinton would look at properties for him now and again. Of course Quinton agreed. It was his forte.

That meeting would directly lead to the purchase of an Australian mine with the highest grade gold in the world and the lowest cost of production.

For much of 2015 Novo kept advancing the Beatons Creek deposit. They completed the preparation work in anticipation of building a mill, but that was never the real plan. It would be the dumbest thing in the world to spend millions of dollars in planning and permissions to build a $100 million mill that was a stone's throw away from another $100 million mill, owned and operated by a company that had never made a dime mining gold.

Gold continued lower. Sooner or later Millennium would fold.

CHAPTER 14
QUINTON DISCOVERS A RICH GOLD MINE

MUCH OF WHAT OCCURS IN LIFE can be put down to pure chance. Stuff happens because other stuff happened. The meeting with Eric Sprott in March of 2015 would have a long-lasting and far-reaching effect that couldn't have been predicted or even recognized at the time. It just occurred, but the reverberations would last for years.

No one in the mining industry is neutral about Quinton Hennigh and his talent. You love him or you hate him. I have never found anyone in between. If you are secure in your own abilities and judgment, and spend five minutes discussing anything about mining or geology with him, you shake your head in wonder because you just learned something simple that you should have known but didn't. And he explained it so clearly that you got it right away, and realized: I could have done that. But you didn't. So you go back for more because he is so interesting to listen to and knows so much. And he wants to share all that he knows with you.

On the other hand, those who have made it in life and the resource industry through a hearty combination of bluster and bullshit hate him. They wish they were that smart or knew that much, and deep in their black hearts realize they come up short. They can't wait to badmouth him or try to cut him off at the knees, knowing they are no comparison. It's amazing how many bullshit artists populate the resource industry. There are almost as many conmen in it as there are in the newsletter business. Those guys hate Quinton because he makes things seem so simple. So simple they just don't get it.

Eric Sprott called Quinton in early 2016. Eric had been

following the Fosterville gold mine in Victoria, Australia. A company called Newmarket Gold owned Fosterville. Sprott had been accumulating Newmarket shares in anticipation of getting it to merge with Kirkland Lake Gold. Eric had a major position in Newmarket and also owned a lot of Kirkland. He felt that Fosterville had potential for a lot more gold. He wanted Quinton to visit the project and tell him if he should buy more.

Quinton arrived at Fosterville on May 12, 2016. Doug Forster, president of Newmarket Gold, greeted him and took him on a tour. Newmarket had been processing the near-surface gold and at the time were milling 6.91 grams of it per ton of rock. What Quinton saw convinced him that they had a lot more potential than that.

The strange thing is that Fosterville had been showing strong sniffs of high-grade gold for a couple of years. That was what convinced Eric it was a rather large deposit (or BFD, in mining parlance). Another Canadian junior, Crocodile Gold, had owned Fosterville until it merged with Newmarket in May of 2015. [42]

In March of 2014 Crocodile reported assay results from the Lower Phoenix structure, [43] later to be named the Swan zone. It showed exceptional grade and thickness, including 122 grams of gold per ton over 6.8 meters and 24 grams per ton over 33.8 meters. Those are home run holes. But Crocodile didn't understand the nature of the system.

In February of 2015 more assay results came out, including 123 grams per ton over 9.4 meters, 51 grams over 6.0 meters, and 294 grams over 1.8 meters. Crocodile may not have understood the nature of the system and what those grades meant, but it didn't pass Eric Sprott by.

By late 2015 you couldn't even give a high-grade gold mine away. Why would anyone want to own gold? What future was there for it?

High-grade gold in veins from the Fosterville mine

This style of mineralization is called epizonal. These deposits are rare but rich. Just prior to Quinton's visit, Newmarket was drilling deeper holes and coming up with higher grade hits than the six or so grams per ton they had been mining. The gold content was spectacular.

This kind of gold is easy and cheap to recover via gravity. When Quinton saw it on the shaker table he was quite impressed.

He sent his report to Eric. He said that there was a lot of high-grade gold potential at Fosterville. It would require deep drilling but the company was well worth buying. Eric continued to add to the shares in Newmarket that he already owned.

Core speckled with tiny gold bits

They drilled in June. It was deep, almost a kilometer below surface. In July, RBC Bank contacted Quinton and asked him to prepare a due diligence report in anticipation of Kirkland Lake merging with Newmarket.

Quinton flew down in August to do the report and by chance ran into Tony Makuch, the president of Kirkland Lake Gold, who had just arrived from Toronto. They shared a taxi to the project. By this time the assays had come back from the deep drilling in June. When they looked at the core it was obvious they had not hit the high-grade zone. The core showed the quartz stockworks, similar to what Quinton had observed during his May visit.

Gold table, for concentrating gold from the Swan zone

Quinton was highly encouraged and told Doug Forster to drill slightly deeper. Tony Makuch told Quinton that he wasn't about to pay $1 billion for Newmarket, and that as far as he was concerned, they had nothing. It was a piece of dog shit.

When Quinton sent his due diligence report to Eric Sprott he told Eric to not worry about the price; the gold was there. Meanwhile, Tony Makuch was pissing on both the project and the company. Eric was as nervous as a long-tailed cat in a room full of old ladies in rocking chairs, but went ahead due to Quinton's encouragement.

In October of 2016 Kirkland Lake made its offer to merge with Newmarket Gold. The offer was valued at about $1 billion. In Toronto no one thought much of the deal and it only went through at Eric Sprott's insistence. The merger closed on December 1, 2016. [44]

Quinton Hennigh did not discover high-grade gold at Fosterville. Crocodile Gold drilled and found it at least two years earlier. Quinton did pinpoint the Swan zone. That deposit would make the fortune of Kirkland Lake Gold. Quinton's secondary contribution was in understanding the nature of the system and how to advance it. His primary contribution was to hold Eric Sprott's hand and tell him it was a project made for Kirkland Lake Gold, in spite of the opposition from Kirkland Lake's management.

On January 17, 2017, just six weeks after the merger took place, Kirkland Lake announced some of the highest-grade assays ever recorded in Australia, and certainly the highest ever at the Fosterville mine. [45] The best was 1,429 grams of gold per ton over 15.1 meters, including an incredible 21,490 grams over 0.6 meters, in hole UDH1817.

It was exactly as Quinton Hennigh had predicted in August. I can't tell you how many people that pissed off.

The Lower Phoenix Footwall was promptly renamed the Swan zone. The head grade for Fosterville went from 6.9 grams of gold per ton to 49.6. It became the richest and lowest-cost gold mine in the world.

At the time of the merger, Kirkland Lake Gold was valued by the market at about $2.4 billion for all its projects. It now carries a value of $16.64 billion. The shares went up over 1,000 percent and made Eric Sprott another one-and-a-half billion dollars.

Tony Makuch and his sidekick Greg Gibson both made millions on their now ultra-valuable options. Neither would forget that Quinton Hennigh made that money for them, and made it look easy.

The relationship between Kirkland Lake and Novo would become more and more interesting, with Kirkland helping Novo at important points and nearly destroying Novo at other times. I seriously doubt that the board or Eric Sprott ever knew about the games being played behind the scenes by Gibson and Makuch.

CHAPTER 15
SNIFFS OF GOLD AT KARRATHA

EVER SINCE QUINTON CAME UP WITH HIS THEORY of how the gold got into the Witwatersrand, he saw the Pilbara region as potentially hosting billions of ounces of gold. He always wanted to go big rather than have just a deposit or two. As he advanced Beatons Creek and Marble Bar he continued to seek out other properties that would work with what he already controlled. He had no intention of breeding and training a one-trick pony.

In August of 2015 Novo announced its intention to purchase what it called the Blue Spec and Gold Spec[46] deposits from Northwest Resources, an Australian company. This would add about 215,000 ounces of high-grade resources to its portfolio. Novo paid with cash and shares. By 2016 the company had begun to devote resources to investigating just what the potential might be at Blue Spec.[47] The two deposits were located about twenty kilometers due east of Beatons Creek, so Novo personnel could easily support geological surveys and work at both projects.

Most of the groundwork necessary to develop a mine and build a mill is boring — way too boring for most investors, who want something interesting, and immediate action. Punters treat a purchase of a junior mining company's shares as they would any other lottery ticket. If the draw will be on Saturday night, they want something happening on Friday and Saturday. But building a mine and a mill can take a decade or more.

Novo advanced Blue Spec and another newly acquired property named Talga Talga[48] during 2016, while moving Beatons Creek towards trial mining. In the background, events began to take place that would have giant implications for Novo and for Quinton.

In the late summer of 2016 Quinton began hearing about the discovery of patches of gold nuggets around the town of Karratha in Western Australia. That's at the west end of the Pilbara Basin.

The best geologists have a snout like a beagle pup. They stick their noses into everything in the hopes of finding something interesting to gnaw on.

Sometime in September, Bill and Kerry Edwards contacted Quinton and reported that someone on a crew digging gravel for some Rio Tinto rail work saw gold nuggets in the gravel. The workers promptly spent $10,000 on a metal detector and returned to the gravel pit on the weekend. That weekend was worth a quick three hundred ounces of gold. I'm certain it was reported to the government and all taxes were paid on it.

Bill and his wife Kerry lived across from the Novo camp at Nullagine. Bill did a lot of small-scale mining and kept his ear to the ground on Quinton's behalf.

Over the past twenty years I have met a lot of geologists. I suppose I have crossed paths with a thousand or more. Most cast no shadow. I'd say that very few of them were much better than shake-and-bake geos who could recite what they were told in class and hadn't had an original thought since their first visit to Lima, where they were trying to figure out where and how to get laid. That's pretty dumb. If you can learn how to fall off a bike you can figure out how to get laid in Lima.

But I have also gotten to meet and talk with the very best in the industry, such as David Lowell, Peter Megaw, Keith Barron, and Quinton. The greats are interested in everything, have a wide range of knowledge about just about anything, and are constantly trying to piece things together.

I've met all these guys and have spent a lot of time with them over the years. All things fascinate them, and each can talk intelligently on dozens of subjects, not just rocks. And they ask questions constantly. They question everything, and perhaps more importantly, they listen. Any opinion might have some value. I've been very lucky to know these guys.

The Pilbara Basin is home to one of the largest and richest banded iron formations in the world. According to Wikipedia, about 59,500 people are employed in iron mining in the Pilbara. Gold mining dates back to 1888 through about 1900, and it would be safe to say that over the years, 100,000 miners of one sort or another have trodden on the Pilbara in search of valuable minerals. It took that rail crew from Rio Tinto, Johnathon Campbell, who I will return to in a moment, and Quinton to put it all together.

The rail crew needed only to notice that the pit they had opened up for fifty truckloads of gravel contained gold, in order to transfer three hundred ounces of it into their pockets over two days. That information told Quinton that he needed to get busy staking, before the whole world realized the potential of the conglomerates. If one pit about two meters deep and perhaps one hundred meters across can generate $500,000 worth of gold in a weekend, how much more gold is still around there?

The point here, and it is an important point, is that the gold discovery at Mallina by Jimmy Withnell took place 129 years before the light switched on in Quinton's brain. Government reports from the early 1900s mentioned gold in the conglomerates and in the eroded gravel. The existence of the gold was established all those years ago, and no one put it together until April of 2017.

On November 16, 2016 a tiny Australian junior named Artemis Resources issued a press release announcing that it had found gold nuggets at surface at its Purdy's Reward project.

Here is an extract, including one of the Figures referred to.

ASX Announcement
16 November 2016
Purdy's Reward Gold Discovery – Karratha

Highlights

> ✓ Visible gold and nuggets exposed in mafic rocks, at Purdy's Reward Project 35 km south–south east of Karratha.
> ✓ Surface gold identified over a potential 800 metre strike.
> ✓ Gearing up of exploration activities

Artemis Resources Limited (ASX: ARV) is pleased to announce that recent exploration activities have confirmed the presence of primary gold mineralization (Figures 1 and 2), with significant free gold, in mafic rocks 35 km SSE of Karratha ("Purdy's Reward Project").

The primary gold mineralization was recently discovered by prospectors in the belief that, because the gold was flat and rounded, it was elluvial in nature. The visible gold actually sits within weathered mafic rock and requires significant hand pick, crow bar and sledge hammer work to liberate. Free gold has now been found over a strike length of 800 metres with widths up to 100 metres within the project area (Figures 3, 4 and 5).

The geology of the project is characterized by Archean felsic and mafic rocks. The only previous exploration work in the area was back in 1971 by Westfield Minerals NL and this exploration programme focused only on base metals exploration with 6 percussion drillholes on the western tenement boundary. These holes were assayed for nickel with assays returning up to 1260 ppm Ni in drillhole 69-SP-07 (Table 1, Figure 5) associated with a chloritized mafic basalt.

Figure 1. Artemis, Purdy's Reward Project - Karratha. Gold nuggets, flat and rounded and up to 13 grams from surface detecting.

All gold mineralization observed and found to date in the

West Pilbara has been associated with quartz reefs. This new style of gold mineralization within mafic hosted rocks increases the potential size of mineralized horizons.

> Ed Mead, Artemis's CEO, commented: "The initial work from Purdy's Reward increases our confidence in this under explored gold region of Western Australia. Off the back of recent results from Silica Hills, the best indication of gold is visual gold at surface, and we certainly seem to be getting that."

Most geologists will confirm that visible gold at surface is indeed a strong indication of the presence of gold. Pedants may wish to discuss the precise meaning of the expression "certainly seem".

For the next couple of years Ed Mead would claim that it was he who discovered gold in conglomerates around Karratha. He also claimed that the Witwatersrand basin contained "watermelon seed gold identical to that found in the Karratha region." Alas, neither claim was true.

In Geological Survey #33, published in 1909 by the government printer in Perth, a geologist named A. Gibb Maitland stated that "nearly all the alluvial gold was of the uniform shape and size of small melon seeds." He was talking about gold from the Egina region, but the gold from Karratha proved to be identical. Another Geological Survey published in 1906 suggested that the source of the gold in the Pilbara was the conglomerates.

Quinton Hennigh realized how significant that was. The distance from Egina to the projects he would pick up at Karratha was about 115 kilometers. And the gold from the two areas was virtually indistinguishable. The gold from the Witwatersrand was in tiny bits, and there was no watermelon seed gold at all.

Credit for any gold rush belongs rightly to the person who made the discovery that people noticed. It was known as early as 1842 that there was gold in California, but it was John Marshall's discovery at Sutter's Mill in January of 1848 that brought about the incredible mass movement of prospectors to the new state of California.

Likewise, in the Yukon, it was George Carmack and Skookum Jim who started the snowball rolling down the mountain after they found gold on August 16, 1896 at Rabbit Creek, later renamed Bonanza Creek, more appropriately. Their discovery began the short-lived Klondike gold stampede of 1897–98. But the natives had known about the gold for centuries. They had no use for it, and much preferred the copper that could also be found there.

The person who actually began the Karratha gold rush was a cattle rancher named Johnathon Campbell. In August of 2016 he and Bruce Woods, a Kiwi helicopter pilot, took a Robinson R-44 chopper up near Karratha to count wayward cattle. Following a creek system downstream, they passed over a camper hidden under some trees. They supposed that it was someone either on the run from the law or prospecting for gold. In Australia that is often the same. As they continued down the creek they saw a number of systematic holes poked in a gully leading from the dried-up creek bed.

Campbell lives near Port Hedland. He placed a notice on Facebook, looking for a gold prospector to work with. A man named Brad Smith answered his ad. They purchased metal detectors and went back to the area that Campbell and Woods had flown over.

Campbell and Smith drove to what is now called Purdy's Reward, pulled out their metal detectors and went to work. They found gold nuggets everywhere they looked. Brad Smith soon caught gold fever and commented that he had never seen anything like it in his life. That comment brings back memories of watching the movie, *The Treasure of the Sierra Madre*, [49] and what happens to people when greed for gold overtakes good sense.

On that first visit, the two intrepid prospectors spent only half a day in the field but took home about $5,000 worth of gold nuggets. Smith was hooked, especially when Campbell promised to split everything they found. On subsequent weekends spent prospecting on their claims they would take home between $5,000 and $20,000 every time.

Australia has an interesting mining right, designed for prospectors and small miners, called a Special Prospecting Licence. [50] It allows prospectors to mine for gold over a mining lease held by someone else, on a limited basis. Johnathon Campbell applied for and was granted SPLs on both Purdy's Reward and Comet Well. Brad Smith applied for a small block further down the creek but soon dropped it, saying it was no good and a waste of money. Campbell picked it up but was mocked as a fool by Smith.

Artemis Resources held the lease on Purdy's Reward, and a man named Peter Gianni owned the lease on Comet Well. The SPL allows mining on a small scale if the primary leaseholder does not object.

Not long after this, on another chopper flight seeking lost cattle, Campbell and his pilot flew over some people operating a small bulldozer to scrape the ground. They were then following the bulldozer with a metal detector and collecting the gold nuggets they found. Campbell and his pilot landed and talked to them. It was Ed Mead and one of his people from Artemis.

Artemis had gotten into hot water with the Department of

Mines and the local native group for disturbing the ground without any permit or agreement with the natives. Claiming they were rehabilitating the ground, the pair was instead inflicting even more damage. They freely agreed they were selling the nuggets to the Perth Mint.

I have looked at Artemis' financial statements. Nothing is obvious about any revenue from illegal gold sales. If it had been following the rules, it would not have been permitted to sell the gold in any case until it had obtained a mining license. Which it didn't have.

Campbell asked Mead who the registered holder was of the "Armada" lease. Mead volunteered that Artemis held the property. Campbell said that he had been trying to contact the company, but no one ever picked up the phone or returned calls. Ed Mead said the company had no money, and that for all purposes, he was the receptionist.

In a neighborly spirit, Campbell told Mead that he had applied for an SPL over the part of the Armada lease which came to be called Purdy's Reward. Ed said he didn't care; he had no interest in the ground other than Artemis controlling it from a legal point of view.

Campbell persisted, and told Mead that it was far better ground than they were working. Mead should at least look at it. Remember, he was claiming to be doing rehabilitation work, and that that was why the bulldozer was there. Artemis had never done a thing with Purdy's Reward, so claiming to be cleaning up past work wasn't just weak, it was impossible. He was using the bulldozer to find gold.

Mead said he didn't care if Campbell worked the SPL over Purdy's Reward, but changed his mind in a hurry once he ran a metal detector over the project and took out a lot of nuggets. He immediately chased down Brad Smith, who was camping nearby, and told him to get off the ground and not to return.

As for Johnathon Campbell, to this day he thinks about Ed Mead as a street he has driven down in Perth. It's marked "ONE WAY." Mead would have never known there was gold at Purdy's Reward, but for Campbell.

Smith and Campbell, through their SPL, still had the right to prospect for and find nuggets at the Comet Well project. One day in late 2016 or early 2017 they talked to a geologist named Rob Jewson about trying to buy the lease on Comet Well from Peter Gianni. Brad Smith was hard to get along with, and people who dealt with him tended not to want to do it twice. Gianni agreed to take $50,000 for the lease but wouldn't deal with Smith.

Then one day Smith called Campbell to say he had found someone who would put up $100,000 for the lease on Comet Well. The three of them would own equal shares of the project. The new investor was named Darren White but everyone involved always referred to him as YT.

Campbell was in agreement and quite willing to sell an interest in the pending lease to White, as was Smith. But when the terms were first discussed in January, Brad Smith refused to put his half-interest in the SPLs on Comet Well into the deal. He told Campbell not to consider adding the SPL to the deal. He could take out thousands of dollars' worth of gold every weekend, so why should he share it with someone new?

But after the sale was concluded in April of 2017, with Darren White now owning half of the Comet Well lease, he demanded Campbell throw his interest in the SPLs into the deal. By that time Smith had flip-flopped, rewarding the guy who had put tens of thousands of dollars into his pocket by joining with White and changing the terms of the deal after it was signed, sealed, and delivered. Campbell reluctantly agreed, but was starting to figure out which way the wind was blowing.

Quinton made his first visit to the Comet Well property, arriving on April 6. What he saw convinced him to get in touch with the leaseholders. YT didn't want Campbell talking to Quinton. Quinton sat down with YT and made him an offer. Fifteen minutes after leaving YT's office, Quinton's phone started ringing. It was Mark Creasy. Quinton didn't answer any of the calls.

Obviously, and as Quinton later confirmed, YT had taken Quinton's offer straight to Mark. Mark made YT an interesting counteroffer. He would buy all of Comet Well for $5 million by paying $50,000 for each percentage point. He would hand over $50,000 to the trio for Comet Well, gaining the right to test it and collect gold without paying one cent more. If it proved valuable, he would buy the additional 99 percent. If not, he would walk.

It was not only a predatory offer. Mark was a Novo insider due to his share position, and there is the interesting question of how he could become involved in negotiations between Novo and a third party, ethically or legally.

YT eventually accepted a higher bid from Novo, but YT talking to Mark Creasy greatly increased the cost to Novo. Dealing with the three owners of Comet Well was difficult because none of them could agree with the other two on how to proceed. Eventually Quinton got Johnathon Campbell to accept an offer for his one-third share. That forced Smith and YT to come to some kind of reasonable terms with Novo. Quinton does not reminisce fondly about dealing with three people who had entirely different views of what would constitute an acceptable deal.

What was interesting to me was that Comet Well was a free ride for Brad Smith, who took out tens of thousands of dollars in gold within just a few weeks after being pulled into the deal by

Johnathon Campbell. YT bought into the project in February and had an agreement with a well-run and well-financed junior exploration company two months later.

The back story of the gold is far more interesting than the above somewhat confusing press release issued by Artemis in November, when Ed Mead began claiming to be the fellow who began the Karratha gold rush. First we must identify some of the more important players.

David Lenigas became executive chairman of Artemis Resources on November 3, 2016. His timing was impeccable. Artemis was a junior with a tiny market cap, no money, no focus, and a zillion shares outstanding. His appointment came with a $5,000 monthly salary and, as an incentive, twenty-five million shares of the company.

All he had to do to make a fortune was to increase the value of the company to an absurd level, sell his shares and ride off into the sunset, leaving the rubble of Artemis in his wake. That is exactly what he would do, over time.

Lenigas was your typical overweight, pompous, beady-eyed blowhard, living in Monaco off the rewards from the other 160 tiny juniors he had churned and burned. In physical appearance and in attitude he resembles Goldfinger, the antagonist of the 1959 novel and the 1964 movie of the same name.

He fancied himself as the reincarnation of Tiny Rowland [51] and indeed there were similarities, including a willingness to stiff shareholders regularly. In mining I have found that some management types mine rocks, and some mine shareholders. Lenigas mined shareholders again and again.

When he became chairman, Artemis was a company without a clue. It claimed to be a copper company and a cobalt company and a nickel company. It intended to mine zinc, lead, platinum, diamonds, obviously gold and silver, and even had a graphite project in Greenland that fit its portfolio perfectly. In other words, management had no idea what they were doing. They would soon spend $3.5 million on a rusted wreck of a processing plant that would forever be about to go into production.

Artemis was perfect for the Lenigas magic. He even had a mysterious investor in the background, Mick Shemesian, [52] dubbed "Many Names" Shemesian by the Australian press due to his refusal to respond using the same spelling of his name to the many lawsuits in which he was involved. Shemesian was the largest shareholder in Artemis.

Allan Ronk, who had worked with Quinton at Beatons Creek, was on an assignment for Artemis in the Karratha region. When he learned about the gold the rail crew had found and saw a February 20, 2017 press release [53] from Artemis, he became the first to connect the dots. Below is a lengthy extract from that press release, plus one of its six Figures.

While Ed Mead may have believed all of the gold in the Pilbara was in lode gold veins, Quinton was well aware of the government reports going back over a century. He was convinced that the key to the gold jewelry box was in the conglomerates. If that was true in the Eastern Pilbara, as it surely was at Beatons Creek, why couldn't it be true in the Western Pilbara, 360 kilometers to the northwest?

ASX / Media Announcement
20 February 2017
New style of gold mineralisation for the West Pilbara discovered at Purdy's Reward gold project – Karratha, Western Australia.

- Archean sediment hosted gold mineralisation identified.

- Typical sediment style deposits of this type are Witswatersrand, Nullagine and Marble Bar which are conglomerate hosted.
- Assays confirm gold mineralisation is typical of this style with fine to coarse gold.
- Geochemical assays from orientation traverse identifies gold anomalies coincident with high levels of prospector activity.
- Free gold continues to be found at surface at Purdy's Reward Gold Project by prospectors.
- Programme of Work for drilling and costeans has been approved by the Government.
- Heritage survey requested and expected in March.

David Lenigas, Artemis's Chairman, commented;

"The results from the new Ionic Leach geochemistry techniques have now been received and they confirm a completely new style of gold mineralisation for this area of the West Pilbara akin to the Witswatersrand style Archean sedimentary hosted deposits. These deposits are similar to those being discovered in the East Pilbara, around Marble Bar and Nullagine, by TSX listed company Novo Corp. Novo's work has identified predominantly fine gold mineralisation, whereas Purdy's Reward not only has fine gold but coarse gold in the form of nuggets. This makes the Purdy's Reward discovery significant for the area and highly prospective. This is the first time that this type of geology and gold mineralisation has been confirmed as Archean sedimentary and associated with conglomerates and fine grained sediments with mafic appearance."

Artemis Resources Limited ("Artemis" or "the Company") (ASX: ARV) is pleased to report the identification of a new gold mineralisation style for the Karratha area at Purdy's Reward, West Pilbara.

Results (Figure 2, 3 and 4) from a geochemical orientation sampling traverse at the Purdy's Reward Gold project, and additional mapping support a model of Archean sedimentary (conglomerate) hosted gold.

The gold occurrence at Purdy's is considered analogous to the conglomerate hosted mineralisation outlined by Novo Resources in their Beaton's Creek Project near Nullagine, but Purdy's Reward is significantly older in age. The style of mineralisation is referred to the Witswatersrand style, after the Witswatersrand gold province in South Africa that has significant gold in Archean sedimentary conglomerates.

The traverse straddles the unconformity between the older Archean basement in the north and the overlying the Mt Roe Basalt to the south.

The samples were collected along a 1 km long traverse; samples for analysis using the ultra-sensitive ALS Global Ionic Leach™ technique were collected at 25 metre intervals. For comparison purposes at every 100 metre sample point an additional sample was collected and analysed using a conventional digest (Supertrace).

Within the central area of the traverse numerous prospector metal detecting pits were present. The area of the detecting pits corresponds to where significant anomalism in both the Ionic and Supertrace results occur. The results are expressed as "Response Ratios" for both techniques. This is to enhance the response to background signal of the data and to perform basic levelling on the results; this allows direct comparison of the results from the different methods. As such these values do not have a unit type.

Figure 1: New Conglomerate Package identified at Purdy's Reward:

[Figure 2 shows Purdy's Reward orientation geochemical sampling traverse.]

The Ionic data (Figure 3) shows a stronger anomaly to background response compared to the Supertrace data. Neither technique (Figure 4) shows significant correlation between the strong gold responses and the typical pathfinder elements: arsenic, silver, bismuth, molybdenum, antimony and tungsten.

154

This lack of pathfinder elements (Figure 4) indicates the gold is not derived from a shear system or from supergene alteration of a shear system. The data suggests that the Purdy's Reward gold is from an alluvial source, albeit Archean in age. The lesser gold responses on the right hand side of Figure 4, which are at the northern end of the traverse (Figure 2) do show coincident gold, silver, arsenic responses indicating there is a secondary zone of mineralisation in the area which is shear system related.

Allan Ronk told Quinton to take a close look at that press release. Allan had of course identified the mineralization as being identical to that of Beatons Creek and Marble Bar, and had provided most of the technical information in the press release.

At this time, literally no one had paid serious attention to the conglomerates for over a hundred years. In the Beatons Creek area, miners in the nineteenth century actually dug caves into the conglomerates to mine for gold, but no one had investigated and then said, "Maybe all the conglomerates hold gold" — except for Quinton Hennigh, and, generations before him, a few government geologists who were largely ignored. Everyone believed that virtually all the gold was from a lode source.

I should mention one confusing detail here. Frankly I don't know why the conglomerates are called "reefs" in the Witwatersrand but "conglomerate layers" in the Pilbara. I like reefs better.

Quinton looked over the press release from Artemis, with its pretty photo of a conglomerate outcrop, and took action at once. Novo's exploration manager at the time was Luke Meter. Quinton told him to pack his bags and go to Karratha and peg every claim around the area where the conglomerates showed at surface.

By April 2017 Novo was down to $1.5 million. Ronan Sabo-Walsh, its financial guy, was sending $100,000 to Perth each week to cover expenses incurred in filing the claims. Quinton couldn't talk about the new properties as it was a work in progress; Novo was still busy staking as much ground as possible. But it would become a great deal for investors within a very short time.

A private placement was being done at sixty-six cents a share, with a full warrant. Initially Novo planned to raise $8 million, but the terms were so attractive that the size of the placement was first increased to $12 million Canadian, and then to $15 million by the time it was closed in April 2017. Those wise enough to participate were in for the ride of their lives by the time the four-month holding period was over.

Novo was still busy on the eastern side of the Pilbara Basin. It had begun a 30,000-ton bulk sampling program in mid-2016 and was busy drilling in the Blue Spec. The results from the bulk sample showed the grade at Beatons Creek to be much higher than the numbers obtained from the drilling and surface samples.

Also in April, Novo announced the deal on Comet Well. All this time, behind the curtain, Novo's geologists were staking land non-stop. On the strength of the Comet Well agreement, Quinton was able to reach agreement with Artemis on a package of just over 1,500 square kilometers, including Purdy's Reward.

Lenigas actually called Quinton to see if Novo was interested in doing a similar deal on Purdy's. Lenigas didn't understand it at the time and perhaps still doesn't, but to Novo, the primary attraction of Purdy's wasn't the quality or size of the project. Artemis had something far more valuable: a drill permit for Purdy's. Getting the same for Comet Well would probably take a year. Dealing with the bureaucrats in Australia is like swimming through cold molasses. It's wet, nasty, and slow.

Mead and Lenigas still believed the bulk of the gold would be in lode gold veins, so they thought they were restricting Novo by making the agreement valid only for the gold in the conglomerates. But that was all Quinton was interested in.

In the press release announcing the binding letter of agreement on Purdy's Reward, Novo mentioned for the first time the staking of over 6,000 square kilometers of ground; over 2,300 square miles. It was a giant position and was news to the market.

In the press release Quinton explained what Novo sought, saying, *"The basis for staking such a large land package is the recent recognition of gold-bearing conglomerates in a previously unexplored sequence of rocks near the base of the 2.7–2.85 billion year old Fortescue Group, a thick pile of sedimentary and volcanic rocks underlying vast portions of the Pilbara region."*

That was the basis of a giant shift in thinking that most of the tiny juniors in the Pilbara region have still not come to grips with. It was a shift from exploration for lode gold sources to a search for gold from the conglomerate structures. Quinton upended the thinking of the gold community. It would take years for the rest of the herd to realize that the easy gold was in the reefs and the gravels extending to the Indian Ocean.

Immediately after completing the deals with the Comet Well threesome and with Artemis Resources for Purdy's Reward, Quinton made a welcome addition to the Novo management team. He is so talented and smart that he tends to want to do things himself. It's not micromanagement but he does take on more than he should at times, because he can often do a better job than anyone else. That's all well and good, but that kind of management style limits a company to what one person can accomplish, no matter how well.

I had been bugging Quinton for months about a matter he had realized himself but hadn't addressed. As an American running a Canadian-listed mining company in Australia, no one was going to go out of their way to do him any favors.

Western Australia still resembles the Wild West days of the United States and Canada. Everyone knows everyone else's business. To a certain extent it's a closed community, even though Quinton had worked for Newcrest, an Australian mining company, and was highly respected in the country. He was still a furriner.

Quinton needed a first-class manager who was Australian. He announced the hiring of Rob Humphryson as CEO. Rob came with a resume as long as your arm and had experience of a broad range of duties over his twenty-five years working in Australia. Best of all, he was Australian; he understood the culture and knew how to get things done.

At the same time, Quinton promoted Ronan Sabo-Walsh to Chief Financial Officer (CFO). I've known the pair now for several years and both are brilliant. Especially given the COVID-19 lockdowns and stupidity, Novo was now well prepared with a widely experienced management team.

A lot of what Quinton does and announces merely flickers through the minds of investors, seen and soon forgotten. However, he is always playing three-dimensional chess and there is always a method to his madness. Investors and outsiders like to throw rocks at him but his moves are always carefully thought out.

One of the biggest dangers that any junior mining company faces is that of succeeding. As soon as a junior makes an announcement of a big find, everyone wants to steal it, including all of their big partners.

In the early days, when capital is almost always the biggest problem, finding investors with deep pockets seems like a good thing. Having Mark Creasy as a major shareholder was good for a lot of reasons. It also could have been the kiss of death to Novo, and in a moment we will see an example of a major investor trying to chop the company into tiny pieces.

Mark had held big positions in other tiny juniors, and when he could hold a company to ransom he was quite prepared to do so. It was like keeping a pet lion in the back yard. It keeps the neighbors relatively honest, but when it is hungry it will eat your leg without a second thought.

So in all of Quinton's dealings with everyone, he kept in the back of his mind that even the biggest shareholders might turn on the company. There were a couple of times when Mark could have done things to put Novo on the map instantly, but he always put his interests as an individual in front of his interests as a shareholder. I thought that was especially dumb, and I like Mark a lot.

The major mining companies can be dangerous too. Everyone wants someone else to do the work but everyone feels they are smarter than management. Or they see a chance to pick up a prime asset cheap. Big companies, small companies, big investors, small investors — everyone is smarter than management.

If you don't believe me, go to Stockhouse or CEO.CA or HotCopper and see all the fools hiding behind their aliases. They know everything about mining, and if you were to turn any company over to them for a month they could sort out all its problems.

Quinton Hennigh did a better job of keeping his potential enemies at arm's length than anyone I have known in mining. He did the same with his friends, for you never know who might turn.

It's a finely tuned balancing act; you don't want to be offending anyone, but you still have to run the business and make progress. If any single investor gets too big a position, it is only natural that they start thinking they should have the majority of the input as well.

Few management teams make it from discovery to production without being sniped at, and often hit and killed. Everyone wants the bonanza deposits with the least effort and expense.

I made a trip to Japan in June of 2017, to Hokkaido, to visit Irving Resources' new discovery there. Quinton serves as an important advisor to the company. He brought the lovely Heather. Brent Cook joined the tour, as well as Maurice Jackson from Proven and Probable.

While we were in Japan, Quinton showed us some samples of the gold Novo was finding on the western edge of the Pilbara region. It is highly unusual. They are gold nuggets but in a hard rock environment.

Gold nugget from Karratha

A week later, Brent joined Quinton and me on a visit to Karratha, to see that new discovery. The three of us met in Perth for drinks. Quinton showed us samples of gold that he had bought on top of a piece of polished conglomerate.

Gold nuggets Quinton picked up on a previous trip

It was one thing to show us a few gold nuggets in a hard rock matrix while we were in Japan, but Quinton was holding out on us. Once in Australia, we began to get an idea of the amount of gold involved.

There were several prospectors with Special Prospecting Licenses who were collecting gold found with metal detectors. We went out into the field and watched a fellow named Rob Beaton detecting and finding lots of nuggets. Rob was an ordinary prospector with a metal detector who had filed several SPLs and was busy making his fortune. We could legally buy gold from him.

Rob Beaton had lots of gold nuggets for sale

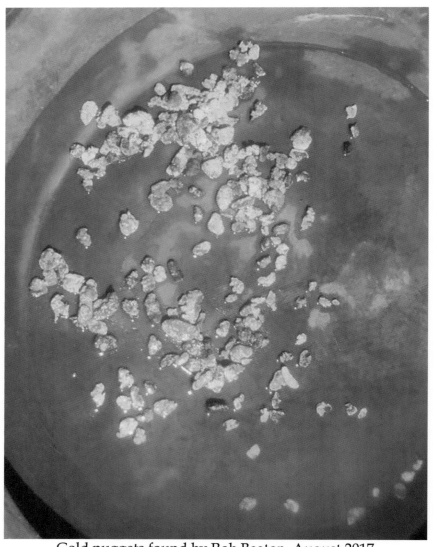

Gold nuggets found by Rob Beaton, August 2017

This was unlike anything I had ever seen. The source of the nuggets was the conglomerate, but unlike Beatons Creek and Marble Bar, the material it was found in was gravel that had been covered in basalt and solidified into hard rock. These were gravel reefs 2.8 billion years old, now metamorphosed into hard rock. Other than the size of the gold, this was very similar to the gold from the Witwatersrand Basin.

I never knew if Rob Beaton was connected in any way to Beatons Creek, some 350 kilometers to the east, but it's a small country and he probably was related.

There were a number of people on the trip. Just after we left the highway we passed the famous gravel pit where the railroad construction crew had made their fortune in one weekend. It wasn't a giant pit, maybe one hundred meters across, but good for three hundred ounces of gold in two days.

Artemis using a small digger to make payroll

We drove out to Comet Well and passed by Ed Mead, who was using a small digger to collect his paycheck for the week. This was the sort of thing that Quinton and Novo wouldn't dream of doing. It was unquestionably illegal but that presented no problem to Ed Mead, Artemis, or David Lenigas.

Novo hadn't progressed to exploration yet. We were visiting Rob and inspecting Novo's ground at Comet Well to see the size of the project. Quinton had his guys stake an incredible total of 7,600 square kilometers, or 2,950 square miles. Gold here at any grade would make this the largest gold find in history.

Rob Beaton detects while Eric Sprott watches

Many of us on the tour bought some gold from Rob. He had a lot of gold.

Quinton and I returned to Karratha and visited the local representative of Minelab metal detectors, reputed to be the best on the market and the most sensitive. At the time of our visit, in the winter of 2017, Minelab's most expensive product cost about $10,000 Australian. There has been so much demand since then, largely because of the incredible stories coming out of Karratha, that the quality of the machines has increased markedly and the price has been chopped in half.

The dealer, located in a dingy hut in a beat-up industrial park, brought out tray after tray of gold. He was so casual about doing so that it was plain he had a lot of gold. We paid a slight premium for what we bought, but it wasn't much. He must have been selling millions of dollars' worth of metal detectors and equipment each year, and taking even more back in purchases of gold nuggets.

He took care of Quinton and me because Novo had brought a lot of welcome attention to the region. He was selling detectors with one hand and filling up the other paw with nuggets, and making a nice margin on both.

He finally brought out a tray loaded with pairs of matching nuggets that he had purchased from a miner with a project near Egina. I would mention to Quinton that the gold from Comet Well that we bought from Rob was identical to that gold from Egina. That implied a giant strike length and a lot of gold.

What I didn't know at the time, and wouldn't learn until much later, was that while Purdy's Reward and Comet Well were both hard rock with embedded nuggets, Egina was an entirely conventional alluvial project. Again, since the Pilbara has either too much water or too little, Egina had been mined off and on for a hundred and twenty years.

The Karratha projects were gold nuggets in conglomerate, while the Egina area was a lot of gold nuggets in gravel. Come back in a few million years and they too would be hard rock.

Gold nuggets from Egina, on display in Karratha

What I also didn't know at the time — because he keeps his cards close to his vest, even from me — was that Quinton had opened discussions with a man named Carl Dorsch, to buy his fully permitted gold project at Egina.

Earlier in this book I mention that years before, I had asked Quinton where the gold at Beatons Creek went when the outcropping conglomerate eroded. He told me that it had been washed into the Indian Ocean.

Well, not all of it had made it all the way to the water's edge. Some of it landed near Egina and was slowly working its way west.

CHAPTER 16
QUINTON FLUBS THE FIRST ASSAYS AT PURDY'S

THE YEAR OF 2017 WAS A LUCKY ONE for Novo and Quinton Hennigh, with a lot of breaks coming their way. Not all of luck is good, however. Quinton flubbed a giant set of assay results and got beat up for years as a result.

When Novo did the placement first announced in March of 2017, it expected to raise $8 million at sixty-six cents a share, with a full warrant at ninety cents. Quinton knew the company's focus was shifting to the Karratha conglomerate gold find but couldn't discuss it while the placement was underway.

Eric Sprott already owned over four million shares and four million warrants, and topped up with another five million of each. Calculated from the total number of shares issued and outstanding, Sprott owned 8.5 percent of Novo. If he were to exercise all his warrants he would own 15.6 percent.

Little did he expect that his position in Novo would soon make him well over $100 million. Remember, he profited by over $1.5 billion at Fosterville, largely because of Quinton's genius in pinpointing the Swan zone and nudging Eric to merge Kirkland Lake with Newmarket Gold. For Eric, it was a great year. For Quinton, not so great.

Quinton signed and sealed the deal with the threesome on Comet Well in April 2017, and the binding letter of agreement with Lenigas on Purdy's Reward in May. In July Novo announced a leap forward: the Sumitomo Corporation had agreed to work with Novo to put Beatons Creek into production.

That agreement didn't look like much because it didn't bind Sumitomo to anything, but anyone who understands how the Japanese do business would recognize just how important it really was. The Japanese look years and years ahead. Sumitomo was in effect becoming engaged with Novo and expected a long and profitable relationship.

Even when Eric Sprott makes a major investment in a company, he rarely has time to hold hands with management. He was now sitting on nearly twenty million Novo shares, which even for Eric was a pretty good size position. He decided he needed someone to represent his interests on the board of directors. In July of 2017 Novo announced it was adding Eric's man Friday, Greg Gibson, to the board.

Naturally Gibson had to be taken care of, so he was awarded half a million options at $1.57. He would sell his shares in early 2019 and clear something like $750,000 to $1 million for less than two years' work. Not all of his work was to Novo's advantage, either. In fact some of it came close to destroying the company.

There was also a side deal between Newmont and Eric Sprott. While it didn't affect Novo at all, it was interesting in that it shows how big finance works.

To recap: Eric had participated in a placement, buying Novo shares for sixty-six cents apiece. The placement called for a full warrant at ninety cents. Quinton knew of things in the works at Karratha, but no information was released to the public or to potential investors. It was a deal of a lifetime for Eric, as those shares would increase in price by 1,400 percent in six months.

But after Novo announced that it had picked up a lot of ground around Karratha, Eric thought it might be nice to own more shares. He asked Quinton if there were any blocks for sale.

Quinton contacted Newmont in Denver. Newmont could reduce the cost of its initial investment in Novo to zero by selling some of its shares. Newmont contacted Eric and agreed to sell him eleven million shares of Novo at $1.56.

Meanwhile, in Perth, the Newmont guys there went out to

Purdy's Reward, saw the trench where the samples had been taken, and became most enthusiastic about Novo's potential. Somehow, the Newmont guys in Perth learned that the Newmont guys in Denver were thinking about selling Novo shares. Perth told Denver that this was inadvisable, as Novo looked to be on to something big in WA.

Eventually Eric Sprott realized that Newmont hadn't invoiced him or made the agreed transfer of shares to him. He asked its Denver office where his shares were. Newmont replied that after careful consideration, they thought they might want to hold on to those shares. Since the first discussion about the sale of the shares, the price had gone up a lot.

Eric had his attorney draft a nastygram to Newmont reminding them that they had agreed in writing to the sale, and to the price. Did they really want a giant nasty public lawsuit on behalf of a billionaire suggesting that Newmont was trying to welsh on a deal?

Newmont made the sale at $1.56, begrudgingly after careful consideration. You could hear the teeth grinding from its office in Perth. So Eric Sprott did the second-best deal he had ever done on Novo shares. Like he needed the money. But on the other hand, principles are principles, especially when money is involved.

The really big news from Novo, and the first real indication of the potential size of the gold deposit, came on July 12, 2017. Quinton wrote the press release [54] while he was on a trip to Japan for Irving Resources. The first paragraph included the following. The emphasis is mine.

Novo [. . .] is pleased to announce that it has found *in situ* gold nuggets up to 4cm long in primary conglomerates from its first trench at its Purdy's Reward prospect and has collected a bulk sample of these gold-bearing conglomerates for analytical test work. The sample originates **from a one meter thick reef near the top of an 11 meter thick stacked sequence of mineralized conglomerate horizons**.

The whole exploration team was present when taking the sample, and there is a video [55] of Barry Rattingan using a portable jackhammer to pry out a great big nugget. Brad Smith had been using the metal detector. He sees the nugget when they turn over the rock and says, "Oh, shit." The jackhammer belonged to YT. He and Brad Smith provided the expertise with the metal detector. Luke Meter and Allan Ronk provided the geological thinking.

They found good gold. The sample was of a reasonable size, at seven hundred kilograms: two meters by two meters, and 5.2 cm thick. The metal detector was used first, to find an area giving a strong signal.

The sample was split into two and sent off to the lab in Perth. But first, Novo ran the material through a Steinert mineral sorting machine. Even at the start Quinton was thinking about how to reduce costs. The sorting machine sorted the first sample to a total mass of only 2.15 percent of the original material. The second sample came in at a tiny 1.82 percent of the original mass weight. After sorting, the material was sent for assay.

Costs in mining are always based on how much material you are working with. If you could get rid of 98 percent of the weight with an inexpensive sorting machine, your total processing cost would really decline.

Quinton's press release included a link to that short video. To the best of my knowledge, it was the first time a video had been included in a press release about finding gold. Since it would be easy to fake results in such circumstances, it surprised me that the exchange permitted it. But investors evidently loved the video as the share price more than doubled, going from $1.49 when it was released to $3.15 on the day when the assay results were published.

Those results arrived on August 8, 2017. One sample graded 87.76 grams of gold per ton (g/t), and the other 46.14 g/t. The average grade of the two samples was 67.08 g/t, and this was the most meaningful figure, as the two assays were from the same sample, just split.

When the results were released, the stock continued to rocket higher. In the first week of October, Novo shares peaked at $8.83, up an incredible 1,700 percent since starting the year at forty-eight cents.

There was one slight technical problem, actually two problems, that could and should have been avoided.

Many people in the peanut gallery [56] have been tossing rocks at Quinton for years, mostly through jealousy. The real issue would cause confusion and anger among shareholders who ended up buying at the top and watching the share price tumble for years.

Above, I highlighted an important sentence in the press release of July 12: *"The sample originates from a one meter thick reef near the top of an 11 meter thick stacked sequence of mineralized conglomerate horizons."*

But it didn't. It came from the bottom.

This wasn't Quinton's mistake. It was a mistake by the geologists on the ground. Quinton was in Japan at the time and didn't catch it; really, he couldn't have at the time. But when the assays came out a month later he should have caught it. He didn't. It remains the only giant mistake I have ever seen Quinton make in all the time I have known him.

A second press release,[57] on August 8, 2017, repeats the error: "*As discussed in Novo's news release dated July 12, 2017, this sample originates from the uppermost horizon of an 11-meter thick sequence of mineralized conglomerate beds.*"

It would take months of ground work and drilling to determine the actual structure of the conglomerate. Eventually Novo realized that the sample had not been taken from the top of an eleven-meter sequence. It was the difference between day and night. The sample came from the bottom of the reef, not the top.

I am the only person so far, writing about Novo, to point out that in any conglomerate sequence, since gold has a specific gravity of about 19 (not being pure), as water flows over the gravel the densest material will sink to the bottom. The richest quantity of gold is always at the bottom of any conglomerate sequence. The error in the press releases made it look as if this was the richest reef ever found. It wasn't. Everyone who believed it was alluvial gold in a reef supposed it would get richer and richer as they went lower in the sequence.

Quinton wasn't on site; he was in Japan when both press releases were issued. But in the Navy, when a ship runs aground, the captain is always court martialed.

This brings me to the second howler.

I return to the Bre-X scandal of 1997. It was a gold project in Indonesia. The geologist salted the samples with shavings of gold from his wedding band at first, and later by buying from the locals some sixty-one kilograms of alluvial gold. He fed small bits of gold into the drill samples being sent to the lab for assay. The project grew from a two million ounce resource to an incredible seventy-one million ounces in eighteen months.

No other gold deposit in history had grown so fast. And neither had the Bre-X project, because it was all fraud, with gold in every drillhole. Literally, it was too good to be true.

When Novo's assay results came out on August 8, 2017 someone should have noticed that there could not possibly be 67 g/t gold in nuggets *at the top* of a conglomerate sequence.

There was one more matter. No one else has ever mentioned it, but it too was and is significant.

They used a metal detector to determine where to take the sample. Of course it was going to be high-grade; it had nuggets in it. That's how metal detectors work.

But miners don't give a shit about cherry-picked samples. They have to have representative samples.

All the work done since 2017 has shown that the grade at Purdy's Reward and Comet Well was 2+ grams of gold per ton over perhaps two meters. The gold is found in the bottom of the sequence, not the top. The sample was not taken from the top of a reef but from the bottom, and it did not represent the average grade of the reef. Much of the reef was correctly barren.

So it wasn't eleven meters of high-grade, with a likelihood of even richer grades deeper in the conglomerate. It was a couple of meters of lower grade.

The mistake was in causing investors to believe the deposit was far richer than it was. You may safely assume Kirkland Lake were pissed after figuring out it wasn't as rich as they thought. They made the same assumption Quinton did, and frankly, it was wrong.

The saying "Only mad dogs and Englishmen go out in the midday sun" originated in the Pilbara region, not in India, as has popularly been supposed. There is crazy, and there is Pilbara crazy. Novo was entering a three-year period in which the nutcases flourished. Pilbara crazy became the rule, not the exception.

Until Brent Cook wrote his first piece on the project in mid-

2017, just as things got interesting, the only newsletter writers who covered Novo were Jay Taylor and me. All of a sudden, the rush was on to claim the discovery of the Pilbara.

John Kaiser swept into the space as if the industry was sitting waiting for his wit and wisdom. According to him, the Novo holdings might contain as much as twenty billion ounces of gold. He did sort of forget to mention that most of his information came from Jay Taylor and me.

I found that to be an interesting number, firstly since it was pretty much based on just two assays, and secondly because only about seven billion ounces has been extracted from the Earth's crust since Abel took his gold pan down to a stream in the Garden of Eden. When, months later, it became obvious that while there was a lot of ground with conglomerate containing gold nuggets, Kaiser said he didn't know if two grams of gold per ton at the surface was economic. That was just as dumb as the projection of twenty billion ounces.

There was an even bigger fruitcake in the pantry named Allan Barry, or Allan Barry Laboucan, depending on the time of day. I had a run-in with him a couple of years earlier. I posted a piece saying I believed we were about to go into a market correction. He came out and attacked me on chat boards, saying I had no idea of what I was talking about, and was wrong.

I wrote him back and suggested that he give it a month to see what transpired. Logically, he couldn't say I was wrong until we waited to see if I was in fact wrong or not. I asked him if he would be similarly quick to apologize if I turned out to be right, and there was a correction.

There was a correction; I was right. He never bothered to admit being wrong in what he said. He was one of those guys who has an opinion about everything, runs his mouth, but can't admit it when he is dead wrong.

He liked David Lenigas a lot, and rather than focus on Novo, Barry sucked up to Lenigas in the hopes of getting him to subscribe to his minor website with its few followers. He also runs a sub-tiny junior, and once issued a press release saying it had drilled into a vein of quartz, and that in that area, gold was sometimes found in quartz. That's perfectly true, but his quartz vein did not contain gold, just as most quartz veins do not contain gold.

Neither Barry nor Kaiser had any real insight into the Pilbara or Novo Resources. They just wanted to pretend they had discovered the company and the project. They didn't do either, and soon faded away into well deserved obscurity.

CHAPTER 17
THERE IS CRAZY AND THERE IS PILBARA CRAZY

ON MAY 26, 2017 NOVO ANNOUNCED a binding letter of agreement with Artemis, adding another 1,536 square kilometers to its already giant land position, now amounting to 6,021 square kilometers in addition to Comet Well. Novo took the first bulk sample from Purdy's Reward on July 12, 2017. Since a metal detector was used they knew the results would be good, perhaps even excellent.

I was married to a Brit for thirty years. She wanted to make it to her seventy-fifth birthday. She came within a week of doing so, but close only counts in horseshoes and hand grenades. I miss Barbara greatly; she was a wonderful wife, mother, and grandmother.

We would argue regularly about the meaning of words, as British and American versions of English often diverge. I think savory means tasty; she thought it meant the opposite of sweet. She spelled it differently, too. Any time you have people from different countries using the same language, there may be fundamental disagreements as to what certain words mean.

David Lenigas signed the binding letter of agreement with Novo in May of 2017. Once the rich sample loaded with gold nuggets was taken from Purdy's Reward, Quinton knew he had to drag Lenigas to the table to sign the definitive final agreement. He chased him by phone and scheduled meetings and tried every way he could to get Lenigas nailed to the floor. Lenigas came up with one excuse after another and canceled one meeting after another. For Quinton it was like trying to pick up a drop of mercury with a pair of tweezers.

The problem was that Lenigas thought of himself as a pirate, literally. He believed that being a robber baron was a great job, for which he was highly experienced. His definition of the word "binding" wasn't in any dictionary. As far as he was concerned, once you make an agreement and sign on the dotted line, that's when you start to negotiate.

Lenigas pulled the slickest piece of legerdemain I have ever seen any conman pull. Quinton was in a bind of his own making. He had an agreement with a pirate who wanted to pump up the value of the twenty-five million shares of Artemis he had been handed for being such a nice guy. Yo-ho-ho, and a bottle of rum.

Quinton had the assays from the sample taken in July. He couldn't have them in his hands and not release them. What he did not have was a definitive and absolutely binding agreement with Artemis.

So Lenigas paid himself, paid his backers, and screwed his own shareholders. On August 4, shortly before Novo would release the bulk sample assays, he announced a private placement for 23.7 million shares at 12.66 cents apiece. That was about $3 million. It all went to Lenigas and his buddies.

Once the first bulk sample results from Karratha were released, on August 8, Lenigas had Novo by the short and curlies. He started going on the HotCopper chat board in Australia to let Quinton know who had the power. Here are five posts, dated August 10 and 15.

werdna007 [58]
10/08/17; 8:25; Post # 26405439 [59]
Hi Doodledog,
I agree with your explanation, well done.
There is no relationship between ARV and NOVO, at present, but subject to certain conditions being satisfied,

there may be in the future. Ed Mead, as a senior geologist and competent person of ARV, would have to be satisfied with the validity of all the statements made by NOVO in that announcement. ASX listing rules must be observed for ARV to release an announcement.
Cheers

werdna007
15/08/17; 17:19; Post #: 26512020 [60]
ARV own 100% of Purdy's. NOVO own 0% at the moment. NOVO's MC has risen more than $400m on ARV's tenement.
Stay tuned.

https://hotcopper.com.au/search/search?type=post&users=werdna007
My estimate of NOVO SP value
werdna007
15/08/17; 18:24; Post #: 26513184 [61]
20–40c, maybe 50c with overvalued Beaton's Creek.
Very little workable real estate for quite some time and little cash to do anything more than that.
No mill to process gold...no native title...no exploration licences....no mining licences.....no money...... no FIRB approval...no TSX approval....no ASX approval...no environmental approvals.....no understanding of the MOA they entered into....almost says it all.

Cheers

werdna007
15/08/17; 19:28; Post #: 26514402 [62]
ARV has NOVO by the short and curlies!

My estimate of NOVO SP value, page-8
werdna007
15/08/17; 22:05; Post #: 26516802 [63]

ZEN, have reconsidered your request and have placed absolute minimum MC's on a few of ARV's tenements over the course of time – minimum values only on (1), (2) and (3)

(1) Purdy's - $1 b mc - wits gold

(2) Oscar - $1b mc - wits gold

(3) Carlow Castle - $1b mc - Cobalt-gold copper

(4) Mt Clement - JV with Black rock - gold mining - $250m mc

(5) other - diamonds - platinum group metals- antimony - whundo copper- silica hills - $100m

(6) Radio Hill Plant refurbished and other fox jorced tenements - $150m

Total $3.5B with 440m shares on expanded capital as at 30/9/17 equals approx $8.00 per share.

With confirmation of wits gold at purdy's and oscar (over the next 6 mths to 3 years) an increase of mc of $108.5B + Carlow Castle (with high grade cobalt/gold/copper) an increase in mc of $20 B to a combined share price of $300.00 per share minimum within 3–5 years (on capital of 440m shares) - profit taking opportunities at $50/$100/$150/$200/250 and $300 a share over that period for around 10–16% of shares per segment.

Please note that all these calculations are based on ideal outcomes and the actual outcomes may vary greatly from these outcomes.

Please also note that these possible outcomes are all based on my own calculations and are not in any way indicative of actual outcomes.

Please also note that these outcomes are based on my own private calculations into the future and the actual results may be wildly different from these.

Please also note that I am just a private investor and none of my forecasts may actually take place.

I am not a paid analyst, nor associated with any Company or Stockbroking firm or any party who has inside information or advance notification of price movements or corporate activity in the field.

I am just like you, who spends 50 hours or more of research on stocks I would like to keep or trade, and will ,in the future, keep or trade , ARV stock at any time I see fit, but not at any time below $10 per share and mostly not below $50 per share, if that ever happens.

Cheers to all prospective SH and traders.

May we all achieve our objectives for massive profit or comfortable living.

For the chairman of a publicly traded company to post something like Lenigas was posting on Hot Copper, under an alias, with a private placement pending and while working with Novo on a definitive agreement on Purdy's Reward, is way beyond shithouse rat crazy, and beyond 10 on the Pilbara Crazy scale. It's totally bonkers and symptomatic of an utter sociopath.

Novo was selling for $4.85 a share on August 15, 2017 when Lenigas posted his estimate that it was worth only 20–40 cents a share, or maybe fifty cents. With all the ups and downs since then, Novo shares are selling for $2.20 as I write this.

But what of Artemis? Artemis was selling for $.215 a share when Lenigas was posting his prediction of $300 minimum within 3–5 years. Artemis is now $.135 a share, and three and a half years have passed. Artemis will have to get cranking soon, if it is to reach even that first profit-taking opportunity, of $50. The share price would have to increase by a percentage in the tens of thousands in the next eighteen months. The $300 fantasy share price is far less likely still.

Looking back, the best part of the prediction from Lenigas was his statement that the Radio Hill rust plant was worth $150 million. How about $1.50 total for the plant? Artemis has been promising for years that it is on the verge of production.

It goes without saying that Lenigas was breaking more rules on full disclosure than I can count. The Australian Securities Exchange knew something funny was going on. They sent him a letter, demanding to know why Artemis did a private placement for insiders when it could reasonably expect the amended earn-in agreement to have a material effect on the price of the stock.

Novo did get Artemis to the table to sign an amended earn-in agreement on August 15, 2017. Lenigas demanded and got an additional four million shares of Novo worth right at $20 million because, as he had posted earlier that day, *"ARV has NOVO by the short and curlies!"*

All the Australian posters on Hot Copper thought Lenigas had pulled the coup of the century. After all, who wouldn't want to screw their partner at the first opportunity? I don't remember anyone suggesting that it might have been a nice chunk of change short-term, but in the long term might not be such a great deal. I looked at it as taking a hooker with you on your honeymoon. While interesting and maybe even fun, it would tend to set the tone for the marriage.

The ASX had a fit, and on August 16, 2017 made its demand for an explanation about the timing of the private placement on August 4, the assay results on August 8, and the corn holing of Novo on August 15 under the entirely different terms for the "binding letter of agreement."

Lenigas was as cute as a cartload of monkeys when he responded to the ASX on August 21. "Artemis was not aware of the Amended Earn-In Agreement prior to conducting the Placement. . . " And, of course, that was perfectly true. Lenigas may well have known he intended to screw Novo, but he hadn't told them yet. He would do the placement knowing full well that the assays were about to become known, and as he said on Hot Copper, *"ARV has NOVO by the short and curlies!"*

But he didn't know on August 4 what the terms would be, so he didn't really lie through his teeth to the ASX. Not really.

Earlier in this book I talked about the games people play. In this particular game Lenigas and his shareholders thought he was playing a zero sum game, where Artemis extracted $20 million worth of shares from Novo, and Novo lost $20 million in value in exchange for Lenigas redefining the term "binding". So Artemis seemed to have won and Novo seemed to have lost.

No one realized it at the time, certainly not the Artemis shareholders. While it was a great deal for Lenigas because all he wanted was to pump up the value of his twenty-five million shares so he could dump them, he screwed his own shareholders and Novo's. The reason is easy to understand.

Novo had 50 percent of Purdy's Reward, after handing over that $20 million in shares. It had 80 percent of Comet Well, that it had already paid for. Novo also had hundreds of square miles of ground surrounding Purdy's Reward and Comet Well, and owned that 100 percent.

The only reason Novo wanted Purdy's Reward was because

Purdy's had permission to drill. In time, permissions would follow for Comet Well and the rest of Novo's ground, but Novo could commence work immediately at Purdy's Reward. Novo didn't really need Purdy's. Quinton needed the right to drill and carry out exploration. He could start at once at Purdy's.

So, as time passed and Novo worked at Purdy's Reward, the question would naturally arise: how did Novo intend to spend its treasury once it had a drill permit for Comet Well or anywhere else on its land around the Karratha area? Would it be spent on the ground it owned 100 percent, or on Comet Well (80 percent owned), or should it continue to pour money into Purdy's Reward (50 percent owned), where Lenigas had just shown them that he would screw them at every opportunity?

At Purdy's Reward, there was even permission to take a bulk sample. That was highly valuable from a technical and geological point of view. It would provide a lot of valuable information. I think it was for twenty thousand tons. For some strange reason Novo didn't take advantage of it. When the drill permit arrived for Comet Well, Novo moved all its equipment there, and left Purdy's Reward to produce dust that could blow away when the winds came up.

It would take a couple of years for a few brave souls on the Hot Copper board to start to ask why Novo wasn't spending its money on Purdy's Reward when it could have. They should have checked the thermometer, because it would be a cold day in Hell before Novo would spend another dime there as long as Artemis owned half of it. Lenigas not only stiffed Novo, he stiffed his own shareholders.

By why should he care? He had all the money from the free shares he had been given.

On August 23, 2017 I wrote a piece blasting Lenigas. While everyone realized he had screwed Novo, few understood that he had stiffed his own shareholders as well. That realization would come later.

In that piece I tried to make the point that Lenigas was shitting in his own lunch bucket. Artemis was partnered with a fully cashed-up Novo and was getting the services of the best geologist in the world, in my opinion. The first thing Lenigas does is to steal from him, for his own benefit.

But at that time I didn't understand his motivation. I didn't know that he was sitting on millions of shares he wanted to blow out as quick as he could, so he could mosey on down the trail to some other scam.

Here is my conclusion to my August 23 piece: [64]

Let's pretend Novo's shares shoot up when the original video came out on YouTube on July 11th and Lenigas puts out a press release from Artemis. "Congratulations to our JV partner with their success at our 50% owned Purdy's Reward. Artemis is thrilled to have gained Novo Resources as a valued associate in this venture. Artemis is overwhelmed to have joined the team of Newmont Mining 3rd largest gold mining company in the world, Sumitomo Corporation, Mark Creasy, the most successful prospector in Australian history and Eric Sprott of Canada, billionaire investor and we get the services of one of the top geologists in the world for free and two million dollars spent advancing our projects. Here's my pen, where do I sign the definitive agreement?"

What do you think the shares would have done that day? Well, they wouldn't have gone to the mythical $8 a share but they would have easily spiked to $1 a share.

Instead, Lenigas gets cute, cuts a deal good only for him and his eastern European backers and stiffs both Novo and his other shareholders.

While Lenigas fully understood what he was saying when he kept issuing the veiled threats to Novo in a number of press releases, he needs to pay a lot of attention to the shot the Australian exchange just fired across his bow. He's playing with the big boys now and the chicken shit little games need to stop at once.

This was not the first time I had embarrassed Quinton; nor would it be the last. Naturally Novo had to issue a press release suggesting I was totally off base. It began as follows:

Novo Resources Corp. [. . .] announces that it has learned of certain recent public media disparaging one of the Company's joint venture partners, Artemis Resources Limited ("Artemis"), and its Chairman, Mr. David Lenigas. Novo wishes to make clear that it does not condone such views.

David Lenigas had just robbed Novo shareholders of $20 million Canadian in shares. He had ensured that Novo would never trust Artemis any further than they could heave him, but you aren't allowed to say such things in a press release. Actually, I think that's exactly what press releases should say, if it's true.

Shortly thereafter Novo announced an investment in Novo by Kirkland Lake at $4 a share, to the tune of fourteen million shares for a total of $56 million, all based on the extraordinary assay results released in early July. It would be the start of an uneasy relationship between the two companies. Novo was

190

looking for a fat treasury. Kirkland Lake wanted a cheap entry into a potential takeover target. As placements go it was rich, at $56 million, and generous with the shares, plus another fourteen million warrants at $6 apiece to Kirkland Lake.

In September of 2017 Novo began serious exploration at Purdy's Reward. The biggest gold show of the year, the invitation-only Denver Gold Forum, was scheduled for later that month. Quinton and Novo were determined to make a splash. They did, but in time it came to look more like a belly flop. Years later, Quinton would suggest that it was probably a mistake to have done it. But at the time it was really cool, and unlike any presentation any of the viewers had ever watched before.

Novo was given a fifteen-minute slot at the Forum, at a time that would be in the middle of the night in Karratha in Western Australia. Novo rigged spotlights and used heavy equipment to clear off a good spot to take a sample. Again, in hindsight, using a metal detector to find gold will only finds gold where it beeps; it doesn't tell you how representative the sample is.

The July assays had created unrealistic expectations. The Denver Gold Forum presentation reinforced those unrealistic expectations. It was a mistake that would take Novo years to get over.

Quinton introduced the company and the project to the watchers at the gold show before handing over to Luke Meter in Karratha; he was Novo's exploration manager at the time. Luke spent a couple of minutes talking about their first trench, where they had marked out gold nuggets found with a metal detector with pink spray paint. There were a lot of nuggets and it had to be impressive. He handed over to Brad Smith, who I mentioned earlier as one of the three owners of Comet Well, and whom Johnathon Campbell had brought in to prospect his SPL at Comet Well.

Brad Smith fires up his metal detector on the video and begins to move around the 30 × 20-meter trench, showing one nugget after another. Brad tended to be a Chatty Cathy and could have shown the same thing in a quarter of the time. I'm not sure what he thought he was supposed to be doing, but in essence he was lecturing on the use of a metal detector.

Finally he hands back to Luke, who had two of his guys prise out some marked nuggets with a jackhammer. John hammers out a tiny melon seed nugget and it comes free from the rock, with Brad working the metal detector.

The Denver Gold Forum is unique in that it is run extremely professionally. Your slot runs for exactly fifteen minutes and not a second more. They will cut your mike off and go on to the next presenter. Quinton was standing at the podium, watching the action in WA and sneaking regular glances at his watch. It looked like Novo's presentation was going to run over time. Finally, Pete and Brad pull out a good size nugget and show it to a suitably impressed crowd in Denver.

What no one understood at the time, in either Colorado or at Purdy's Reward, was that if you clean the conglomerate level with an excavator down to where the nuggets are, you will find a lot of nuggets. But unlike the advertised eleven meters of conglomerate, it was a lot more like two meters containing all the nuggets. Instead of the sixty-seven grams of gold per tonne that was hoped for, it was a lot more like two grams.

It was a mistake, and one that would cost Quinton Hennigh and Novo a lot of credibility. The mistake was in initially assuming they were testing the top of the conglomerate section, when in fact they were testing the bottom, where the richest gold had to be.

The way in which the gold industry tends to work is that everyone makes mistakes, but also everyone wants to cover their mistakes like a cat in a litter box. The difference between two grams and sixty-seven grams of gold per tonne was so great that it demanded an explanation, but the market didn't get it. So the Novo naysayers beat up on the company for years.

In my view, someone should have come out and said, "We screwed up. We don't have sixty-seven gram rock but we do have hundreds of square miles of two-gram material at or near surface."

CHAPTER 18
THE PEAK, AND THEN THE LONG DECLINE

ON THE STRENGTH OF THE INITIAL ASSAYS from Purdy's Reward, the injection of $56 million by Kirkland Lake Gold, and the video show at the Denver Gold Forum, Novo's share price peaked in the fall of 2017. It was a heady time for the company, rising from twenty cents a share in 2010 to $8.83 on October 1, 2017.

A return of 4,300 percent is what every junior resource investor hopes to find once or twice in a lifetime. But holding onto that kind of return can be difficult. For the next three years Novo would struggle to regain credibility, with a cast of thousands of keyboard commandos in the peanut gallery throwing rocks with every new press release.

Shares of Artemis would peak one month later, in November, at just over forty Aussie cents, before starting a long and painful decline to a low of $.025 by December of 2019. When Lenigas signed the original "binding" letter of agreement with Novo in March of 2017, Artemis shares were about eleven cents. When he got his twenty-five million shares in late 2016, they were worth two cents apiece. They went up twenty-fold and came right back down to where they started.

I can't see exactly where Lenigas dumped his shares. The Artemis annual report in June of 2017 shows him owning all twenty-five million. A year later there is no record of him owning any reportable shares. The shareholders didn't realize it but he had pocketed something up to $10 million for a year's work. Once he unloaded his shares, he just coasted. He had accomplished what he wanted to accomplish.

One important issue that I don't recall anyone mentioning

about the conglomerate find at Karratha was just how unusual it was. The deposit was virtually identical to that of a high-grade alluvial gold project, but since it had been covered with basalt for 2.7 billion years it was now hard rock — hard rock with the characteristics of an alluvial deposit. Beatons Creek was similar in nature but it contained loosely consolidated material that was perfectly suitable for gravity recovery.

For the time being, Beatons Creek and Nullagine were on the back burner. Millennium Minerals had never made a profit and clearly never would, but kept limping along. The price of gold was tending higher and the Aussie dollar was declining, and that was enough to keep the corpse warm.

Novo had the easy-to-mill gold. Millennium had a mill and a lot of uneconomic rock. Millennium wouldn't consider doing any sort of deal with Novo, even if it might keep Millennium running for a bit longer. That was dumb but you can't fix stupid.

Quinton was just about the only person in mining who understood that much of the gold in the Pilbara came from the conglomerate reefs. Beatons Creek was pretty close to being conventional in nature, with smaller gold, which was at least somewhat measurable in its quantity and grade in the reef. Purdy's Reward and Comet Well were another kettle of fish.

In Quinton's first attempt to measure the gold, he used a large-diameter drill rig with a large bore for digging a big hole. It was designed to drill water wells. His theory was that if he picked up enough material he could get a reasonable idea of grade.

At the same time, Novo's geological team used a diamond core rig to determine the structure of the reef. It worked great, and provided a clear three-dimensional picture of the reef. By November of 2017 Novo had completed over sixty short core holes indicating the structure of the conglomerate reef.

The large-bore water drilling rig, on the other hand, was an abject failure. It would grind up the material, but since gold is about nineteen times heavier than water, any nuggets in the hole just dropped to the bottom. You could pull the drill out and vacuum out the cuttings and nuggets, but you had no idea where in the reef they came from. The reefs ranged from a couple of meters thick up to twenty meters, so knowing where the gold was would be key to mining at a profit.

For the first time, Novo acknowledged that most of the nuggets found by the metal detectors were located near the bottom of the conglomerate sequence. If the geologists had ignored the fact that they were dealing with hard rock and pretended it was nothing more than a gravel bed, similar to any alluvial project, it would have been obvious to them from the start that the lower part of the sequence would be the rich area.

That's a failing throughout the gold mining industry; geologists trained for hard rock exploration look down on placer miners and don't want to learn what makes those deposits, or their nature. Placer miners know placer deposits, and hard rock miners know hard rock deposits. Rarely do they mix. That's just dumb.

To obtain a mining permit in Australia it is necessary to prove to the mining department the grade and quantity of the gold. But in a large nugget system, such as Novo faced, it was almost impossible to measure to government standards. Novo couldn't use a conventional core or RC rig, as the nuggets were too big and too widely spread out. The water well drill rig left all the gold at the bottom of the hole.

Figuratively speaking, since not a great deal of it remained, Quinton was pulling his hair out. Without measuring the gold to Australian standards he would never get a mining permit.

My attitude, as I said a number of times when I wrote about

Novo, was that you can't measure nugget gold; you can only mine it. The Romans wouldn't have had any problem mining the gold-bearing reefs. The Spanish would have grabbed their picks and started digging. Hell, the *garimpeiro* miners in Brazil today would have mixed some diesel fuel and ammonium nitrate and started blasting holes in the reef to pluck out gold nuggets until they made enough money to feed their family. None of them would ever get permitted in Australia.

Quinton and I talked dozens of times. I kept repeating that you can't measure nuggety gold, you have to mine it. He insisted the Department of Mines would never permit it until it had been measured. In the end he began to take bulk samples to determine grade. As far as I can tell, taking a bulk sample is the exact same thing as mining.

I cannot over-emphasize how difficult a project the Karratha conglomerate gold story was. There was one simple reason why over a hundred and thirty years had gone by and tens of thousands of miners had walked right past the gold in the conglomerates. They didn't understand it and couldn't figure out how to mine it. Quinton understood it but hadn't quite gotten around to figuring out how to mine it.

One Novo decision from early in the program was to make use of sorting machines. Quinton believed he would find small, finely disseminated gold that could easily be drilled to determine grade and quantity. It would never happen.

He chanced on the idea of the sorting machines. They can sort just about anything, based on color or size or density or material. They are commonly used to sort trash, especially trash containing metal. Their use is skyrocketing in the mining industry. The technology is advancing rapidly. Right from the gitgo, Quinton included representatives from two companies making the sorting machines. They will revolutionize the industry in time.

Australia depends on the mining of coal and iron ore for a major portion of the taxes paid to the state and federal governments. With billions pouring in from China each year, Aussie miners are the highest-paid miners in the world. A truck driver can make $200,000 a year. The salaries are amazing. If you visit Perth airport, all you will see is thousands of miners waiting to fly off to a mine site or just returning. There seems always to be a labor shortage in the country.

As a result, the crews can be picky. Australia is a socialistic country. The workers believe everything should revolve around what they want. As such, starting up a major mining project is similar to herding cats. The workers go where they want, when they want. Management is fairly low on the power pole.

Beatons Creek had been held in abeyance since it made no sense to expand the resource further. Quinton had outlined a resource. He had done test mining and processing and was now stuck until Millennium could be budged from TDC. Beatons Creek was important to Novo but was clearly not a company-making property unless and until it could be put into production. But the Karratha collection of projects, taken together, clearly were a company-making proposition.

Quinton had brought on Leo Karabelas in 2011 to handle corporate communication for Novo. Leo would do the same thing for a whole slew of companies with a Hennigh connection. Bringing on Rob Humphryson as CEO in June of 2017 was a major achievement. As in every other country, Australians tend to prefer working with their own kind. No Canadian company was going to march into Australia and start telling them how the mining game works. Rob took a lot of the burden off Quinton, and ever since the COVID-19 nonsense began he has had to shoulder most of the work, keeping exploration moving to get into production.

The exploration staff at Purdy's Reward and the Karratha area was more problematical. Everyone had their own idea of how the exploration should be conducted. Getting everyone to march in the same direction was difficult, to put it diplomatically. People came and went. One senior fellow simply didn't show up one day, or ever again; he waltzed off and no one had any idea of where he was or what his problem was.

The management team was excellent, and worked smoothly together. Ronan Sabo-Walsh kept track of the money and did an excellent job of it.

Since the hard rock conglomerate deposit was close to being unique, no one had any experience of dealing with it. Over time, this showed. There were a lot of mistakes made and then instantly pointed out on various chat boards.

People afraid of making mistakes will never succeed at anything. Everything is a process of trial and error. Even flight manuals approved by the FAA will talk about straight and level flight as a series of turns and banks, with both climbs and glides around a central direction. If you aren't making mistakes, you aren't making enough decisions. That said, while it's only human to err, if the erasers on your pencils always wear out before the lead, God is trying to tell you something.

As the chat room warriors noticed immediately, Quinton made a number of mistakes in 2017 and 2018. But it's OK to make mistakes when you are doing something for the first time; you are bound to do so. It's OK as long as you don't make the same mistake again and again. Quinton made mistakes but rarely made the same mistake twice.

Late 2017 was an exciting time, as major progress looked to be under way. As a major shareholder, Eric Sprott joined the board in November. Greg Gibson was already a board member.

A heritage agreement was penned with the local Ngarluma Aboriginal Corporation. That was necessary, to get permission to begin exploration at Comet Well and the other surrounding tenements. Until Novo could explore and drill Comet Well, Artemis had the company by the short and curlies, as David Lenigas so kindly put it, as Novo was obliged to work at Purdy's Reward.

Artemis did announce that the Department of Mines had granted a permit to take a 20,000-ton bulk sample at Purdy's Reward. If anyone other than David Lenigas had been running Artemis, that permit would have been valuable to both companies, because for the first time a real mining sample could have been taken. That would have saved a year of fits and starts and disappointed shareholders demanding instant gratification. But anyone who has ever been bitten by a rattlesnake knows you don't put your hand down to pet the next one to wander by.

By mid-December 2017 the Department of Mines had granted the tenement for Comet Well to Novo, which meant exploration could begin. In late December Novo issued a press release talking about results at Purdy's Reward, where some sixty-nine diamond core holes had outlined the conglomerate reef across a width of five hundred meters and a length of over fifteen hundred meters. By now Novo had picked up over 12,000 square kilometers or over 4,600 square miles of property. Novo controlled one of the largest land positions in Western Australia.

Trench samples from Purdy's came in at 15.7 and 17.7 grams of gold per ton, from bulk samples of over three hundred kilograms. Novo was realizing that to obtain an accurate estimate of the gold grade, the sample size had to be larger, perhaps much larger. There remained the question of where in the reef the gold came from. If you were to test only the lowest part of the reef, you would get exceptional grade. If testing the entire conglomerate sequence, it would be much lower.

Kas De Luca joined Novo in early 2018 as exploration manager, from Newcrest. Getting everyone on the exploration team to work in the same direction was still difficult. She was an important addition to the Novo team.

CHAPTER 19
NOVO SET THE BAR TOO HIGH AT KARRATHA

THE DISCOVERY OF GOLD in the conglomerates in the Karratha area by Quinton and his team was a sort of good news–bad news story. Thousands of prospectors and miners had worked the region for over a century. Weekend warriors with metal detectors took millions of dollars' worth of gold nuggets out of the area for several years. Until Quinton came along, no one put it all together. Dozens of tiny Australian junior mining companies were all searching for the mother lode; that is to say, lode gold. Quinton didn't care, thinking that after all, gold is gold. Who cares whether it is in hard rock vein systems or in nuggets in the conglomerates?

If you want to make a monster discovery — and that was the first item on Quinton's to-do list — you must start with a good theory. His theory of gold precipitation out of brine and salt water 2.7 or 2.8 billion years ago inferred a lot of gold. It just didn't suggest where exactly it was.

Beatons Creek was actually a sort of slam-dunk. The project was close to being conventional. While the nugget effect made the gold hard to measure with accuracy, that same feature made it easy to process. Although the gold in the conglomerates had long been known about and mined, at least on a small scale, without the theory in hand, miners didn't realize that where there were conglomerates, there was gold. Since this sort of deposit is almost unique in the world, it was overlooked by prospectors and the management of junior gold lottery tickets in their search for lode gold.

Beatons Creek was so close to conventional that it was no stretch for Quinton to put together a feasible project. There was the one small issue of having no suitable mill, but Millennium was trying hard to go out of business, after which Novo could use their mill.

The grade of the first two samples from Purdy's Reward set shareholders' expectations so high that Quinton was under the gun for years, both from several respected geologists writing about the company and from a bunch of clowns writing about a project they had never visited and didn't understand. It can't be fun standing there dodging cobbles all day.

Nobody in history ever discovered gold by finding an exceptionally high-grade sample at the first attempt. Whatever you come up with, some samples will be higher, most lower. But by announcing two samples averaging sixty-seven grams per ton, Novo elevated the bar right to the moon. Stating incorrectly that the sample was taken from the top of an eleven-meter segment of reef just made it worse.

Sampling and small bulk sampling continued into 2018, with work being done at Purdy's Reward and, for the first time, at Comet Well. Due to the nugget effect, no two samples ever showed similar grades. At Purdy's, Novo mapped the conglomerate structure very well with a large number of short core holes. Bulk samples from the conglomerate showed gold nuggets over a two-meter interval, in the range of two to three grams per ton, all located near the bottom of the gravel sequence. That would be economic anywhere in the world. With Novo controlling around 12,000 square kilometers of land in the Karratha area, that is potentially a lot of gold. But at Comet Well there were actually several reefs, all near each other, all carrying various grades of gold.

You can't mine it until you have measured it, and you can't measure it.

I thought the whole issue was stupid. The gold was

economic and near surface. This discovery would have been put right into production at any point in history prior to the Bre-X scam. So in the name of protecting investors, more regulations and rules would somehow make investing safer. But this was Australia. Welcome to modern professional mining.

Throughout 2018, more and more samples were sent to the lab for assay. Some personnel issues popped up, with blame due on both sides, and with the net effect that results took far longer to be published than impatient investors would tolerate. In my mind the issue went all the way back to those first two assays. Investors couldn't come to grips with the fact that those results simply were not reflective of the nature of the deposit.

Using a metal detector is a great way to find nuggets but a terrible way to sample areas. You can find gold but you could never mine it that way. Of course the samples will grade high. But those samples do not reflect anything other than the tiny area they came from. The initial samples went only centimeters below surface. They didn't mean anything in terms of the overall project.

Communication would have helped, but many of the problems became obvious only well after the fact. There is a lot of trial and error in any mining project, and this was the most unusual gold deposit in the world. Karratha provided a sufficient supply of trials and a selection of errors.

Quinton was working on a couple of problems at the same time. It was obvious to the market that the use of mechanical sorting machines was a high priority for him, since the first discovery at Purdy's Reward. Novo would spend the next three years working with the technical teams of two companies to find a way to process crushed rock to concentrate the gold in the smallest amount possible. I've been to Perth and seen such a machine in action. When a machine can pick up pinhead-size gold and separate it from the gunge, that's quite remarkable. Novo has been able to reduce the total mass down to less than one percent of the volume it starts with.

At Artemis, David Lenigas had managed to dump all his shares and was busy blowing the remainder of the company's cash. Artemis still had the four million Novo shares he had extracted in mid-2017. In May of 2018 Kirkland Lake agreed to buy them, to add to the twenty-one million it already owned. When Novo issued the shares to Artemis in August of 2017, one condition of the deal was a twelve-month hold. Novo agreed to release Artemis and Kirkland Lake from that condition. Kirkland Lake paid $5 a share and sent the money to Artemis in May.

The transaction made Quinton slightly nervous. Whenever any party becomes a large shareholder in another company, they tend to want to start making decisions for the company. They feel they should have authority but don't want the responsibility. It's always handy to have management nearby, to blame for anything that goes wrong.

Eric Sprott had put a lot of money into a company called Pacton Gold, another Canadian junior lottery ticket. Greg Gibson was a director of both Novo and Pacton.

Quinton was working on picking up land in the central Pilbara region. The proposed acquisition was discussed with all the directors, and then to Quinton's great surprise, Pacton suddenly announced it was doing a deal on the same parcels. It could have been coincidence, and a stroke of pure luck for Pacton. Alternatively, it could have been double-dealing on the part of Greg Gibson. With my customary diplomacy, to both parties, I may say that the incident did not leave Quinton with enormous confidence in the business ethics of Greg Gibson. It looked as if someone had been telling tales out of school.

But behind the scenes Quinton had been doing some serious

wheeling and dealing for months, in order to pick up a couple of projects at Egina from local prospectors. (It is pronounced *Ed Ja Na*, by the way.) We had bought gold nuggets from Egina, from the metal detector dealer in Karratha. We compared them to the Purdy's Reward and Comet Well nuggets, and agreed they were virtually identical.

That was important. Egina was alluvial in nature; simply gravel, not hard rock. It would be easy and cheap to process, if you had water. Egina certainly had water during the monsoon season. It had so much water, you couldn't even move around the project. The other nine months it was as dry as a bone.

No one ever said Quinton's job was easy.

Novo was now sitting on the Beatons Creek property while advancing its far larger land position near Karratha. It was that phase in the lifespan of a junior resource company where progress is being made but, quite bluntly, it's dull progress. Gradually it was becoming obvious that there would be no more assays of sixty-seven grams of gold per ton gold, just a boatload of assays at two grams. Two-gram rock was a giant success because of the vast land position, but it paled in comparison to the early results.

On September 17, 2018 Novo announced two property transactions in the Egina region. [65] Only in hindsight would it become evident how important this was. Novo was paying $8 million in shares and cash to a company that held a mining license at Egina and had been mining gold nuggets from its alluvial gravels for years. Sincerely, I think I was the only person who really understood it. Clearly it conflicted with the direction that Kirkland Lake and Eric Sprott wanted Novo to take.

All investors want instant gratification, no matter if it is mom and pop with a hundred shares or a major with ten million shares. Novo set the bar so high with the first assays that everyone tended to ignore the considerable progress being made. Then, all of a sudden, it appeared that Quinton was off on a tangent with the Egina purchase.

Novo closed the deal two weeks later. [66] It would take a long time for the market to understand just how important the move into a large land position at Egina was.

Of course, this being Australia, the deposit there came with its own unique set of problems to be solved before any money could be made.

CHAPTER 20
KIRKLAND LAKE ATTEMPTS A COUP

I'M UNSURE HOW MANY VISITS I have made to Novo's Pilbara projects. My first was in June of 2009, even before its deal with Millennium Minerals on Beatons Creek. Since then I've been back almost every year until 2020, when the COVID-19 nonsense shut down the world's economy.

Since Quinton Hennigh is my best friend and we chat almost every week, I suspect I know more about Novo and about him than anyone else outside of the company. I also know more about Novo than many of the people in the company. And so I should, because this is a project I have been thinking about for a dozen years.

In late September of 2018 Eric Sprott, Tony Makuch, Greg Gibson and Quinton flew a plane Eric had chartered across the Pacific to Australia for an inspection trip on behalf of Kirkland Lake Gold (as one of the biggest shareholders). Eric is the Toronto billionaire who is also a big shareholder in Novo and probably the biggest shareholder in Kirkland.

They were going to take a look at Fosterville, and to determine where Novo stood and why not much seemed to be happening. Again, the initial assay results set the bar so high that almost nothing the company did would seem very impressive. To say that the atmosphere in the cabin was tense would be to understate it. If Quinton felt he was being ganged up on, he was probably about right.

Fosterville was looking more and more like a giant home run on Kirkland's part. Even though its merger with Newmarket had taken place less than two years before, Kirkland's share price had gone from about $5 to over $21 by this time.

Eric later spoke at a mining conference in Melborne on October 3rd. Tony, Greg, and Quinton sat right at the front, in the first row. Eric talked about owning shares in Newmarket and Kirkland. He believed the two companies would become a powerhouse if they joined together. The market obviously agreed, with the shares up 300 percent. He talked about the discovery of the Swan zone at Fosterville, and how important it was in hindsight to the finances of the company.

I like Eric Sprott a lot. I have met him only once, and chatted for a couple of hours at best, but without question he has done more for Canadian mining than anyone I know over the last twenty years.

Because he's a billionaire, and they think differently, he made a giant mistake in his talk. He gave full credit to Quinton for the discovery of the Swan zone. Also for convincing Eric to proceed with the merger. He did this in front of hundreds of movers and shakers in the Australian mining sector in front of Tony Makuch and Greg Gibson. They hadn't made the discovery; they had thought it a waste of time and money and opposed the merger.

Tony Makuch and Greg Gibson demonstrated how no good deed goes unpunished. They were furious that Quinton got the credit for the work he did and they didn't do. They were determined to get even, and if it destroyed Novo that was just tough shit.

Eric would make somewhere up to $1.5 billion out of the merger. He takes care of the people who take care of him. He told Greg to put Quinton on the payroll and pay him a nominal but regular salary. Greg did. It was six months after the merger between Newmarket and Kirkland Lake before Quinton was paid anything. Tony Makuch promised in writing to give Quinton options on sixty thousand shares, and in the end gave him fifty-three thousand.

After Kirkland Lake injected $56 million into Novo, Quinton realized he now had a conflict of interests. If he was president of Novo and was receiving a monthly salary from Kirkland Lake, just whom did he work for? He told Greg to stop the monthly stipend.

Quinton made a little money from his work for Eric on Fosterville, but given the ten billion dollar or so increase in the value of Kirkland Lake, it was pitiful. It was like tossing peanuts to the monkeys at the zoo. It was all because two functionaries were green with jealousy.

Quinton is the man in the arena, doing battle with everyone at the same time, including those who had made millions of dollars from his efforts and his skills. In a speech made at the Sorbonne in Paris in 1910, Teddy Roosevelt made an interesting observation about the man in the arena.

It is not the critic who counts; not the man who points out how the strong man stumbles, or where the doer of deeds could have done them better. The credit belongs to the man who is actually in the arena, whose face is marred by dust and sweat and blood; who strives valiantly; who errs, who comes short again and again, because there is no effort without error and shortcoming; but who does actually strive to do the deeds; who knows great enthusiasms, the great devotions; who spends himself in a worthy cause; who at the best knows in the end the triumph of high achievement, and who at the worst, if he fails, at least fails while daring greatly, so that his place shall never be with those cold and timid souls who neither know victory nor defeat.

I was in Australia a month later, for another visit. While I liked the Beatons Creek story and was so-so about Karratha, I loved the Egina story from the beginning.

Beatons Creek was almost conventional. Karratha was anything but conventional, but it was obvious that it was a work in progress, albeit slow progress. Karratha's millions of ounces of gold will take from fifty to a hundred years to mine and process. The thousands of miners that all walked past the conglomerates over the course of a century had their very valid reasons for passing over the deposits.

Egina was perfectly conventional as a placer mine of sorts, other than the fact that there was always either too much water or too little. That's one of those niggling little facts that gets in the way of a good story. Not much was written about Egina because people simply didn't understand it. I went there in 2018 and saw the trenches that Carl Dorsch had put in, but beyond the fact that the pay streak was near surface, you couldn't tell much.

Carl had set up a simple plant near a hard rock shaft where gold had been produced decades before. There was some water to be had from the decline. He had holding ponds set up, but summers in Egina are hot. The ponds would quickly dry up until they could be replenished with the typhoon season's rains.

Quinton's sortie into the world of mechanical sorting would prove valuable. If sorting machines could separate gold from rock without the use of water, Egina could literally become a gold mine.

I've owned placer projects, and you can mine fairly low-grade material at a profit because you don't have to crush and pulverize it. Novo was on its way to picking up about 1,200 square kilometers of gravels. That's over 460 square miles. If all the gravel had gold in it, even at a low grade, there would be a lot of gold and a lot of profit.

Quinton and I visited Egina together. Carl Dorsch had left the fines from his small operation in 55-gallon drums. When we were bored we would pan the material. Unlike Karratha, there were a lot of fines. Our panning was always profitable.

Fine gold panned at Egina, November 2018

We made a chopper tour of some ground belonging to Kairos Minerals. George Merhi showed us around the property. George, you will remember, was the chief geologist for Mark Creasy, with whom we had swilled $750-a-bottle wine on our tour of Beatons Creek back in 2009. George was now doing the exploration for Kairos, who had a project a little to the southeast of Novo's Egina tenements.

A few days before we walked the property, George and a mate had run metal detectors over an area of no more than two hundred square meters, and taken out over ten ounces of gold in three hours.

I cannot fathom why investors and commentators cannot come to grips with the remarkable potential of the Pilbara, which Quinton knew about twenty-five years ago. Where else on the surface of the Earth could two blokes make money so quickly? If any weekend warrior (I don't mean George) could do that with simple and relatively cheap equipment, what could a well-financed and well-managed mining company produce?

Quinton made some unsettling discoveries while we were at the Egina camp. Evidently, Greg Gibson had returned to Australia after the Diggers and Dealers gold show and, behind Quinton's back, was stirring up trouble. It appeared that he was telling people that Kirkland Lake would fire most of the Novo board and essentially would be running the company after its annual meeting, scheduled for the first week of December. Quinton was aware that Gibson was talking to Novo personnel, and it was probably not a good solution he had in mind for Novo.

John Youngson was running the exploration program at Egina and almost certainly would have been one of the people Gibson would be courting. Quinton confronted John, and I then understood what the "deer in the headlights look" actually meant. John protested and said that he was entirely innocent and would never stab Quinton in the back.

George Merhi's three-hour gold haul

I had met John and his partner Sue Attwood on the trip to the South Island in 2009. I had a lot of time for her. John's genius was mostly in his mind, although he had all the paper credentials of a PhD.

After my visit to New Zealand in 2009 I paid him and Sue to visit two of my placer projects, one in British Columbia and the other in Tanzania. I began to form the view that John lacked any feel for what he was doing. He could write a paper using all the proper words but he couldn't tell you where the gold was or how to recover it.

But I paid him a couple of hundred thousand dollars to source a placer plant and ship it to Tanzania for my operation there. The plant he provided would have been perfect for New Zealand, with its coarse gold, no clay, and lots of fresh water. Alas, I had fly-speck gold, lots of clay, and very little water. The plant was unsuitable. When I complained, John told me he hadn't designed it, just ordered it. But the purpose of having them visit the project before supplying the plant was to show them the conditions under which I had to operate. I wasn't a big fan.

When rot sets in, you must cut out that part of the structure. All Greg accomplished was to create ill-will among the people Quinton had spent years forming into a team. A number of people lost their jobs because they were confused about who was behind their paychecks. I felt bad for Sue as she had far more sense than to fall for a trick like Gibson's, but it appeared that John fell for it, hook, line and sinker. It cost him the greatest opportunity he would ever see.

After our quick trip to the Kairos ground with George, the chopper dropped us off at Port Hedland airport. Quinton and I returned to Perth. The next day, he and I and Rob Humphryson and Ronan Sabo-Walsh met in the Novo conference room to discuss what we thought was going on.

Quinton knew Tony Makuch and Gibson had it in for most of his board. Eric Sprott and Kirkland Lake had been busy in the area. They had put cash into Pacton Gold, of which Greg was a director. Pacton held ground in the Egina area that Novo had wanted to secure.

In addition, Kirkland Lake had put money into Artemis. As far as I was concerned, that proved that Makuch and Gibson were clueless. How dumb is it to deal with a company run by a pirate who makes it clear that he intends to stiff you at the first opportunity? The Boobsy twins weren't befriending the best and the brightest in Western Australia. They were climbing into bed with the gang of fools.

There had been a heated Novo directors' meeting in mid-November, and it seemed that Kirkland Lake wanted either to have Pacton take over Novo or to strip Novo's assets and hand them to Pacton.

These games go on all the time when a mid-tier company puts money into a junior. The tendency is always for the moneyed partner to believe itself to be smarter than the junior, and better able to make decisions. Almost always, that means better for the funding partner, not for the management or shareholders of the junior.

Quinton had been around long enough to have seen this game played before. Ever since starting Novo he had done his best to balance power between the potential vultures. And while Gibson and Makuch were quite impressed with themselves, I wasn't much impressed.

Eric Sprott had stakes in so many junior resource companies that he had to have someone representing his interests. When making cash injections he would often insist on having a seat on the board. That seat would be filled by Greg Gibson. This was a bad idea, for several reasons. Gibson was supposed to be running the Jerritt Canyon mine in Nevada but was also on something like twenty boards. I doubt he could have named half of them, much less provided management expertise. Eric's reach was exceeding his grasp.

When Eric came up with the idea of merging Kirkland Lake with Newmarket to achieve economies of scale, he needed someone senior to run the new, much larger Kirkland Lake. Greg Gibson nominated Tony Makuch, so there was always going to be a brotherhood of sorts between the two. Neither favored the merger, and made that clear. So there was conflict with Quinton, who fully approved of the proposed merger, foreseeing the new Kirkland Lake as a powerhouse. In the end he was proved correct, and neither Gibson nor Makuch has ever forgiven him.

I mentioned that tense flight from the U.S. to Australia. Progress appeared to be slow at Novo, and like all investors, Eric Sprott wanted the share price to be going up right now. Next, he was unhappy with Greg Gibson because Jerritt Canyon was turning into a money pit. And Makuch had been slow off the mark after taking over Newmarket.

The Fosterville mine was surrounded by a number of tenements important to the company. Australia has a loose way of controlling land. The properties were under the control of Newmarket but the expiry date on the tenements was approaching. Makuch should have made contact with the Department of Mines immediately and secured the land. He didn't, and the Department snatched the ground back and put it out for application.

When Rob, Quinton, Ronan and I met in Perth, another item for discussion was Novo's forthcoming annual meeting. Clearly something was going to happen, but what?

I am a neophyte when it comes to board meetings and annual meetings. I know they take place, but as with the manufacture of sausage, I just don't care to know the details. Ronan said something about shareholders abstaining. I had never heard the term before.

It seems that a shareholder can vote his shares either for a proposal or a director, or against, or can abstain. If more shareholders abstain on the matter of a director's election than vote in favor, it apparently means the director should resign.

When I found that out, I realized that even though Quinton and his people on the Novo board of directors could outvote Eric Sprott and Greg Gibson on issues because there were more of them, they could be booted from the board. In that event, Eric and Greg could then do as they wanted since they would become the majority.

I realized that Eric, Greg and Tony planned a backdoor coup. All they had to do was vote against the other Novo directors and the company was theirs to plunder.

Without a word to either Rob or Quinton, I returned to my hotel and wrote a fairly scathing piece about the Toronto Mafia. I suggested that all Novo shareholders abstain from voting on the matter of Greg Gibson as a director. I didn't suggest they did the same with Eric Sprott. I didn't want to attack him directly, only to fire a shot across his bow. I knew all the senior management at Novo and thought I understood their opinions of Novo's land position and Quinton's management skills. I thought Makuch and Gibson wanted to run everyone off the board except Quinton, Eric, and Gibson himself, thus giving all the power to Eric and ultimately to Pacton.

I pointed out in my piece that if the coup were to succeed and they canned Rob Humphryson, Quinton would leave as well. In fact the entire staff would walk out, leaving the keys to the office in an envelope taped to the door for the new management team. Jingle mail [67] works for companies as well as for homeowners. And just to let everyone know there was a price to be paid, I would lay the responsibility for the demise of Novo at Eric Sprott's feet.

When someone has a couple of billion dollars, money becomes almost meaningless. It's an accounting entry, not something real. How many houses, boats, airplanes and trophy wives can you keep track of, anyway? But believe it or not, billionaires are concerned about their legacies. If this triumvirate was determined to sink Novo, I wanted the world to know who to blame.

The vote took place in December of 2018. Makuch waited until the very last moment to vote the Kirkland Lake shares. He not only wanted to abstain on the three directors recruited by Quinton; he wanted Quinton gone as well. As I had recommended, there were many abstentions on Greg Gibson's election. Not enough to can him but enough to wake him up. Quinton twisted Eric's arm a little, pointing out that as a director he had at least a moral obligation to vote in line with management recommendations.

I am certain that my warning shot did remind Eric that while his vote could fire the entire board, the responsibility would be on his record. Discretion being the better part of valor, he voted along with the rest of the management team.

My piece may not have been the deciding factor in saving Novo but it did shed some light on a furtive attempted coup d'état. And I'm dead certain it made me many friends in the Toronto Mafia.

CHAPTER 21
TAKING ADVICE FROM KEYBOARD COMMANDOS

LEGEND HAS IT THAT STEVE JOBS of Apple Computer once said that the most wonderful thing about the Internet is that it gives everyone a voice. Alas, he continued, the Internet's biggest flaw is also that it gives everyone a voice.

Who knows if he did say it or not? He's not around to vouch for it but if he didn't say it, perhaps he should have.

We tend to forget that as little as twenty years ago there was not much to be found on the Web on junior resource companies. I began 321gold.com because I realized there was a demand for free information. The junior lottery ticket companies wanted their stories known, and investors wanted a choice. It was a perfect fit for us. Nova Gold (that's Nova, not Novo) was the biggest gainer on the Canadian stock market in 2001, and few realize the reason why. It was this: Nova was telling its story in a compelling way on the Web. I helped.

Doug Casey was still posting expensive paper newsletters with his stock recommendations to his subscribers. They got the information a week or so after we would have posted it on 321gold. Between the low in gold in August of 1999 and 2008, communication on the Web aided the creation of hundreds of new junior mining companies. They could raise money and attract investors because they now had a cheap, swift way to communicate. For years it seemed that every former taxi driver in Vancouver or drill crew supervisor was setting up his own company. Of course, given the 95 percent failure rate of mining projects, companies came and went regularly.

Over time, more and more investors would migrate to websites such as HotCopper and Stockhouse and CEO.CA. There they could read articles about new and upcoming mining companies and projects. In addition, many sites offered chatboards. There the punter was free to express his opinion, and all too often to attack anyone who disagreed with it.

Chatboards seem like a waste of time to me. I always wonder how many people would run their mouths the way they do if they didn't have an alias to hide behind. If they are so proud of their opinions, why won't they identify themselves?

When a guy has a keyboard and an opinion about something, and when the keyboard gives him access to a cast of perhaps thousands of readers, it seems only natural to start believing people actually care about your opinion. But keyboards and opinions are a dangerous mix. Having an opinion is not quite the same as having an opinion that anyone gives a shit about. Opinions are a lot like assholes. Everyone has one and most of them stink.

Quinton Hennigh thinks unconventionally. That is what makes him a genius. Unlike most geologists, he is not regurgitating the stuff that was fed to everyone in Geology 101. He thinks for himself. That's both rare and valuable. Miners and prospectors have crisscrossed the Pilbara for generations. No one put it all together until Quinton started Novo. Mark Creasy had a clue, and for most of the last thirty years was the biggest landholder in the basin. He didn't know exactly where the gold was but he pretty much knew there was a lot of it.

It should go without saying that the vast majority of people, and especially people possessed of a keyboard and an opinion, do not think outside the box. When they encounter someone who does, they have difficulty with it. Obviously, because they don't get it, then whatever is being done must be wrong.

Novo Resources is followed by a fair number of stock lovers who get it. It also has its full share of stock haters who clearly do not get it.

One of the most valuable posters in my view is on Stockhouse, and calls himself TX Rogers. His posts are thoughtful and intelligent. He understands the concept and largely agrees with the direction Quinton is taking. When he writes he adds insight to what can be fairly complex issues. He has no agenda, unlike other posters.

Another poster calls himself Rhino10. His real name is Ken Watson. He was a minor player in the Australian resource field and he hates Quinton. You see, when the managing director of Millennium Minerals made the deal with Novo on Beatons Creek, he didn't mention that he had also done a deal on the ground with Ken Watson and his partner for the alluvial rights. Since the reefs are basically unconsolidated gravel, there wasn't a world of difference between the rights Millennium sold to Novo and the deal Millennium did with Ken Watson and his partner.

For some reason Ken Watson and his partner broke up, leaving the alluvial rights to Beatons Creek in a sort of limbo. He blamed Quinton, and has spent the last nine years criticizing Quinton on chat boards, sometimes a dozen times a day saying the same thing again and again. But Quinton wasn't any part of the problems between Watson and his partner going their separate ways. Watson is so goofy that he came back in 2020 and wanted Novo to do a deal with him on some ground. That was pretty weird; you spend years badmouthing someone on a chat board and then ask them if they want to do a deal with you. I don't think that's going to happen.

It seems odd to me that Watson would have such hatred for Quinton and Novo. But he has posted hundreds, perhaps thousands of nasty grams on HotCopper and Stockhouse pissing on Novo's parade. And because he had sent me half a dozen meaningless emails ten years ago, he included me in his venom.

There should be some sort of check made, to prevent anyone from posting more than (say) five hundred emails all saying the same thing, to make certain the person posting isn't shithouse rat crazy. Watson wouldn't come close to passing. He is shithouse rat crazy. Saying so will probably set off a chain of another thousand stupid posts, all saying exactly the same thing.

Another very strange poster was someone named Taylor Dart who writes on occasion for Seeking Alpha as anyone can do. He wrote a pretty biased piece on Novo. That's not all that unusual. It happens all the time from those unable to think outside the box. But Taylor Dart then went to the CEO.CA chat board and proceeded to write one scathing post after another about what a lousy investment Novo was. That also happens. But he posted under a semi-alias of Trad. Someone noticed and pointed out that Trad is nothing more than Dart spelled backwards.

Trad had a fit and insisted he was not Dart. But if anyone posted anything aimed at Dart, Trad responded because he is Dart, just unwilling to admit it. How many times does a person have to lie before you realize he is a liar. Dart/Trad is a liar and that pretty much takes his credibility and flushes it down the toilet. He has an agenda and nothing more.

One change on the web over the last twenty years is the disappearance of paid subscription newsletter writers. There used to be half a dozen or so, and most have now retired or moved on to other ventures. Barb and I ran a computer website twenty-five years ago and we soon learned that the way to make something valuable on the Web was to give it away. Paid newsletter writers are caught in a bind. They pretty much have to tell people what they want to hear, or they will go elsewhere. Also, why should someone pay $500 a year for information available for free elsewhere?

One scribe based in South America had an interesting solution. Mark Turner is a self-hating Jew who supports terrorists in both Israel and the U.S. He instantly attacks anyone who even mentions the word Israel. He goes out of his way to make Jews look as evil and devious as possible. He whines and cries and pretends he is some sort of victim.

I caught him pumping and dumping a stock in March of 2009. I pointed out to him that pumping a stock under his own name for Hallgarten while trashing it on his other website was illegal. He lies all the time. He uses projection, accusing others of the self-dealing that he constantly engages in.

He is probably the worst financial forecaster I have ever read. Earlier in 2020, when silver dropped below $12 an ounce, he mocked everyone who bought or owned it. He continued his tirade as it rose all the way to nearly $30. He missed a 150 percent gain and was making fun of people on the other side of the trade the entire time. But he hates everyone, including the entire mining industry. People go to his admittedly meaningless website just to see Dr. Pimple Popper in action.

Mark Turner lives in Lima. I don't think he is there for the beaches. Lima has the ugliest beaches in the world. The nice beaches are on the other side of South America, in Rio. But Lima is a great place to hide out from the law.

Now I don't know if it is the same man, but the authorities in England are looking for a Mark Turner accused of child molesting. Almost certainly it is some other Mark Turner. And the authorities in Peru have a complaint of wife-beating from a former spouse of a Mark Turner. No doubt it is some other Mark Turner, that being one of the most common names in Peru.

Mark Turner would like to have an impact on the mining business. Perhaps he does have some influence. He makes the industry look as if it is run by and for evil people, with him being the leader of the pack. He is vicious, cowardly, and a stone cold liar. Everything he writes sounds as if it is coming from the mouth of either a drunk or someone on drugs. Clearly he has a chemical issue as well as being evil.

Naturally he is jealous as hell of Quinton Hennigh, who is everything Turner is not.

When I pointed out to him in 2009 that pumping and dumping, while momentarily profitable, was illegal, he threatened to sue me. I'm still waiting.

Mark: Please, please sue me. Because if you do, I will be able to take your deposition, and two minutes later it will be in the hands of the SEC and TSX, who will run your ass out of the industry.

CHAPTER 22
MOVING FORWARD TOWARDS PRODUCTION

THE ATTEMPTED COUP on the part of Greg Gibson and Tony Makuch did no permanent damage to Novo but did require some personnel changes. It didn't make sense to retain those who weren't entirely loyal.

Gibson's back door maneuvering left a bad taste in everyone's mouth and it was only a matter of time before he would be shown the door. Pacton Gold, the heir apparent, realized it now had no game plan. It began pulling up roots in WA and moved its operations and projects to Canada.

Kirkland Lake Gold achieved the highest grade of any gold mine in the world, without a shred of appreciation ever being shown to the guy that Eric Sprott credited with the success of the merger with Newmarket Gold and development of the Swan zone at Fosterville. It was as if there wasn't enough attention on offer for Tony to be able to share any with the man who handed him a gold mine on the proverbial silver platter.

Novo's stable of projects now had three distinct and different targets. Beatons Creek was near conventional. Since Quinton was waiting for Millennium Minerals to realize it had no way forward, it didn't make much sense to do anything other than publish an updated 43-101. That came out in April, showing just over 900,000 ounces of gold at an average grade of just over 2.6 grams per ton.

Novo had carried out test mining and had run material through a small plant, so was prepared if and when Millennium bit the dust. That 900,000 ounces was sufficient for five or six years' production, and in reality, how much sense does it make to spend your scarce money to define production material decades into the future?

The tough nut to crack was always going to be Karratha. Purdy's Reward and Comet Well were both permitted, but as long as David Lenigas was around, Novo wasn't going to advance Purdy's Reward an inch.

The problem remained how to define ounces to the satisfaction of the Department of Mines. Quinton made it clear in a press release on May 19, 2019.

> It is well understood that delivering a conventional mineral resource at the Karratha Gold Project is challenging owing to the extremely nuggety nature of the deposit. Accordingly, Novo has worked closely with independent experts (Mr Ian Glacken, Director of Geology at Optiro Ltd and sampling and geometall-urgical expert, Dr Simon Dominy) to ensure the Company's QA/QC processes and sample collection methodologies are sufficiently robust to underpin this mineralization report. Development of this minerali-zation report has become a guiding discipline to ensure Novo can pursue a mining lease application as well as update its NI 43-101 technical report for the Karratha Gold Project.

Quinton was juggling two balls in the air. The more difficult was determining how to measure the unmeasurable. Novo had already extended the length of the conglomerate reef for over ten kilometers, including both Purdy's Reward and Comet Well. Once the Department was on side with the process — drilling for the structure of the conglomerate reefs, and taking bulk samples to determine grade and distribution — Novo would hopefully qualify for a mining lease.

Quinton then needed some cheap, swift method of mining and processing the hard rock.

Ever since the first samples were taken at Purdy's Reward, some two years earlier, Quinton had been working with Steinert, a German company, on the use of its mechanical sorters.

The Steinert XSS T mechanical sorting machine used X-rays to identify the particularly dense gold in rocks. When it encountered a rock containing gold, a puff of air would pop it into the ore bin for later processing. The process was highly successful but Novo spent a lot of time deciding just how to crush the particularly hard rock conglomerate, and to what size.

One of the original samples, taken in 2018, was for some reason crushed to 2 mm size and sent for processing. That is a pretty small rock, and it overloaded the capacity of the machine to sort. By the time someone realized the error, tons of bulk sample material had already been crushed to 2 mm. It's easy to make mistakes when you are working on something entirely new, attempting something that has never been done successfully before. But investors are impatient with any error, and raked both Quinton and Novo over the coals.

Initial tests of the Steinert machine showed it could concentrate the rock down to less than two percent of its original mass. That was a game changer for Novo and for the entire mining industry, as it meant that projects previously assessed as uneconomic could be economic now. Much of the cost of mining is incurred in moving around tons of material, again and again. If the crushing and sorting could be carried out near the mine, and if only two percent of the mass need be moved around, costs would be far lower.

But even the Steinert machine had limits, and it didn't like all the material being of 2 mm size. Rocks that small would blank the sensors. If it worked at all, it was like molasses in winter.

Later in 2018, Novo began to test mechanical sorting machines made by Tomra, a company formed in Norway decades ago but now based in Germany. Novo had realized that the nugget effect caused a giant problem with the accuracy of assays, since the nuggets were distributed at random, both horizontally and vertically. However, that same issue was also an opportunity, if they could pick the correct machine for mechanical sorting. Nuggets were hard to measure but easy to process.

The Tomra machine used a combination of X-ray transmission (XRT) and electromagnetic (EM) induction. The former could detect high-density material while the latter was in effect a metal detector. Gold nuggets have a specific gravity of about 19, depending on what other minerals they contain, so were perfect for both XRT and EM induction.

Novo had been testing the Steinert machine since work began at Purdy's Reward in 2017. They were quite satisfied with the 98 percent reduction in mass, but with the Tomra device, even the initial tests on material of 10–63 mm size showed a 99.7 percent reduction in mass. From a technical point of view, that was a home run. For every 2,214 pounds of input (one tonne), the sorted gold was about 6.6 pounds, or three kilos.

Novo continued to test both machines until well into 2020, with wonderful progress. The vendors of both the Steinert and Tomra devices worked closely with Novo. As experience grew, both showed constant improvement in the potential of their machines, both in capturing as much of the gold as possible and in reducing the sorted mass to the lowest fraction of the original amount.

The gold from the Farno deposit at Egina almost demanded the use of mechanical sorting, if the necessary economies of scale were to be achieved. Egina and Karratha material of various sizes were sent to both Steinert and Tomra for testing. They soon found that rocks larger than 63 mm were simply too large for the machines, while rocks smaller than 6–10mm were too small for the machines to work well.

The new projects in the Egina area were being largely ignored by Novo investors. They were different, and as difficult in their own ways as the Karratha story.

Quinton had done a deal to acquire the Farno McMahon tenements as early as September of 2018. He had been working behind the scenes for a year to pick up the tenements, realizing that the size and shape of the gold nuggets were remarkably similar, from Purdy's Reward to Egina. But Karratha was hard rock and Egina was fairly loose gravel.

Carl Dorsch had been mining at Egina and selling the gold to the metal detector guy in Karratha. That is where Quinton and I bought gold in mid-2017.

Carl was doing the most primitive mining possible. He stripped off half a meter to a meter of sand and other material at the surface, then worked in a layer of gravel sediments 1–2 meters thick that contained the gold. Using a metal detector, he would have someone mark the hot spots and they would dig up the nuggets. It was a cheap method, and more or less effective.

Quinton needed to find a way to test various areas to see if they were prospective for gold nuggets, and find a low cost method of processing the gold-bearing gravel.

Again — and you will hear me say this a lot — Quinton's ability and willingness to think unconventionally meant that it might take some time to identify a good way forward, but it would work and it would be profitable. Naturally, most of the people in the industry would mock him, but that was because they weren't smart enough to figure out the problems, analyze them, and devise a practical solution.

Carl Dorsch had been mining at Egina for years

What is most remarkable about the Pilbara region is the long history of gold mining but the limited production of gold. Many people had worked the Pilbara; they just hadn't thought outside the box.

Carl had a small fleet of trucks to carry the unconsolidated gravel to a small plant he had back at his camp, for further processing. It was not efficient, as most of the gold had been located and removed as nuggets. But if you looked at his costs and the amount of gold produced, it was easy to see how much more efficient the process could made to be.

The camp was built at the site of a former underground gold mine. The adit contained water after the monsoon season. As long as the water lasted, Carl would process material.

Quinton sought a better way.

The first thing he had to do was to get Heritage Clearance from the natives on the areas outside the Farno tenements. That has become a slow process. Prior operators had reached agreements with the two local native groups who controlled the region, but then failed to do what they said they would do. As a result, the locals weren't very trusting or easy to work with.

The process could take years. Eventually the natives will realize the financial potential. There is a lot of gravel-bearing ground, from Egina to the Indian Ocean. A fair and reasonable royalty would change their lives for the better, for a very long time. It hadn't happened in 135 years, but with Novo and Quinton Hennigh it would.

Without gaining Heritage Clearance, Novo could not take samples or disturb the ground. But it could test the two pieces of tenement that had been granted mining rights, and interpolate those results to the rest of its Egina properties.

One instant solution that did not require a Heritage Agreement to be in place was the use of ground-penetrating radar, to determine the very slight differences between the half-meter of overburden, the gold-bearing reef of 1–2 meters of loosely consolidated gravel, and bedrock. The radar didn't disturb the ground at all. It made quick work of determining which areas should be the most favorable for mining.

The next step was to determine grade. In 2019 I visited Egina for the second time. Keith Barron was on the same tour. We also brought Erik Wetterling with us, a young Swedish writer who had been covering Novo for a couple of years.

Keith has a sapphire mine in Montana so he knows full well what it costs to process a cubic meter of alluvial material. I've had several placer mines from British Columbia to Ghana, Tanzania, and Sonora, Mexico.

When you talk about the costs of mining alluvial material, it is industry practice to use cubic meters rather than tons. The yellow gear that is used to move the gravel around is all denominated in cubic meters — the capacity of a shovel or bucket or dump truck. The weight of the material is meaningless. Keith and I sorta agreed that costs would be in the range of $10 to perhaps $15. But when we later saw the progress made with the latest and greatest sorting machine, in Perth, and then went on to Egina and saw how close to the surface the gold-bearing gravel was, we felt that $5 to $6 U.S. per cubic meter was a good target number for Egina.

On December 20, 2018 Novo announced mechanical sorting results from the Tomra machine, using ore from Comet Well. The concentrations were as little as 03 percent of the original mass. That's the two kilograms per ton result from rocks in the 6–63 mm size range.

In addition, but missed by almost all investors, were the alluvial tests from the Egina gravels showing 107.88 grams of raw gold from 95 cubic meters of gravel. Guessing a purity of about 93 percent fine gold, that works out to 1.06 grams per cubic meter.

At the time a gram of gold was worth about $50. So you have a basic mining and processing cost of perhaps $5 to $15 a cubic meter to make $50.

When Quinton and I visited the South Island of New Zealand in 2009 we looked at projects where owners were delighted to mine 0.3 grams per cubic meter. There were areas at Egina that were returning over three times as much. One gram per cubic meter would be an excellent result for any alluvial mine.

Quinton was thrilled to get numbers as high as his team achieved at Egina. They used the ground-penetrating radar to find the swales where the highest-grade material should be. I was of a mind that they really didn't need to do that. My preference would be to test what should be the lowest-grade gravels. We knew that there was gravel pretty much from Nullagine to the Indian Ocean. From Egina, a straight line to the ocean at its closest point was seventy kilometers long. It was virtually all gravel and mostly flat, with a few hills with conglomerate reefs outcropping.

Novo's outright land position was of about 1,200 square kilometers, or over 460 square miles. Adding the ground belonging to De Grey Mining, on which Novo had an option, took the total to more like 2,500 square kilometers or nearly 1,000 square miles.

If a square kilometer contained a one-meter layer of gravel bearing 0.3 grams of gold per cubic meter, that would equate to over 9,000 ounces of gold per square kilometer.

Of course, gravel layers will not be found beneath the entire 2,500 square kilometers, and some of the land is hilly. Neither is there any evidence that all the gravel will be as high-grade as 0.3 g/m^3, but my guess is that a lot of it would be. Obviously the +1 gram material will make the most money, but even 0.3-gram material would make a 50–66 percent margin. Every mine manager in the world would love that.

All investors should be required to memorize the Lassonde Curve before being allowed to make their first stake in a junior resource company. All stocks go up and all stocks go down, but resource stocks tend to follow a fairly predictable pattern that the remarkable Pierre Lassonde first identified thirty years ago.

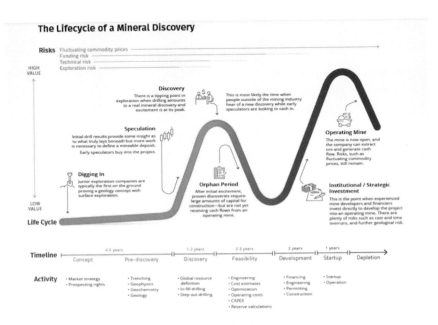

The Lassonde Curve

It took from 2009 until 2017 for Novo to get traction with the discovery of conglomerate reefs bearing gold nuggets in the Karratha region of Western Australia. After its share price had meandered between around fifty cents and $2 between 2011 and 2017, Novo shareholders hit the jackpot in 2017 when the first assays were released from Purdy's Reward. The shares peaked at $8.83 in late 2017 before tumbling by as much as 85 percent into the crash of March 2020, when investors finally understood that the world really was coming to an end.

But it didn't, and Novo has since been in the boring period of the cycle, after a peak powered by speculation and a later realization that maybe it wouldn't continue to rocket higher. All juniors do the same thing. From 2017 until well into 2020 the stock price climbed at times and declined at times, just like every other stock.

From 2018 into 2020 there was progress in defining grade and tonnage at Beatons Creek and Egina. Constant progress with the mechanical sorting machines from both Steinert and Tomra improved their ability to pick up smaller and smaller gold, along with a most dramatic reduction in total mass. At least for Purdy's Reward and Comet Well and the surrounding ground, profitable mining was possible only if the gold could be removed from the dross at an early stage. Finding it would remain an issue, but clearly, the use of mechanical sorting machines made profitable mining possible there for the first time.

We reached the first peak in the Lassonde Curve and are now in the "orphan period" prior to the start of production. While many investors moan and groan constantly about the lack of important news, and other investors with an agenda will constantly bash Novo, actually Novo offers a gift to investors every year. That's equally true for the bulls and the bears.

I have never noticed any other newsletter writer or commentator mention this, but the junior resource market gives small investors an edge that exists in no other market. Small investors in resources have a great advantage over the large funds and the deep-pocketed investors. At the worst of times and the best of times, the tiny traders can get an execution. Now it's true that at market tops there is total liquidity and at bottoms it's hard to get a fill. But it's still possible to buy or sell $5,000 worth of any stock at any time. The only time a large investor can get a fill on, say, a one million share block of anything is at market tops.

The juniors have a trading range that almost all investors ignore. It's something I discuss in my books. It is common for the price of a junior stock to cover a 300–1,000 percent range in a single year. (I mean the difference between its lowest and highest prices in the course of a calendar year.)

Novo has a smaller range, about 300 percent usually, but that means that if an investor were to buy Novo near a yearly low and was willing to hold for a 50 percent gain, it would happen every year. Likewise, if an investor went short near a yearly top in the shares, he could make a 50 percent profit every year. Where the vast majority of the investors I am aware of go wrong is to hold for a 1,000 percent gain.

It's true that they happen. I said in 2012 that Novo would be a 10–100 bagger. So far it has gone up only 3,600 percent. But anyone, bear or bull, could have made 50 percent on Novo every year, and those wins add up over time.

The real advantage of Novo over other juniors is that it has great liquidity. Many juniors trade only by appointment at lows. You can still make the same profitable trade. I do it by putting in stink bids at ten percent under the lowest price of the day before. No stock really loses ten percent of its value in a day without announcing terrible news, but at lows the weak hands love to dump the shares they bought at the highs.

Novo is in that part of the Lassonde Curve where serious investors make the most money with the least risk. It would be worth printing any of the variations of the curve you can find with Google, and keeping it handy.

In early 2019 Novo made an interesting announcement, [70] saying that Sumitomo Corporation had agreed to provide up to five billion yen (about $46 million at that time) on a mutually agreeable project, and as a result would have a right of first refusal (ROFR) once it had made an investment, should a third-party investor make an offer for all or part.

While it was true that the press release didn't actually commit either side to a particular deal, it did indicate an interest, on the part of one of the foremost mining and metals companies in the world, to be a part of something that Novo had. The Japanese tend to take a long-term approach to dealing with other companies. They do not invest in the flavor of the month, and much internal discussion and research will precede an approach to another company. It was significant, for several reasons that will take years to fully understand.

In April of 2019 Novo announced a 30 percent increase in the resource at Beatons Creek, including just over 750,000 ounces in oxide and fresh rock mineralization.[71] The oxide portion, of 325,000 ounces, was especially suitable for the Millennium Minerals mill, when that migh become available.

May of 2019 brought the ultra-important environmental approvals to mine the Beatons Creek deposit.[72] In addition, all Native Title and tenure documents were in place in anticipation of mining. These agreements and permissions are what often delay junior mining companies in moving forward. Novo was ready to mine. All that was needed now was a mill.

June of 2019 saw an agreement signed whereby Sumitomo would spend up to $30 million U.S. over three years at Egina to earn a 40 percent interest in Novo's project there, including the Farno tenements and the ground included in the joint venture between Pioneer Resources and the De Grey Mining.[73]

Late in June came a binding letter of intent with De Grey, with Novo to explore the lag gravel deposits on ground held by De Grey. De Grey concentrates on hard rock underground deposits in the region, so was happy for Novo to pick up rights to the 1,100 square kilometers of gold-bearing lag gravels.

In August of 2019 a Novo press release announced just over one gram of gold per cubic meter of gravel at Egina, from four bulk samples totaling 282 cubic meters.[74] A purity of 89–95 percent would equate to 0.9–0.95 grams of pure gold per cubic meter; excellent numbers, and potentially very profitable.

Novo took a batch of nuggets from Egina and sent them to Germany, for Steinert to test on a machine using a new technology called an eddy current separator. The tests were announced in a press release in September of 2019.[75] Novo fully understands that if it is to succeed in mass production at Egina, it must use some form of dry processing with a mechanical sorting machine. Novo has worked closely with Steinert for over three years now.

October of 2019 saw more bulk sample results from Egina: 222 cubic meters of gravel produced 337 grams of gold nuggets when processed, or just over 1.5 grams per cubic meter.[76] Guessing a purity of 90 percent would mean a solid 1.35 grams of pure gold per cubic meter, worth about $67 U.S. and costing perhaps $10 to recover, even if Novo does a really rotten job of mining and processing.

Investors continued to ignore these incredible results. That would be a world-class number in any placer operation anywhere. Given that the Egina area demands dry processing, Quinton has done a series of deals on over 1,100 square kilometers of gravel-bearing ground with exceptional results. My view is that Egina will in time become the jewelry box for Novo.

Keith Barron and Erik Wetterling had visited Novo at its Karratha projects, but neither had been to Egina or to Nullagine. We joined Rob Humphryson for a short tour of all three of Novo's major gold deposits in October of 2019.

By this time Karratha was on a sort of hold, since a fair bit of money would be needed to advance it further. We knew where the gold was. We knew how to separate it from the country rock. Novo needed a mining permit, and that was in the works. But any further progress was going to eat money.

Quinton was at home in Colorado, awaiting the birth of his second grandchild. He had strict instructions to stay there until she arrived.

Keith Barron and Rob Humphryson at Egina, 2019

I looked forward to this trip with my friends, partly because I wanted to take the measure of Rob Humphryson. We were close to the point at which his work would become more important. Wherever in the world a mineral discovery is made, someone has to put it into production. Quinton had done 90 percent of the hard geological thinking, but someone now had to put the pedal to the metal to advance matters from theory to practice.

You have to have a few beers with someone over a period of a week or so and share some lies before you start to understand what they are and are not capable of. Rob was a great guy to go on a tour with. He told lies with the best of us. If he would just learn to hold his beer a little better he could be a good companion on a site visit.

Erik Wetterling had begun following Novo a couple of years before. On his own, with no prompting, he had written a series of short pieces talking about the company, and why he saw it as a great investment.

I had previously suggested to Quinton that he bring Erik over to Western Australia. Jay Taylor and I were about the only other writers covering Novo. Brent Cook pretty much wrote off Novo after the disappointing assay results in late 2017, but he had never visited Egina. Quinton hadn't picked up any of that ground when Brent washed his hands of Karratha.

Erik brought a new pair of eyes to the story, and a different way of looking at the company.

Novo brought its IGR3000 test gravity plant from Beatons Creek to Egina and fastened it to the plant operated by Carl Dorsch. It worked well, and recovered even the tiny gold not picked up by the original plant. The gold from Egina was quite remarkable, and at times Novo recovered nuggets larger than one hundred grams.

Erik Wetterling made his first visit to Egina

One thing I want readers to think about is the soil attached to the nuggets, and its solid red color. That is all iron staining. Later in the book I will explain why, if you understand how the iron got there, you will also understand at once the giant scale of the gold deposit in the Pilbara Basin.

Egina test plant with IGR300 for fine gold

Since most of the gold at Egina was in the form of nuggets of decent size, either the Steinert or the Tomra sorting machine would pick up all of it. Tests had also shown that the machines would even pick up gold of near pinhead size.

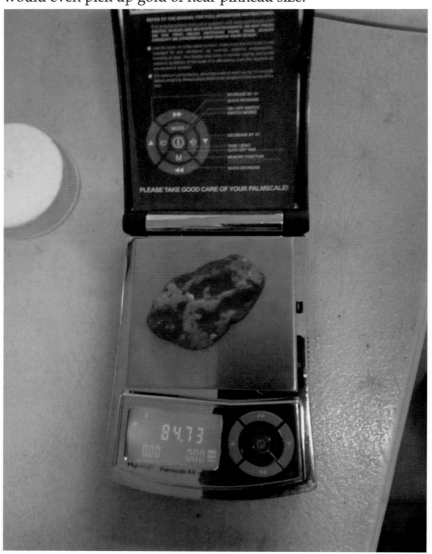

A not uncommon large gold nugget from Egina

Of the three projects, Egina impressed both Erik and Keith the most. Without seeing how Novo prepared the ground for mining, and without seeing an actual cut showing the nugget-bearing layer of gravel, it wouldn't mean much. Since Keith mines alluvial sapphires, he realized how inexpensive the mining could be.

Typical nuggets found with a metal detector

First, you strip off the thin layer of overburden. Novo intends to literally slide a steel plate under a section of overburden and move it to the side. Then, process the meter or two of gold-bearing gravel (marked "GV" in the photo). The loose gravel is run through either a Steinert or a Tomra sorter and put right back where it just came from, minus the gold picked up by the machine, which is fed into a container of its own, never to be touched by human hands.

At the end of the process the overburden and vegetation is put back in place on top of the gravel. It will cost more in operating expenses to reinstate the ground than it will cost to mine and process the gold.

There is a clear difference between the gold-bearing gravel and the basement rock. While the reef does pinch and swell, some small amount of basement will be processed. Nuggets can be found into a few centimeters of the basement.

Rob does enjoy showing visiting firemen around Egina, since it cannot fail to impress. The exploration crew were out with metal detectors, scanning every level of the reef and carefully measuring what they found for their records. Rob had us give the metal detector a go. We each found a nugget. We could hardy fail to, there being so many of them.

While the grade wasn't as good as in the reefs at Karratha, it was remarkable to me how similar the stripped ground looked when the nuggets had been mapped. Except that Karratha was hard rock and Egina was unconsolidated gravel, easy to process.

Overburden has been stripped; "GV" is gravel with gold

Karratha and its many kilometers of outcropping conglomerate reefs will contain far more gold than in the area from Egina to the Indian Ocean, but Egina will be the lowest-cost gold mining operation in the world, and the easiest to process.

Keith Barron was quite comfortable with the estimate of $5 to $6 a cubic meter. Most of that cost will be in stripping and after-mining remediation. The worst case looks something like $500 to produce one ounce of gold.

Bob Moriarty metal detecting at Egina

I am exceptionally willing to suggest that Quinton and Rob have opened a whole new world of low-cost mining though the use of effective and cheap mechanical sorting machines. For that alone they both deserve the thanks of the mining industry.

After we had visited all three projects with Rob, we returned to Perth for a demonstration of the Steinert machine in operation. It was remarkable. It could pick up pinhead-size gold. Since the tiny gold is only a small fraction of the Karratha conglomerate story, that will not be a significant contributor to project economics, but in the Beatons Creek conglomerates there is a lot of small gold.

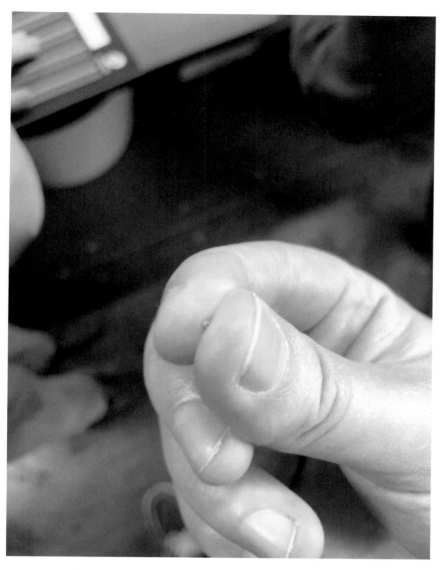

Pinhead gold that the Steinert machine could detect

Steinert machine in Perth, October 2019

We, all three of us, found the technical capability of the machine quite incredible. But when you delved into the operating costs it became even more interesting. The biggest expense was operating the air compressor that popped the gold into the ore bin. (A machine costs about $750,000 in Australian money.)

The rock, crushed to the correct size, ran up a conveyor belt, over the rotating sensor wheel and into the sorted material bin — except for the gold, which was popped right into the gold bin. It would be untouched until the sealed and weighed lockbox was delivered to the refinery. The cost per ton of sorted material was going to be tiny.

THE NEWS THAT QUINTON had been waiting patiently for arrived a month early for Christmas in 2019. It was a wonderful present nevertheless. He had known for years that it was coming, but the price of gold would inch higher and give Millennium just a bit more breathing room. Finally it arrived. [77]

Millennium Minerals moves into voluntary administration

© November 25, 2019   🖿 News   ♠ Vanessa Zhou

This meant that Novo had to shift direction at once. Work would continue at Egina in determining the grade and extent of the gold-bearing gravel. Karratha would remain on the back burner, and Novo would move towards the purchase of Millennium's mill and going into production at Beatons Creek.

Millennium's bankruptcy was interesting because of the legal structure. IMC, a Singaporean company, had advanced about $69 million to Millennium, using the mill as security. When Millennium went into bankruptcy IMC took over the company.

Novo had to come to an agreement with IMC if it was to take over the mill. But IMC wanted to sell the entire company, which had possible advantages for Novo as there were numerous land packages included.

Quinton spent from November of 2019 until August 4, 2020 finalizing an agreement. The mill had cost about $100 million to build but had of course been used, and poorly maintained, with the management of Millennium just trying to keep the wolf from the door. In a way it was a Mexican standoff, where two knife fighters tie their left wrists together before starting to battle. If Novo didn't buy the mill, IMC would have a pile of rust to get rid of, but Novo would have no quick route to production.

IMC actually made out like a bandit but could have done far better still. They should have taken the proceeds in Novo stock but chose to take $60 million Australian in cash. I'm always baffled by the actions of those running big companies. It seems to me that the bigger the company, the dumber the decisions. Millennium Minerals had never been anything more than a supplier of jobs to mining and mill personnel. Why would IMC be happy to keep throwing money at Millennium, but prefer cash to Novo shares?

Now Quinton could have the production he had been trying to get for so long. Building a mill would have taken five years to permit and construct and would have cost well over $100 million.

Running a production company is a whole different kettle of fish than operating a junior resource exploration company. As an exploration company you can never fail. You can always whine that all you need is a few more bucks for another drill program which is sure to hit this time. As a production company, it's either deliver or get fired. It's a lot like the difference between ham and eggs. Now the chicken, she's involved. But the pig, he's committed.

Quinton had to consolidate the company and make whatever changes he had to make before production started. The COVID-19 virus that arrived early in 2020 actually helped Novo. Rob Humphryson, an Australian in Australia, could oversee the refurbishment of the plant and the hiring of people and assembling of equipment. Since Quinton and everyone outside Australia were essentially locked out of the country for the best part of a year, Rob could do the job he had been hired to do without someone looking over his shoulder and offering advice. It was his baby to screw up, or he could look like a hero. I have a lot of confidence in him and believe that hiring him was one of the best decisions Quinton ever made. But if he screws this up, I'd fire his ass.

So while Quinton was operating behind the curtains, working with IMC on a plan and a price for the mill, he started spreading his wings in other directions.

Karratha was under control, but would remain on the back burner until Novo had the cash flow generated by going into production at Nullagine with the Beatons Creek material. At Egina, various areas were being tested with the ground-penetrating radar. Research continued with the Steinert and Tomra sorting machines.

As for the Heritage Agreements for the other ground, Novo has still to reach agreement with the two native corporations involved. Since the advent of COVID-19 they have hunkered down, and little progress will be made until the parties get together.

But Quinton always wanted to have at least a mid-tier gold mining company, and maybe even a major. In early January of 2020 Novo announced the purchase of ten million shares of Australian-listed Kalamazoo Resources.[78] Eric Sprott took a similarly-sized chunk of shares. The private placement was done at forty cents a share, with a full warrant at eighty cents for eighteen months.

As I write, the shares are up about 50 percent, but hitting the warrant price in another six months might be a bit of a struggle. Kalamazoo looks as if it has its hands on another Fosterville. Time will tell.

February 2020 brought yet another brilliant bulk sample from Egina in the gold-rich swales. [79] This one showed 562.25 grams of gold and nuggets from a 413.6 cubic meters. Alas, only Novo, Keith Barron, and I understand how rich a test result of 1.36 grams per cubic meter is. Figure maybe 1.25 g/m^3 of pure gold, and today's gold price of $60 U.S. a gram, with costs of $5 or $6 a cubic meter for a 90 percent margin.

Getting Egina into production is a priority, but Novo must work within the confines of what the natives will allow. One day soon we can hope they will realize it would be nice to have steady dividend checks coming in monthly to each corporation.

The results of testing the Tomra machine in Sydney with Egina material showed 100 percent recovery of the gold in rocks in the 6–18 mm range, and 100 percent recovery from rocks of 18–50mm size. [80] The Tomra sorter really likes the 6–18 mm material, with 100 percent recovery while rejecting 99.7 percent of the waste rock. They could run the 6–18mm material at a rate of about ten cubic meters an hour, and the 18–50mm material at about twenty cubic meters an hour.

In March of 2020 Novo made a major investment in a tiny junior in Newfoundland, trading 6.94 million of its shares for fifteen million shares of New Found Gold. [81] New Found later went public, and as I write in January of 2021, those same Novo shares are now worth $16.53 million, while the shares they picked up in New Found are worth an incredible $61 million.

It took Artemis only two and a half years to wake up to the fact that by stiffing Novo in 2017 and demanding $20 million in hostage money for the "binding" letter agreement signed by David Lenigas for Purdy's Reward, Artemis had only ensured that Novo wouldn't drop a dime into the property as long as Artemis held it.

In March of 2020 Artemis announced it would sell its half of Purdy's Reward to Novo, as well as the 47K property, for two million Novo shares and $1 million Australian. The Artemis–Novo joint venture was now a thing of the past. Novo would now control 100 percent of Purdy's Reward. [82] David Lenigas was long gone from Artemis, having sold all his shares at the top, but his legacy still left a stench.

In late March of 2020 Novo announced it had taken a stake of nine million shares in Australian-listed GBM Resources, with an option to earn up to a 60 percent interest in GBM's Malmsbury project, located in the famous Bendigo gold belt in Victoria province. [83] Novo would earn in with a combination of shares and gold exploration expenses.

With the acquisition of the Millennium mill under discussion, Quinton wanted to consolidate as much potential ground in the Nullagine area as possible. In June Novo announced it would acquire nineteen square kilometers of ground in three exploration licenses from the vendors of what was called the Mt. Elsie project, about 75 kilometers from Nullagine. [84] It cost Novo 324,506 shares and $100,000 in cash. Novo now controlled a total of about 13,750 square kilometers of ground in the Pilbara Basin. That's over 5,300 square miles; nearly one Connecticut or two Delawares, but hopefully containing a lot more gold than either of those states.

Novo's technical team at Egina began to use a new device in 2020, called a mobile alluvial Knudson or MAK.

The difficulty with such a giant land position is how to home in on the highest grade, or the material that is easiest to test, rather than just using Kentucky Windage to guess where the gold might be. The MAK unit takes grab samples of approximately one ton or 0.4 cubic meters of gravel at regular intervals from a pit 1–2 meters deep, and counts the gold grains. In July of 2020 Novo announced it had taken 342 such samples, with another 750 scheduled. [85]

The purpose of the MAK sampling is to get a feel for the gold in a certain area, and then follow up with 120 bulk samples of 30–50 cubic meters of gravel for detailed processing through the IGR3000.

Due to the nugget effect, the MAK samples will provide only a rough idea of grade. The use of the IGR3000 will tend to correlate where the MAK is most effective after the larger bulk samples are processed through the plant at the camp. The MAK processing is done in the field. The IGR3000 bulk sample gives an exact amount of gold in a bulk sample, but takes a lot longer to schedule and to make arrangements to scoop out the gravel and deliver it for processing.

By far the most important press release in Novo's history came out on August 4, 2020. [86] It announced the agreement to acquire the mill of the defunct Millennium Minerals and its 230-man camp for $44 million in Novo units priced at $3.25 a share and a half warrant at $4.40 good for three years. Following Novo's acquisition of the Millennium shares, Millennium would repay IMC $43.3 million in cash and 6.5 million Novo units. Novo was providing Millennium the difference between what Novo is paying for the shares and what Millennium must repay to IMC for its secured debt.

Novo needed cash for the transaction, and so raised a total of $56 million in private placements and secured up to $60 million in debt financing with Sprott Lending, in two parts. The sum of $35 million was to be advanced at closing, with an additional $25 million available up until March 31, 2021 contingent on Novo completing a pre-feasibility study on Beatons Creek that is acceptable to Sprott. The debt bears an eight per cent coupon plus LIBOR or one per cent, whichever is greater. Repayment begins twenty-four months after closing, in equal quarterly installments.

To put this in terms investors can understand, Novo is on the hook to Millennium and IMC for about $65 million, plus whatever it takes to put the mill back into operating condition. It is also on the hook to Sprott Lending for an additional $35 million.

The mill has averaged 1.88 million tonnes of throughput per year for the past five years. I've made it clear that I didn't think much of Millennium's management. I do have a high estimation of Rob Humphryson and his crew. If all Novo does is to equal Millennium's throughput, feeding in rock averaging two grams of gold per ton, that will result in the production of about 120,000 ounces of gold a year. I should be sorely perturbed if Rob cannot beat that, and in such an event I will not buy him any more beer. I think the all-in cost of production will be in the $700 to $900 U.S. range. But what do I know?

In October of 2020 Novo announced progress in Perth, with the use of a Steinert KSS sorting machine for use in bulk samples taken from both Purdy's Reward and Comet Well. [87] This would start in March or April 2021, after the worst of the cyclone season. The machine had been ordered and was due to arrive in Perth in November. It would then be necessary for Steinert personnel to assemble and finish the machine according to Novo specifications before it is taken to Karratha and put into use.

As of early January 2021, as I write, Novo is still on track to do the first pour in about mid-February. I plan to release this book on the day of the first pour.

Investors wondering about what to do next might want to look again at the Lassonde Curve for a hint.

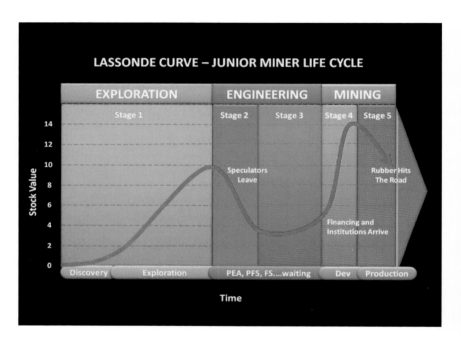

CHAPTER 24
IT'S NOT ROCKET SCIENCE

NO DOUBT THE NAYSAYERS will be out in force, telling everyone just how stupid Bob Moriarty was for calling this the greatest gold discovery in history. But then, unlike me, they weren't saying that Novo was going to go up 10-fold or 100-fold when its shares were forty-five cents, so maybe I'm not quite as stupid as they will say. I did get old but I didn't get stupid.

You see, I am unencumbered by what some geology professor told me in Geology 101. I never took Geology 101, or any other chemistry or mining course. I had to learn to think and to see what was right in front of me. If you can develop that ability you will see the same thing.

I was actually part of the Novo story as early as 1976, long before Quinton came up with his theory. I didn't know it at the time; I thought I was just a ferry pilot (do get the spelling of that word correct) taking a Rockwell 685 from Southern California to Melbourne for Lang Hancock. [88]

Hancock was the guy who jumpstarted the entire Western Australia iron boom. While he claimed to have discovered the world's largest iron deposit, it was actually Harry Page Woodward who did so. [89] In 1890, in the *Annual General Report of the Government Geologist*, he wrote:

"This is essentially an iron ore country. There is enough iron ore to supply the whole world, should the present sources be worked out."

I listened to Lang Hancock's son-in-law talk about how rich the Pilbara iron deposits were, for the seventy-five hours he and

I were in the air together. Official reports today say that Western Australia possesses 29 percent of the world's iron reserves. It's a banded iron formation. The iron precipitated out of salt water over two and a half billion years ago.

So when Quinton started talking to me about how he believed the source of the gold in the Witwatersrand was its precipitation in the presence of oxygen produced by single-cell bacterial colonies, I became a believer. If the iron precipitated out of salt water in the presence of oxygen, and the gold precipitated out of salt water in the presence of oxygen, you cannot have one without the other.

Fifty years from now, after tens of millions of ounces of gold have been produced from the Pilbara, it will be as obvious to everyone as it was to me in 2008.

You see, it's not rocket science.

Novo has made their first gold pour at Nullagine as I release this book for sale.

To be continued...

REFERENCES

1. https://en.wikipedia.org/wiki/Bre-X

2. https://en.wikipedia.org/wiki/4-H

3. https://en.wikipedia.org/wiki/Roebourne%2C_Western_Australia

4. https://fremantlestuff.info/people/angelo.html

5. https://en.wikipedia.org/wiki/Emma_Withnell

6. https://en.wikipedia.org/wiki/Lang_Hancock

7. https://www.thefreedictionary.com/pommy

8. http://y20australia.com/rab-drilling-mining-explained/

9. https://en.wikipedia.org/wiki/De_re_metallica

10. https://en.wikipedia.org/wiki/Caribou,_Colorado

11. https://paperity.org/p/4026623/witwatersrand-gold-deposits

12. https://www.mountvernon.org/library/digitalhistory/digital-encyclopedia/article/first-in-war-first-in-peace-and-first-in-the-hearts-of-his-countrymen/

13. https://www.history.com/news/the-day-skylab-crashed-to-earth-facts-about-the-first-u-s-space-stations-re-entry

14. (directed to FT subscription page)

15. https://www.proactiveinvestors.com.au/companies/news/201

791/legend-mining-continues-to-draw-comparisons-with-nova-bollinger-deposit-201791.html

16. https://en.wikipedia.org/wiki/Dead_centre_(engineering)

17. http://www.321gold.com/editorials/moriarty/moriarty081512.html

18. https://www.novoresources.com/_resources/news/2012-08-13.pdf

19. http://www.321gold.com/editorials/moriarty/moriarty082712.html

20. https://en.wikipedia.org/wiki/National_Instrument_43-101

21. https://www.voiceamerica.com/Show/1501

22. https://www.northernminer.com/news/newmont-takes-35-7-stake-in-novo-resources/1002578301/

23. https://en.wikipedia.org/wiki/Bulk_leach_extractable_gold

24. https://www.novoresources.com/_resources/news/2013-12-17.pdf

25. https://www.novoresources.com/_resources/news/2014-03-04.pdf

26. https://www.novoresources.com/_resources/news/2014-03-19.pdf

27. https://www.novoresources.com/_resources/news/2014-07-24.pdf

28. https://www.novoresources.com/_resources/news/2014-08-28.pdf

29. https://www.novoresources.com/_resources/news/2014-12-10.pdf

30. https://www.businesswire.com/news/home/20140512006751/en/Newmont-Signs-Agreement-to-Sell-Jundee-Underground-Gold-Mine-in-Australia

31. https://www.amazon.com/Games-People-Play-Psychology-Relationships/dp/B0007DYNTE/ref=tmm_hrd_swatch_0?_encoding=UTF8&qid=1602437696&sr=1-1

32. https://en.wikipedia.org/wiki/Volcanogenic_massive_sulfide_ore_deposit

33. https://www.geographie.uni-wuerzburg.de/fileadmin/04140600/WR_BKGR/Frimmel_Wits_SEG_SP18_2014.pdf

34. https://www.springer.com/journal/126

35. https://www.researchgate.net/publication/273278727_First_whiffs_of_atmospheric_oxygen_triggered_onset_of_crustal_gold_cycle

36. (directed to Northern Star subscription page)

37. https://www.hecla-mining.com/midas/

38. https://www.novoresources.com/_resources/news/2014-12-10.pdf

39. https://www.novoresources.com/_resources/news/2015-02-09.pdf

40. https://www.novoresources.com/_resources/news/2015-03-26.pdf

41. https://www.novoresources.com/_resources/news/2015-04-02.pdf

42. https://www.novoresources.com/_resources/news/2015-05-26.pdf

43. https://www.theglobeandmail.com/report-on-business/investor-eric-sprott-stepping-down-as-chairman-of-sprott-inc/article34630762/

44. https://www.globenewswire.com/news-release/2015/05/11/1279060/0/en/Newmarket-Gold-and-Crocodile-Gold-Merge-to-Establish-a-New-Platform-for-Gold-Asset-Consolidation.html

45. https://finance.yahoo.com/news/crocodile-gold-reports-drilling-results-220000972.html

46. https://www.spglobal.com/marketintelligence/en/news-insights/trending/3s62jiinj9202_qd-e0a0w2

47. https://www.kl.gold/news-and-media/news-releases-archive/default.aspx

48. https://www.novoresources.com/_resources/news/2015-08-17.pdf

49. https://www.novoresources.com/projects/pilbara/blue-spec/

50. https://www.novoresources.com/_resources/news/2016-08-25.pdf

51. https://en.wikipedia.org/wiki/The_Treasure_of_the_Sierra_Madre

52. https://www.dmp.wa.gov.au/Documents/Minerals/132298_Mining_Notice_Special_Pros_Licence.pdf

53. https://en.wikipedia.org/wiki/Tiny_Rowland

54. https://www.perthnow.com.au/business/gold/meet-mick-shemesian-the-mystery-man-behind-the-pilbara-gold-scrambleng-b88662343z

55. https://hotcopper.com.au/threads/ann-new-style-of-gold-mineralisation-for-west-pilbara-at-purdys.3237395/

56. https://www.novoresources.com/news-media/news/display/index.php?content_id=232

57. https://www.youtube.com/watch?v=SiAGlWyjRq8

58. https://en.wikipedia.org/wiki/Peanut_gallery

59. https://www.novoresources.com/news-media/news/display/index.php?content_id=232

60. https://hotcopper.com.au/search/34225434/?q=%2A&t=post&o=relevance&c%5Bvisible%5D=true&c%5Buser%5D%5B0%5D=194432

61. https://hotcopper.com.au/threads/why-no-announcement-from-arv.3598161/page-66

62. https://hotcopper.com.au/threads/possession-is-9-10-ths-of-the-law.3610566/

63. https://hotcopper.com.au/threads/my-estimate-of-novo-sp-

fair-value.3610662/

64. https://hotcopper.com.au/threads/possession-is-9-10-ths-of-the-law.3610566/page-9

65. https://hotcopper.com.au/threads/my-estimate-of-novo-sp-fair-value.3610662/page-8

66. http://www.321gold.com/editorials/moriarty/moriarty082317.html

67. https://www.novoresources.com/news-media/news/display/index.php?content_id=319

68. https://www.novoresources.com/news-media/news/display/index.php?content_id=324

69. https://www.investopedia.com/terms/s/strategic-default.asp

70. https://www.novoresources.com/news-media/news/display/index.php?content_id=339

71. https://www.novoresources.com/news-media/news/display/index.php?content_id=346

72. https://www.novoresources.com/news-media/news/display/index.php?content_id=352

73. https://www.novoresources.com/news-media/news/display/index.php?content_id=355

74. https://www.novoresources.com/news-media/news/display/index.php?content_id=361

75. https://www.novoresources.com/news-

media/news/display/index.php?content_id=363

76. https://www.novoresources.com/news-media/news/display/index.php?content_id=366

77. https://www.australianmining.com.au/news/millennium-minerals-moves-into-voluntary-administration/

78. https://www.novoresources.com/news-media/news/display/index.php?content_id=376

79. https://www.novoresources.com/news-media/news/display/index.php?content_id=379

80. https://www.novoresources.com/news-media/news/display/index.php?content_id=380

81. https://www.novoresources.com/news-media/news/display/index.php?content_id=381

82. https://www.novoresources.com/news-media/news/display/index.php?content_id=384

83. https://www.novoresources.com/news-media/news/display/index.php?content_id=387

84. https://www.novoresources.com/news-media/news/display/index.php?content_id=394

85. https://www.novoresources.com/news-media/news/display/index.php?content_id=399

86. https://www.novoresources.com/news-media/news/display/index.php?content_id=402

87. https://www.novoresources.com/news-media/news/display/index.php?content_id=419

88. https://en.wikipedia.org/wiki/Lang_Hancock

89. https://en.wikipedia.org/wiki/Harry_Page_Woodward

Printed in Great Britain
by Amazon